Nanotechnologies:
The Physics of Nanomaterials
Volume II

Physical Properties of Nanostructured
Materials and Their Applications

Nanotechnologies:
The Physics of Nanomaterials
Volume II

Physical Properties of Nanostructured
Materials and Their Applications

David S. Schmool

APPLE
ACADEMIC
PRESS

First edition published [2021]

Apple Academic Press Inc.
1265 Goldenrod Circle, NE,
Palm Bay, FL 32905 USA

4164 Lakeshore Road, Burlington,
ON, L7L 1A4 Canada

CRC Press
6000 Broken Sound Parkway NW,
Suite 300, Boca Raton, FL 33487-2742 USA

2 Park Square, Milton Park,
Abingdon, Oxon, OX14 4RN UK

First issued in paperback 2021

© 2021 Apple Academic Press, Inc.

Apple Academic Press exclusively co-publishes with CRC Press, an imprint of Taylor & Francis Group, LLC

Library and Archives Canada Cataloguing in Publication

Title: Nanotechnologies : the physics of nanomaterials, volume II : physical properties of nanostructured materials and their applications / David S. Schmool.

Names: Schmool, D. S. (David S.), author.

Description: Includes bibliographical references and indexes. | Contents: Volume II. Physical properties of nanostructured materials and their applications.

Identifiers: Canadiana (print) 20200297066 | Canadiana (ebook) 20200297120 | ISBN 9781771889476 (set) | ISBN 9781771889490 (v. II ; hardcover) | ISBN 9781003100225 (v. II ; ebook)

Subjects: LCSH: Nanotechnology. | LCSH: Nanostructured materials. | LCSH: Nanoscience. | LCSH: Physics.

Classification: LCC QC176.8.N35 S36 2021 | DDC 620/.5—dc23

Library of Congress Cataloging-in-Publication Data

....

CIP data on file with US Library of Congress

....

Nanotechnologies: The Physics of Nanomaterials, 2 Volume (Set)
ISBN: 978-1-77188-947-6 (hbk)
ISBN: 978-1-00310-019-5 (ebk)

Nanotechnologies: The Physics of Nanomaterials, Volume II
ISBN: 978-1-77188-949-0 (hbk)
ISBN: 978-1-77463-967-2 (pbk)
ISBN: 978-1-00310-022-5 (ebk)

Dedication

For Ulysse

About the Author

David Schmool, PhD, is a former Director of the Groupe d'Etude de la Matière Condensée GEMaC at CNRS (UMR 8635) and is currently Professor in Physics at the Université de Versailles/Saint-Quentin-en-Yvelines and Université Paris-Saclay, France. Prior to this he has held teaching and research positions at the University of Perpignan/Laboratoire PROMES – CNRS, Perpignan, France; University of Porto, Portugal, University of the Basque Country, Bilbao, Spain; Istituto IMEM, Parma, Italy; University of Exeter, UK and University of Liverpool, UK. David Schmool has also been a visiting fellow to several institutions, including Simon Fraser University (Canada), the University of Versailles (France), the University of Duisburg-Essen (Germany), and the University of Glasgow (UK). He obtained his DPhil in Physics from the University of York in 1994. In addition to his research experience, he has lectured on physics since 2000 on a broad range of subjects. He has over 25 years of research and teaching experience in areas related to nanosciences and nanotechnologies. He has also developed master's and PhD level courses in nanotechnologies and related subjects, which he has also taught. He has published widely, including over 80 journal papers, 12 book chapters, a textbook and has given many conference presentations, including 20 invited talks and over 20 invited seminars.

Contents

Preface

This two-volume book stems from a master's level course in *Nanotechnologies* that I originally gave at the University of Porto in Portugal from 2008 to 2013. It was morphed in various forms to cover subjects in parts of courses I gave in *Nanomedicine* (2011–2013) (at the Faculty of Engineering of the University of Porto), *Characterization of Nanomaterials* and *Innovative Materials* (2015) at the University of Perpignan, France. More recently, I have used some of these chapters in the preparation of a master's course on *Nanomaterials and Characterization* at the University of Versailles/the University of Paris-Saclay. I have also adapted the course to a doctoral level taught course, which I also gave at the University of Porto. This was a shorter more focused module that aimed at an introduction for first-year graduates. Since leaving the University of Porto, my colleague Dr. André Pereira has taken up the role of teaching the course that I initially designed. He has extensively used my lecture notes. For the most part, he has maintained the structure of the course and has incorporated some innovations and adaptations of his own. I am very grateful to Dr. Pereira for his useful suggestions in the compilation of this two-volume book. In addition to this, I have greatly benefited from feedback from students who have taken my courses, with some helpful suggestions, particularly in its early development.

In recent years, there has been a rapid expansion of the literature in all areas *nano*. This ranges from the fabrication techniques to specialized texts in nanomagnetism and nanoelectronics, for example. There are also many textbooks on general introductions to nanotechnology. In this two-volume textbook, my main objective has been to provide a reasonably in-depth account of the subject, aimed principally at university students at the postgraduate level. It is intended for students of physics and engineering-based courses, though it may also serve as a general introduction to the expert newcomer.

As will be noted from the Contents, I have tried to start at the beginning. Or at least where I think the beginning should be. Chapter 1 is meant to provide a gentle introduction to the subject. I have opted to provide a general overview of why nanotechnologies are of such importance in modern technologies and today the groundwork for the context in modern science and technologies. As with the whole book, the emphasis being on Physics. I feel a little uncomfortable discussing biological issues as it is not an area I specialize in. That said, I am well aware that biological issues are extremely important in modern developments in nanoscience. However, this is intended to be a book on the physics of nanotechnologies.

I have chosen to divide the book into three parts and in two volumes. The first part of Volume 1 aims to introduce some of the basic physics required for understanding where nanotechnologies came from. After discussing vacuum physics in Chapter 2, much of the rest of the following chapters in Part I are dedicated to surface physics. Here I introduce the main concepts regarding the nature of surfaces and in particular the modification of the surface crystalline structure due to symmetry changes, which leads to the formation of surface reconstructions. The chapter on thin films is a basic introduction to the main concepts of film growth and methods of preparation. I make no pretense at being exhaustive. My main aim is to provide enough basic physics that the reader can use these concepts to understand other techniques when she/he is confronted with them. Chapter 5 aims at providing an introduction to a broad spectrum of methods used in the analysis of the characterization of surfaces. Many of these methods are widely used in the study of nanomaterials. I have partitioned this into sections of the study of structures via diffraction methods, electron spectroscopies for chemical analysis and microscopies.

The fabrication of nanostructured materials is a large and important topic in nanotechnologies. I have used Part II of Volume 1 to introduce some of the main methods employed to prepare nanostructured materials. Chapter 6 looks at one of the main classes of methods; lithographies. There are a number of techniques available and these are discussed in some detail. In more recent times other techniques have come to the fore, such as the replication methods, this is the subject of Chapter 7. Nanoparticles and nanowires are of huge importance and particularly in terms of their applications, notably in biomedical sciences. In Chapter 8, I have provided an overview of the main methods of fabrication. Chapter 9 brings together other specialized subjects that didn't fit into the previous chapters of Part II. Here we discuss the important class of carbon allotropes, such as the fullerenes and carbon nanotubes (we also discuss graphene, but more in terms of its physical properties). Self-assembly is a fascinating method for the preparation of nanostructures and comprises many approaches. This chapter also provides a brief introduction to some of these methods.

The last part of this two-volume book, i.e., Volume 2, is dedicated to an in-depth study of the physical properties of nanomaterials. I have selected to treat these as the main physical properties of materials: mechanical properties, electronic properties, optical properties, and magnetic properties. The approach I have used in general is to provide a basic introduction to the subject at a level which anyone with a general grasp of solid-state physics should have no problem understanding and is meant as a revision of the main issues. This provides the context for understanding the specific properties of materials with reduced physical dimensions. I have tried to provide ample examples and applications throughout these chapters. It is inevitable that I have missed some topics.

As I mention in the opening paragraph, this book emerged from my preparations for a master's level course in physics on the subject of nanotechnologies. In the preparation of my notes, I used a number of texts on all areas covered in the book. Since turning those notes into the book form, I have expanded and extended many of the chapters. I have also added some subjects that were missing. The book maybe a little too long for one single course on nanotechnologies. My intention is not to provide an exhaustive compendium on the subject, nor is it meant to be a definitive course. I see it more as a guide to the basic principles of the subject. As such I think the book can be used in a number of ways. Principally as a guide to the main subjects. The course tutor/lecturer can choose those topics she/he finds the most pertinent to their approach. They may also wish to provide more in-depth coverage of certain topics. The student can use the book in a similar way, by picking and choosing the subjects of interest and those required in their course. The general reader may wish to delve into specific topics related to their interests. The book may also provide the basis for specialized courses. The chapters on Volume 2, for example, could be used as the basis for specialized topics. In which case, the course provider may use the chapter as a guide upon which further material can be added and treated in more detail. In all of these chapters, I have attempted to provide a general introduction to the principal physical properties of solids. I then attempt to show how the physical properties are modified by the size of the objects. Often we will consider the characteristic length scales which determine at what size of object these modifications will occur. Where possible I have added examples taken from recent research literature which should help demonstrate the main physics involved as well as to provide some examples of how these properties can be harnessed in the plethora of applications and devices that researchers the world over have dedicated their research. Part II of Volume 1 can also be seen as an overview of fabrication techniques, which while not being exhaustive, should provide a solid basis upon which a course on this subject can be prepared.

The use of mathematics is essential for our understanding of the physical properties of materials. I have tried to keep this to the essential. Anyone with undergraduate level mathematics will have no problem following the mathematics in this book. In each chapter, I have tried to end with a brief summary of the subject as well as giving some problems at the end for the student to test their knowledge and understanding. I also list the main references and some further reading at the end of each chapter.

I must admit that when I started the writing of this book, I had some romantic notions of what I would like the book to look like, etc. Reality hits fast! One of the hardest tasks for anyone writing on subject matters which are as vast as those of nanotechnologies is what to include and importantly, where to stop! I can't claim that I have made a good job at this, but at some point, you have to stop. Nanotechnologies are a broad-ranging subject and indeed each chapter could be turned into a specialized book. The literature is extensive and seemingly infinite. A choice must be made. I have opted to try to cover the main topics and provide some pointers for further reading. Despite these difficulties, I have to say that I really enjoyed writing this book. First of all, it is a subject I have come to love and am always discovering some new aspect or recent application. It is indeed an amazing area of science. I hope you will enjoy it as much as I do. I am frequently struck by the ingenuity of scientists and the clever tricks they can do.

I would like to thank the editors of this book for their patience, particularly with respect to the extensions to the extensions of deadlines... I have had to balance the writing of this book with a number of tasks over recent years. As well as the ever urgency of teaching duties, I have also been occupied by the demands of the administrative tasks of being the Director of a research laboratory. Inevitably there will be some errors and mistakes. I am hugely indebted to the dedication and assistance of my colleague Dr. Daniel Markó, who has generously dedicated a large amount of time to read the entire draft of the book, making many corrections and providing countless useful comments and suggestions. I am sincerely grateful for his careful and extensive corrections. Thanks, Daniel! I am grateful to my family for their love and support, especially Mike and Debbie. Thanks are also due to Louise, for being Louise. I cannot begin to estimate the debt I owe to my partner Virginie, she has suffered my extended absences with good humor and my life would infinitely more difficult without her love and support. I would also like to mention Ulysse, our son, who has provided us with so much joy over the last two years and to whom I dedicate this two-volume book.

—David Schmool
Versailles, France, 2019

PART III

THE PHYSICAL PROPERTIES OF NANOSTRUCTURED MATERIALS AND THEIR APPLICATIONS

Chapter 1

Mechanical Properties of Micro- and Nanostructures

In the consideration of the mechanical properties of micro- and nanometric devices, we are essentially discussing the mechanical effects produced in miniaturized systems. Frequently, this will mean micro-electromechanical systems or MEMS as well as their nano-counterparts (NEMS – nanoelectromechanical systems).

The mechanical properties of solids, as with other physical properties, will be significantly modified by the size reduction of an object. To be able to exploit such properties, it is important to be able to classify them, as well as to be able to measure them and of course, understand them. The exploitation of the physical properties of solids is indeed one of the main reasons why we study them in the first place. A vast area of study has been given over to the development and application of reduced or miniaturized objects, which are designed to perform very specific functions. An important class of such devices is known as *sensors*, with a counterpart called *actuators*. Sensors and actuators are designed to detect some physical stimulus or to perform a specific task in response to say an applied electrical signal for example. An example of a sensor may be a temperature measurement, while an actuator could be a motor. Often, these systems require the conversion of one physical property into another. In the examples given here, the first could be the conversion of heat energy into an electrical signal, while the latter concerns the conversion of electric energy into mechanical energy or motion.

Besides the traditional micro-fabricated sensors and actuator, the study of MEMS covers a broad range of applications, in which micro-mechanical components and systems are integrated or micro-assembled with electrons on the same substrate, achieving high-performance functional systems. Such devices can play key roles in many important areas such as transportation, communications, automated manufacturing, environmental monitoring, health care, defense systems, space travel and a wide variety of consumer products.

MEMS are inherently small, thus offering attractive characteristics, such as reduced size, lightweight, and low power dissipation as well as increased speeds and precision when compared to their macroscopic counterparts. The development of MEMS devices requires suitable fabrication technologies, enabling precise and reliable object definition on a very small scale, precise dimension control, design flexibility, interfacing with microelectronics, high yield, repeatability, and low cost. Integrated circuit manufacturing meet all of these requirements and has been the principal fabrication technology for MEMS, with some appropriate modifications. Advances in both technology and methodology have been spectacular over recent decades. As we have advanced in our knowledge and capabilities, much like with the progress in the microfabrication of ICs, NEMS is also becoming more and more realizable and will become a major area of future technology. Nanoscale devices, integrated with nanoelectronics will further the development of new devices and is the subject of vast areas of research in science and engineering, with a huge array of applications, ranging from electronics to health care and life sciences.

1.1   Scaling Laws

Scaling theory is a valuable guide as to how the physical properties of a system may respond to miniaturization. In such a consideration, we aim to understand how physical phenomena behave and change as their scale size changes and thus gain insight into the best strategies for attaining the desired properties and responses of a particular system in reduced dimensions. In a general manner, we can consider three scale sizes: (i) Astronomical objects, (ii) normal day-to-day objects or macro-objects, and (iii) very small or micro-objects. Things that are effective on one of these scales are frequently insignificant on another. Let us consider the following example: the motion of our planets in the solar system is governed by gravitational forces, on the macro scale of objects on my desk, these gravitational forces between two objects, such as a book and my coffee cup, are insignificant. Often, what is obvious on the astronomical scale or macroscale is not so on the microscale.

As we shall discover in this chapter, the field of micro- and nano-electromechanical devices is very broad. It encompasses many traditional science and engineering disciplines on a small scale. As the size, or scale, of a system changes by several orders of magnitude, the system tends to behave differently. Let us take the case of a container of water. For a glass of water, of say 5 cm diameter, consider what happens when we pour the water onto a table. We first observe the water flowing from the glass and pouring over the edges of the table onto the floor. Let us reduce the size of this container by two orders of magnitude; i.e., a container of 0.5 mm. You can intuitively imagine what is likely to occur. Our observation of pouring the water onto the table will be very different. Most likely the water will remain as a drop, held together by the cohesive intermolecular forces, which are responsible for surface tension and capillarity. Let us decrease the size by another two orders of magnitude; our container is now just 5 m. If we can manage to extract the water from the container, it is unlikely to fall onto the table. It will probably be carried away by a current of air. Now, surface tension will be the dominant force acting on the drop. In this simple example, we can see how different forces can come into play at different length scales. As we shall see, also in the following chapters, length scales are an extremely important consideration of the physical properties of all systems on the nanoscale. These play a crucial role in determining many of the physical properties of systems; this is true for electronics, optics, and magnetism.

In Figure 1.1, we show the full range of sizes available and to which Physics as a whole engages to understand. This goes from the Planck length (10^{-34} m) to the size of the Universe (10^{26} m). According to current estimates, the observable universe is roughly 8.8×10^{26} m in diameter. With atoms being towards the lower scale of observable scales, with a size of around 1 Å, this means that the realm of observable Physics is around 37 orders of magnitude!

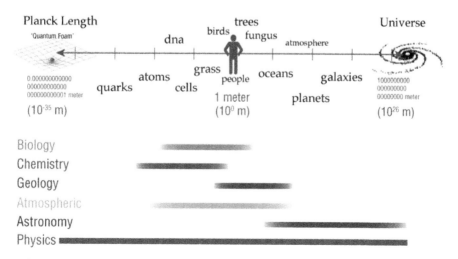

FIGURE 1.1   Logarithmic plot of mechanical systems and sizes in science.

In nanotechnologies, we are only concerned with a small subregion of length scales, which we can generously give as say $10^{-3} - 10^{-10}$ m, which is still a large range of sizes, seven orders of magnitude. On the large end, we can consider an object in the mm range for large MEMS devices, while at the lower end, we consider the atom as the limit of interest.

In mechanical systems, we can think of road systems as the largest mechanical object. However, buildings and aircraft carriers are probably the largest self-contained mechanical system that man will construct in the near future. In the microscopic region, we are capable of manipulating objects on the atomic scale. However, the construction of objects atom by atoms is not a feasible method for making anything larger than a few tens of atoms at most. As was discussed in Part II of this book, there are many different techniques available to us for the fabrication of a wide range of objects using many different materials for a whole host of applications. This is not to say that there are no longer any challenges and obstacles.

As the size of a system changes, its physical parameters often change in a dramatic way. To understand how these parameters change, let us consider the scaling factor, S. This scaling factor is similar to the notation used for the scale on drawings and maps, for example, 1:10 means that the actual object is ten times the size of the drawing. When the size scale changes, all the dimensions of the object change by the same factor S, such as we have 1:S.

We can use the scaling factor to describe how physical phenomena change. All the lengths of the drawing scale by the same factor S, but other parameters, such as the volume, can scale differently. Since the volume is the product of the length, width, and height, $V = lwh$, when the scale factor changes by say a factor of 1/1000, for example, the volume will scale as $1/(100)^3$, or 1/1,000,000, i.e., the volume scales as S^3. The forces due to surface tension scale as S^1, the forces due to electrostatics scale as S^2, magnetic forces scale as S^3 and gravitational forces as S^4. If we consider the reduction from 1 m to 1 mm, i.e., $S = 1/1000$, this change of dimension will give the following scaling factors: surface tension, $S^1 = 1/1000 = 10^{-3}$; electrostatic forces, $S^2 = 10^{-6}$; magnetic forces $S^2 = 10^{-9}$; gravitational forces $S^2 = 10^{-12}$. Thus we see that a change in the size of a mechanical system will alter which forces are of importance.

Knowledge of how a physical force or phenomena changes, as the power of S, will guide our understanding of how to design small micromechanical systems. Let us consider the example of a water insect. The weight of the insect will scale with volume, i.e., as S^3, while the force that supports it on the water, surface tension, scales as S^1 times the distance around its feet, S^2. So when the size scales S decreases, the weight factor decreases more rapidly than the surface tension forces.

We can use these scaling arguments as a guide as to how miniaturization will affect the workings of small systems, though it does not provide an exact solution to the specifics. Each case must be considered on its own merits and particular characteristics. We can express the different scaling laws in a notation, which shows all the scaling laws at the same time to see what happens to the different terms and parameters of an equation as the scale size changes. Let us consider the force laws, which we outlined above, each with their scaling law; S^1, S^2, S^3, S^4. These can be expressed as:

$$F = \begin{bmatrix} S^1 \\ S^2 \\ S^3 \\ S^4 \end{bmatrix} \tag{1.1}$$

This form of expression is referred to as a vertical Trimmer bracket. The topmost element of the bracket refers to the case of the force scaling as S^1, the next element down refers to the forces which scale as S^2, etc. We can use this representation to illustrate how other physical quantities can be represented. For the work done, W, which is a force, F, times a distance, D, we have:

$$W = FD = \begin{bmatrix} S^1 \\ S^2 \\ S^3 \\ S^4 \end{bmatrix} \begin{bmatrix} S^1 \\ S^1 \\ S^1 \\ S^1 \end{bmatrix} = \begin{bmatrix} S^2 \\ S^3 \\ S^4 \\ S^5 \end{bmatrix} \tag{1.2}$$

Here we note that distance always scales as S^1 and its bracket will only consist of elements of S^1. Let us now consider a more complex quantity, acceleration. We can evaluate how these scales in the following manner, where we exploit Newton's second law:

$$a = \frac{F}{m} = Fm^{-1} \tag{1.3}$$

Since the mass scales as S^3, the inverse of the mass, m^{-1} will scale as S^{-3}. We now express the scaling law as:

$$a = \begin{bmatrix} S^1 \\ S^2 \\ S^3 \\ S^4 \end{bmatrix} \begin{bmatrix} S^3 \\ S^3 \\ S^3 \\ S^3 \end{bmatrix}^{-1} = \begin{bmatrix} S^{-2} \\ S^{-1} \\ S^0 \\ S^1 \end{bmatrix} \tag{1.4}$$

This shows that when the force scales as S^1, the acceleration will scale as S^{-2}. Therefore, if the size of a system decreases by a factor of 100, the acceleration will increase by a factor of $(1/100)^{-2} = 10000 = 10^4$. So as the system gets smaller, the acceleration increases. A majority of forces used in the microdomain scale as S^2, for which the acceleration scales as S^{-1}, so a decrease of size by a factor 100 will be accompanied by an acceleration increase of 100. In general, small systems tend to accelerate very rapidly. Where the force scales as S^3, acceleration will remain constant, while for forces scaling as S^4, it will decrease.

A more comprehensive list of forces and their scaling factors can be expressed as:

$$F = \begin{bmatrix} S^1 \\ S^2 \\ S^3 \\ S^4 \end{bmatrix} = \begin{bmatrix} \text{Surface Tension} \\ \text{Electrostatic, Pressure, Biological, Magnetic } (J = S^{-1}) \\ \text{Magnetic } (J = S^{-0.5}) \\ \text{Magnetic } (J = S^0) \end{bmatrix}$$

$$(1.5)$$

The surface tension scales as S^{-1}, thus increases rapidly relative to other forces as a system reduces in size. However, changing the surface tension requires a change in temperature, adding a surfactant or altering some other parameter that is usually difficult to control. Most forces currently used by micro-designers scale as S^2. These include electrostatic forces, forces generated by pressure and biological forces, usually related to the cross-section of the body. Magnetic forces scale in a number of ways depending on the system. For example, if the current density, J, in a coil remains constant (S^0), the magnetic force between two coils scales as S^4. In this case, the magnetic forces become weaker in the microdomain. However, we can remove heat more efficiently from a small body, so the current density of a microcoil can be much higher than in a large coil. If the current density scales as S^{-1}, when the system decreases in size by a factor of 10, the current density increase by a factor of 10. In this case, the coil has much higher resistive losses, but the force scales much more advantageously, as S^2.

1.2   Micro- and Nano-Mechanical Properties

The mechanical properties of solids are a well-established area of Materials Science and form a core subject in the study of materials by physicists and engineers. There are standard properties and tests that are used to provide a reliable system for the comparison of mechanical behavior by different materials. Material properties are based on a microscopic interpretation of the interatomic forces between atoms and will depend greatly on the nature of the bonding mechanism between atoms and the crystalline and microstructure of the materials. In bulk materials, this forms the underlying basis of our understanding of mechanical properties. This remains true for micromechanical systems. When considering objects on the nanoscale, this becomes even more critical, since the object will have a significant proportion of its atoms located at or near the surface of the body. In accordance with the calculation of the proportion of atoms on the surface to atoms in the bulk, the importance of the surface will be more or less dominant. This was discussed in Chapter 1 of Volume 1, where we considered the proportion of surface atoms to bulk atoms as a function of the size of the nano-object.

1.2.1 Mechanical Properties

New technologies tend to demand new approaches and, frequently, new materials along with fabrication processes. This is indeed the case with the development of MEMS technologies. The emphasis over the first decade or so was on new materials and manufacturing of new microdevices. The technological advances were paralleled by an increasing interest in the mechanical testing of the materials used in MEMS devices. Typically, the mechanical properties of interest were the elastic and inelastic behavior as well as the strength of these materials. Such properties are of vital importance to device designers, since it is important to know how a material of a specific shape will react to an applied force. If elastic, what is the deflection obtained for a certain force? If inelastic, or ductile, how much deformation will a specific force produce? Also, the material strength is important for defining the allowable operating limits for the device. These are clearly complex problems that require careful and reliable testing procedures.

Materials scientists and engineers are well aware that the mechanical properties of material should be studied independently of the size of the structure. This is certainly the case for bulk materials and engineering structures. However, for MEMS materials, this may not be the case, adding a further level of complexity of the study. Furthermore, fabrication processes for micro- and nanostructures are very different from those used in bulk materials. It is therefore important to ask whether specimen size needs to be taken into account for the specific device/material being considered. These are again complex issues that often require empirical testing. Given the nature of the materials, it is also important that test methods are developed, which are sufficiently sensitive and reproducible to differentiate material behavior.

The measurement of the mechanical properties of MEMS structures is problematic for a number of reasons. To perform such measurements, we need to (i) obtain and mount the sample; (ii) measure its dimensions; (iii) apply a force or displacement to deform it; (iv) measure that forces; and (v) measure the displacement or preferably, measure the strain. All of these steps are well developed and standardized for common structural materials, where typically the minimum dimensions of a gauge section will be around 2 mm. Certain tests have been developed for metallic wires as thin as 100 m, though other dimensions are large and adapted to mounting on a large scale apparatus.

There is a range of mechanical properties that are of interest to designers and developers of structures, which are essential to take into account to provide a reliable structure, which maintains its structural integrity for the functions it is designed to perform. The principal mechanical properties are expressed as the *modulus of elasticity*, also commonly known as *Young's modulus*, the *bulk modulus* and the *shear modulus*. These three properties define how a material responds to different applied forces. It is useful to introduce these physical quantities one by one.

1.2.1.1   Young's Modulus and Elastic Behavior

The most common mechanical property is probably the modulus of elasticity, which defines the relationship between stress (force per unit area) and strain (proportional deformation) of a material in the linear elasticity regime. A solid material will undergo an elastic deformation, when a small load is applied to it in compression or extension. Elastic deformation is by definition reversible, so once the load has been removed, the material will return to its original form, exactly. The definition given above can be expressed mathematically as:

$$E = \frac{\text{stress}}{\text{strain}} = \frac{F/A}{\delta l/l} = \frac{Fl}{\delta lA} \tag{1.6}$$

where F is the axially applied force, A the cross-sectional area of the sample, δl the amount by which the length of the object changes and l is the original length of the object. More correctly, we state that the Young's modulus is the slope of the stress-strain curve, which is at least approximately linear in the elastic regime. We can make a microscopic evaluation of the Young's modulus. We will consider a section of a crystal, which for simplicity should have a simple cubic structure and equilibrium atomic separation (lattice parameter) of r_0 and a length $l = Nr_0$, where N (or more correctly $N+1$) is the number of atoms along the length of the crystal. We apply a force F to the crystal, such that the force on each atom will be F/n, with n being the number of rows of atoms the force is acting on and the cross-sectional area will be $A = nr_0^2$, see Figure 1.2. Therefore, the force acting along each row of atoms along the length of our specimen can be expressed as $F/n = Fr_0^2/A$. The length of the crystal will be $L = l + \delta l$ and $\delta l = N\delta r$. The strain will take the form:

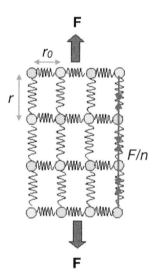

FIGURE 1.2 Application of a force, F, to a crystal, a force of F/n will act along each row of atoms, where n is the number of rows of atoms in a cross-section of area, A.

$$\varepsilon = \frac{\delta l}{l} = \frac{N\delta r}{Nr_0} = \frac{\delta r}{r_0} \tag{1.7}$$

For small strains, we can write $\delta r = [-dr/dF(r)]\delta F$. The negative sign is required, since the force is acting in the opposite sense to the interatomic forces $F(r)$. Here, we have $\delta F = Fr_0^2/A$, such that Eq. (1.7) will now take the form:

$$\varepsilon = -\left(\frac{dr}{dF(r)}\right)_{r=r_0} \frac{Fr_0^2}{Ar_0} = -\left(\frac{dr}{dF(r)}\right)_{r=r_0} \frac{Fr_0}{A} \qquad (1.8)$$

We can re-arrange this expression to obtain the Young's modulus in the form:

$$E = -\left(\frac{dF(r)}{dr}\right)_{r=r_0} \frac{1}{r_0} \qquad (1.9)$$

Now, since $F(r) = -d\mathcal{V}/dr$, we can write:

$$E = -\left(\frac{d^2\mathcal{V}(r)}{dr^2}\right)_{r=r_0} \frac{1}{r_0} \qquad (1.10)$$

where $\mathcal{V}$ denotes the potential. We can use a generic Mie potential of the form:

$$\mathcal{V}(r) = \frac{A}{r_0^m}\left[-\left(\frac{r_0}{r}\right)^m + \frac{m}{n}\left(\frac{r_0}{r}\right)^n\right] \qquad (1.11)$$

If we consider a NaCl crystal, with $A = e^2/4\pi\varepsilon_0, m = 1$ and $n = 9$, we can write:

$$\mathcal{V}(r) = \frac{e^2}{4\pi\varepsilon_0 r_0}\left(-\frac{r_0}{r} + \frac{1}{9}\frac{r_0^9}{r^9}\right) \qquad (1.12)$$

from which we obtain:

$$\frac{d^2\mathcal{V}(r)}{dr^2} = \frac{e^2}{4\pi\varepsilon_0 r_0}\left(-\frac{2r_0}{r^3} + \frac{10r_0^9}{r^{11}}\right) \qquad (1.13)$$

We can substitute this into Eq. (1.10) along with $r = r_0$, giving a Young's modulus expression of:

$$E = \frac{8e^2}{4\pi\varepsilon_0 r_0^4} \qquad (1.14)$$

This provides a reasonable approximation of the value of the Young's modulus for NaCl, where for $r_0 = 2.81$ Å, we obtain a value of $E = 2.95 \times 10^{11}$ Pa, which compares to a value at low temperature of $E = 5.8 \times 10^{11}$ Pa (Walton, 1987). The above model does not take into account the cross-linking in the crystal, which explains the discrepancy.

One experimental method of evaluating Young's modulus is via a measurement of the speed of sound. It can be shown that the relation between these two parameters can be expressed as:

$$v = \left(\frac{E}{\rho}\right)^{1/2} \tag{1.15}$$

where ρ denotes the density of the solid. For the case of copper, where $\rho = 8.9 \times 10^3$ kgm^{-3} and $E = 1.3 \times 10^{11}$ Pa, we obtain a speed of sound of $v = 3.8^3$ m s^{-1}, which is in close agreement with low frequency measurements for copper.

We can make a direct observation from the nature of the stress–strain relationship and its link to the law of Hooke. As the tensile force applied to a body is proportional to the extension in its length in that direction, Hooke's law is expressed as:

$$F = k\delta l \tag{1.16}$$

Dividing both sides by the section multiplied by the original length of the sample, Al, gives:

$$\frac{F}{Al} = \frac{k\delta l}{Al} \tag{1.17}$$

we recognize the quantities of the stress, $\sigma = F/A$, and the strain, $\varepsilon = \delta l/l$, from which we write:

$$\frac{\sigma}{l} = \frac{k\varepsilon}{A} \tag{1.18}$$

We can now write:

$$E = \frac{\sigma}{\varepsilon} = \frac{kl}{A} \tag{1.19}$$

This then illustrates that the stress–strain behavior is actually just a manifestation of Hooke's law, which concerns the linear region of the stress vs. strain curve, in the elastic limit. A full stress-strain curve is illustrated in Figure 1.3, showing the elastic region of mechanical behavior in solids. Briefly, this curve shows some of the common regimes for a ductile material. Beyond the elastic limit, the materials enter the plastic regime, where some form of permanent deformation takes place. Releasing the load or force at such a point means that the unloading curve is slightly displaced with respect to the loading line in the elastic region. This results in a small onset, which can be associated with the permanent stretching (deformation) of the solid. The further loading of the sample leads to further deformation up to a maximum or ultimate stress point. In this region, the material hardens (work hardening) due to the entanglement of dislocation lines, which are introduced from the

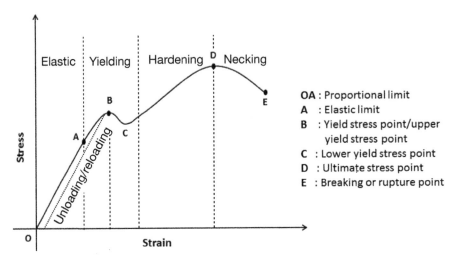

FIGURE 1.3   Stress–strain curve for a ductile material. This schematic behavior is typical of many metals, showing an elastic region, a plastic region, where the material yields to the applied forces, causing permanent deformation. Once the ultimate stress point has been passed, the material will start to neck, before the material breaks.

yield point. After loading beyond the ultimate stress point, the sample starts *necking*, which basically means that a permanent narrowing of the cross-section occurs. The process ends with rupture. The position of each of these critical points reveals something about the sample's hardness, ductility, and strength. For example, a material, which shows no plastic behavior, is said to be brittle, with no deformation occurring beyond the elastic limit. The material breaks very abruptly. The mechanical behavior is intimately related to the interatomic bonding between the constituent atoms of the material.

1.2.1.2   Bulk Modulus

The bulk modulus is another important bulk mechanical property of solids and assumes a hydrostatic pressure applied to the sample, such as obtained when the sample is placed in a fluid. Assuming a constant temperature, the *isothermal bulk modulus*, K_T, is given by:

$$K_T = \frac{dp}{(-dV/V)} = -V_m \frac{dp}{dV_m} \tag{1.20}$$

The negative sign is incorporated so as to make the value of K_T positive. We can also relate this quantity to the interatomic potential energy of the constituent atoms. In this case, we start by using the first law of thermodynamics and then relate the change in the internal energy to the variation of the inter-atomic potential energy. We further note that the inverse of the bulk modulus is referred to as the *compressibility*, β, while the quantity $-dV/V$ is called the *volume strain*.

If we consider the molar volume V_m of a solid being subject to a hydro-static pressure, p, which causes a small change in its volume by an amount dV_m, then the first law of thermodynamics tells us that the internal energy of the solid changes by:

$$dU_m = -p\,dV_m \tag{1.21}$$

In the act of compression, the change of volume will be negative, this will mean that the internal energy will increase. Rewriting this expression in the form:

$$p = -\frac{dU_m}{dV_m} \tag{1.22}$$

We now substitute this into Eq. (1.20) to obtain:

$$K_T = -V_{m0}\left(\frac{d^2U_m}{dV_m^2}\right)_{V_m=V_{m0}} \tag{1.23}$$

where the evaluation of the isothermal bulk modulus is made at the equilibrium molar volume, V_{m0}. This is actually an approximation, since Eq. (1.20) defines the isothermal bulk modulus, while Eq. (1.22) assumes that the compression is adiabatic. This means that Eq. (1.23) should involve the adiabatic bulk modulus, K_S. However, at low temperature K_S and K_T approach a common value.

The internal energy, U_m, is intimately related to the interatomic potential, which allows us to model this quantity. Before doing so, it is useful to express the differentials of the internal energy with the nearest neighbor distance between atoms, a. This can be done as follows:

$$\frac{dU_m}{dV_m} = \frac{dU_m}{da}\frac{da}{dV_m} \tag{1.24}$$

and

$$\frac{d^2U_m}{dV_m^2} = \frac{d}{dV_m}\left(\frac{dU_m}{da}\frac{da}{dV_m}\right) \tag{1.25}$$

It is a simple matter to show that this yields:

$$\frac{d^2U_m}{dV_m^2} = \frac{dU_m}{da}\frac{d^2a}{dV_m^2} + \left(\frac{da}{dV_m}\right)^2 \frac{d^2U_m}{da^2} \tag{1.26}$$

We can express the molar volume in a generic form as:

$$V_m = cN_A a^3 \tag{1.27}$$

which leads to:

$$\frac{dV_m}{da} = 3cN_A a^2 \tag{1.28}$$

Here, c denotes a constant which depends on the crystalline structure. It is possible now to express the second derivative of the internal energy with respect to the molar volume as:

$$\frac{d^2U_m}{dV_m^2} = \frac{1}{9c^2 N_A^2 a_0^4}\left(\frac{d^2U_m}{da^2}\right)_{a=a_0} \tag{1.29}$$

Substituting the above into Eq. (1.23) allows us to write:

$$K_T = \frac{1}{9c^2 N_A^2 a_0}\left(\frac{d^2U_m}{da^2}\right)_{a=a_0} \tag{1.30}$$

To make an estimate, we need to model the molar internal energy, $U_m(a)$. A Mie potential can be used for the case of an ionic crystal such as NaCl, which we can express in the form:

$$U_m(a) = N_A\left(-\frac{e^2 A}{4\pi\varepsilon_0 a} + \frac{B}{a^n}\right) = -\frac{N_A e^2 A}{4\pi\varepsilon_0 a}\left(1 - \frac{1}{n}\right) \tag{1.31}$$

where we use $B = (Ae^2/4\pi\varepsilon_0 n)a_0^{n-1}$ and evaluate at $a = a_0$, to give:

$$U_m(a_0) = -\frac{N_A e^2 A}{4\pi\varepsilon_0 a_0}\left(1 - \frac{1}{n}\right) \tag{1.32}$$

We now obtain the relation:

$$K_T = \frac{1}{4\pi\varepsilon_0}\frac{Ae^2(n-1)}{9ca_0^4} \tag{1.33}$$

So for NaCl, where $A = 1.748, n = 9, a_0 = 2.81$ Å, and $c = 2$, we obtain $K_T = 2.87 \times 10^{10}$ Pa, which is in good agreement with the measured value of $K_T = 3.0 \times 10^{10}$ Pa.

1.2.1.3 Shear Modulus

Forces, which act along the direction of the plane of a surface, or area, A, are referred to as *shear* forces, subjecting the solid to a shear stress, defined in the same way as the elastic or tensile stress, as F/A, where we account for the direction of the applied force. This is shown in Figure 1.4. The shear strain

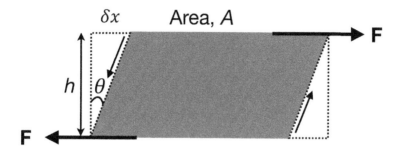

FIGURE 1.4   The application of a force, F, to a crystal, along the direction parallel to its plane of area, A. This results in a shear stress of F/A, and a shear strain of $\delta x/h$.

is defined as the ratio $\delta x/h$, forming a small deformation angle of $\theta \simeq \delta x/h$. The ratio of the shear stress to shear strain defines the shear modulus:

$$G = \frac{F/A}{\delta x/h} = \frac{Fh}{\delta x A} = \frac{F}{\theta A} \qquad (1.34)$$

The shear modulus is also referred to as the *modulus of rigidity*. It will be noted from Figure 1.4 that to prevent the solid body from moving, a system of forces is required: the top and bottom forces are equal and opposite, while an equal and opposite couple is also applied to prevent the rotation of the solid.

Inside a crystal, the shearing force attempts to move two crystal planes of atoms past one another. By considering the force necessary to displace one plane with respect to another, we can formulate a model for the displacement of the planes by one interatomic separation, from equilibrium position to the next, a distance b. These positions are characterized by a zero restoring force, which will also be the case at the mid-point. In the first half of the displacement, a restoring force will act to move the atom plane back to its starting position, while after the mid-point, this will act to push the plane to the new (nearest) equilibrium position. It is thus possible to deduce a simple displacement model for the shear stress as a function of the position x, as follows:

$$\sigma_s = C \sin \left(\frac{2\pi x}{b} \right) \tag{1.35}$$

where C is a constant, whose value is determined by noting that for small values of x, the above expression can be written as:

$$\sigma_s = C \frac{2\pi x}{b} = C \frac{2\pi a}{b} \left(\frac{x}{a} \right) \tag{1.36}$$

where a is the intercolumnar separation. Since the strain can be expressed as x/a, we can write the shear modulus as $G = C 2\pi a/b$, this allows us to write the constant as $C = Gb/2\pi a$. Substituting into Eq. (1.35) yields:

$$\sigma_s = \frac{Gb}{2\pi a} \sin \left(\frac{2\pi x}{b} \right) \tag{1.37}$$

Defining the critical point of the shear stress as that when the atoms of one plane slip over the adjacent plane of atoms, $x = b/2$, we can then write the critical shear stress as the maximum value of Eq. (1.37):

$$\sigma_{sc} = \frac{Gb}{2\pi a} \tag{1.38}$$

Given that a and b are almost equal, the fracture of a crystal should occur, when the shear stress $G/2\pi$ is reached. This is an approximation and somewhat overestimates the fracture point in most solids. This is because we have assumed a sinusoidal variation of the shear force. Also, this model does not account for the effects of defects in the solid as well as grains, etc. All such defects will usually reduce the critical value of shear stress.

Indeed, the application of mechanical forces usually generates defects in the solid, typically in the form of slip planes and dislocations. We will not discuss these issues here. A number of textbooks on Solid State Physics and Materials Science deal with these questions.

1.2.1.4 Poisson's Ratio

Another quantity, that is frequently used in the characterization of the mechanical properties of solids, is the *Poisson ratio*, which is expressed as the ratio of the contractile strain with the linear, or tensile strain:

$$v = -\frac{\varepsilon_t}{\varepsilon_l} \tag{1.39}$$

This accounts for the deformation of the body both along the direction of the applied force as well as that in the transverse direction. Under a tensile force, for example, the body will lengthen, leading to a positive linear strain ε_l. Frequently, this extension in the body is accompanied by a small reduction of the dimension across the body, leading to an effective reduction in the cross-sectional area. For most solids, v has a value somewhere in the range of 0.2 to 0.4. For the case where $v = 0.5$, there will be no change in volume in tension or compression. The case of rubber is very close to this situation. There are rare exceptions to the normal behavior. However, materials which, when stretched, become thicker perpendicular to the applied force have a negative value for the Poisson ratio. This occurs due to their particular internal structure and the way this deforms when the sample is uniaxially loaded. Such materials are called *auxetics*. The term auxetic derives from the Greek word $\alpha \upsilon \xi \eta \tau \iota \kappa o \varsigma$ (auxetikos), which means "that which tends to increase" and has its root in the word $\alpha \upsilon \xi \eta \sigma \iota \varsigma$ or auxesis, meaning "increase" (noun).

It is often useful to express the different moduli with respect to each other and this can be achieved using Poisson's ratio. For isotropic materials, these can be expressed as:

$$G = \frac{E}{2(1 + v)} \tag{1.40}$$

and

$$K_T = \frac{E}{3(1 - 2v)} \tag{1.41}$$

For the normal range of values for v, we can see that the values of K_T and E are comparable. In Table 1.1, we indicate some of the mechanical properties of common solids. We note that, in general, materials display anisotropic mechanical properties, which means that their properties depend on the direction of measurement with respect to the crystalline axes. In some cases, the range of values reflect this variation.

1.2.1.5   Testing Methods

The methods of measuring the mechanical properties of solids are indeed numerous and we will not discuss them in any detail here. Interested readers should have no problem in finding such information in the many textbooks and webpages available. Suffice it to say that there are various techniques available. For example, in what concerns Young's modulus, this can be measured directly, for example using Searle's method. Other techniques available

TABLE 1.1 Mechanical Properties of Selected Solids (*Graphene has been reported to exhibit auxetic properties)

Material (at 293 K)	K_T $(\times 10^9)$ Pa	E $(\times 10^9)$ Pa	G $(\times 10^9)$ Pa	v
aluminum	74	71	26	0.34
Copper	138	130	48	0.33
Iron	170	211	82	0.21–0.26
Lead	46	17	5.5	0.44
Zinc	72	109	43	0.25
Brass	112	100	37	0.33
Mild steel	169	212	82	0.27–0.31
Glass (crown)	40	70	30	0.18–0.3
Silicon		129–187	64.1	0.22–0.28
Silicon carbide	100–176	90–137	32–51	0.35–0.37
Diamond	442–590	910–1250	442	0.1–0.29
Graphene		~1000		*
SWNT	191	971	436–478	0.28

include the loading of a static beam, torsion measurements, a resonant beam and bulge tests using a pressurized membrane made of the test material. Inherent in such methods is a measurement of the forces applied and the static and dynamic response of the specimen under test. This requires a measurement of the dimensions of the sample or a resonance frequency etc. The bulk modulus requires a homogeneously applied hydrostatic pressure and the measurement of the sample dimensions. Alternatively, the sample can be made into a tube form and subject to internal pressure, but an external measurement of its dimensions is still required and the evaluation of the bulk modulus must take into account the geometrical shape of the tubular sample, such as interior and exterior radii. The shear modulus will typically require some form of torsion measurement

In addition to the mechanical properties given above, there are a number of other mechanical properties, which classify the hardness, toughness, resistance, and creep. All of which can be important factors in whether a particular material is suited to a specific structural task. There are a number of test methods, which are used to evaluate these properties, in addition to the loading tests indicated above. Bending, scratching and indentation tests are common in materials testing experiments.

For micro- and nanostructures, the testing methods are more complex, as we noted earlier, due to the practical problems associated with making physical contact with the object and observing its response. It is useful to distinguish between on-chip test methods and property tests. Testing methods, therefore, require much planning and test structures can be built along with the testing structure on the same chip. One of the most useful structures is the cantilever, which provides a model system for many testing methods, as well as being the basis of the structure of many scanning probe methods. We will discuss this in the following section.

1.2.2   The Cantilever and Its Applications

The cantilever is a diving board like structure and has a particular interest in the study of the mechanical properties of solids. The cantilever has been traditionally used as a model structure for the determination of the mechanical properties of materials, and in particular Young's modulus. In the case of nanostructures and microstructures, the cantilever has found an enormous interest linked to its use in the construction of the tips used for the various types of scanning probe microscopy. For this reason, we will take a look in more detail at some of the physical properties that can be extracted from this type of structure. At the micro- and nanoscale, very small physical factors such as surface stresses are enough to substantially bend the cantilever structure. This bending of cantilevers can be further measured to determine surface activity. In Figure 1.5, we show some examples of SEM images of cantilever systems used in MEMS and NEMS applications.

As we noted above, the cantilever has some important applications, with the AFM and other SPM tips being one of the most important in recent times. In Section 5.3 of Volume 1, we outline some of these techniques. We noted that the AFM has several modes of operation, which depend on either a static deflection of the tip, due to the tip-sample interaction, or a dynamic mode, where the tip is driven at or near its natural resonance frequency. These methods show that both static and dynamic modes of operation are extremely sensitive and allow us to infer the various properties of a sample surface just by scanning a tip mounted on a cantilever over the surface.

In the static mode, the bending of the cantilever under the force applied can be studied from classical mechanics. For example, the deflection of the end of a cantilever, of length L, with width w and thickness t, subject to a force F applied at the tip, or end of the cantilever, can be expressed as:

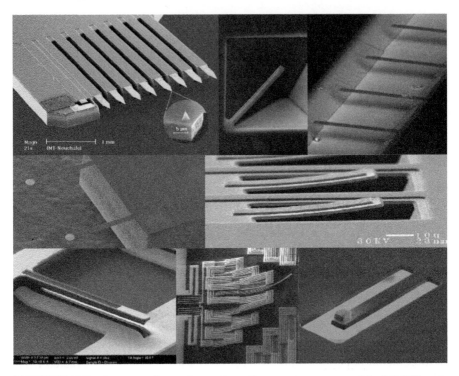

FIGURE 1.5 Images of a variety of cantilever structures used in MEMS and NEMS applications.

$$\delta = \frac{4FL^3}{Ewt^3} \tag{1.42}$$

Other formulae apply if the loading of the cantilever is anywhere else. If we substitute for Hooke's law, $F = kx = k\delta$ in this expression, then we can relate the spring constant, k with Young's modulus as:

$$k = \frac{Ewt^3}{4L^3} \tag{1.43}$$

The deflection of the cantilever beam can be approximated as the arc of a circle, from which the radius of curvature, R, is given as:

$$\frac{1}{R} = \frac{6(1-v)}{Et^2} \tag{1.44}$$

This allows then the surface stress change to be expressed in the form:

$$\Delta\sigma = \frac{Et^2}{6R(1-v)} \tag{1.45}$$

The relation between the cantilever deflection and the stress can be expressed as:

$$\delta = \frac{3\sigma(1-v)L^2}{Et^2} \tag{1.46}$$

This equation is known as Stoney's formula. Since the deflection of the cantilever is an easily measurable quantity, this provides a reliable method for the determination of the elastic properties of the cantilever material in the microscopic domain. The detection of deflection can be performed in the same way that deflection is detected in the AFM for example. This means the deviation of a laser beam reflected from the surface of the cantilever beam is made using a quadrant detector for example. There are a number of different types of cantilevers that can be used to perform sensing and actuating functions. Sensors can be made in a number of ways. For example, the diffusion of molecules onto a polymer layer can lead to swelling in the polymer and eventually a bending of the cantilever beam. Coating one side of the cantilever with specific receptors of biomolecules will also change the surface stress of the coated side. Once again, this will result in the bending of the beam. If the cantilever is fabricated from a bimetallic layer, the difference in thermal expansion coefficients will mean that it is very sensitive to changes in temperature and can be used as a heat sensor. Alternatively, the heating of the cantilever can also be exploited as an actuator, bending the cantilever under the application of heat.

In MEMS and NEMS cantilever structures, the use of piezoresistive materials enables some very interesting possibilities. The upward to the downward bending of the cantilever will exhibit a change of electrical resistance, thus providing a very simple measurement of the beam deflection and hence the stresses and strains in the structure. Piezoresistive MEMS and NEMS cantilevers can be functionalized, in the same way as nanoparticles, and used as sensors for various applications. The key is to selectively activate one surface of the cantilever by coating it with a compound, which shows specific affinity to the target molecules also called analyte molecules. The opposing side of the cantilever can be passivated with chemically passive compounds, so that it does not react with the analyte. In the presence of analyte molecules, the cantilever is subjected to differential surface stress due to the active and passive sides and bends accordingly. This bending causes a change in the resistance of the cantilever structure, which is electrically measured and recorded. This provides a strong basis for cantilever sensors for the detection, analysis, and characterization of surface chemical interactions. Arrays of cantilever sensors can be fabricated to detect a number of different compounds in a controlled environment. This is schematically illustrated in Figure 1.6.

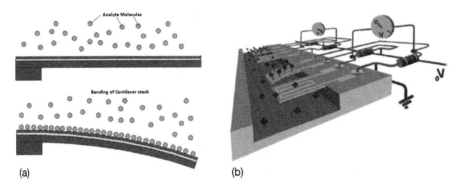

(a)   (b)

FIGURE 1.6   Schematic illustration of the detection of analyte molecules on piezoresistive cantilever structures. (a) The adsorption of the analyte molecules on one surface of the cantilever induces a differential surface stress causing the cantilever to bend. (b) The bending causes a resistance change in the piezoresistive material allowing the detection of the analyte to be measured electrically.

In dynamic mode, we will drive the cantilever to resonance at its natural frequency or one of its harmonics. Once again, the cantilever provides an excellent basis for sensing applications, which often rely on mass changes due to the adsorption of molecules. In the case of the AFM, we saw that the tapping mode relies on its sensitivity to the changes in the effective spring constant of the cantilever due to the tip-surface interaction. Such effects will cause a shift in the resonance frequency of the cantilever.

The derivation of the frequencies of the natural modes or resonance for a cantilever beam, fixed at one extremity, can be deduced using the Euler–Bernoulli beam theory or classical beam theory. This approach is valid for small deflections of a beam which is subject to a lateral load and was first enunciated in the mid-eighteenth century. This equation can be expressed as:

$$EI\frac{\partial^4 u}{\partial x^4} + \mu\frac{\partial^2 u}{\partial t^2} = \mathscr{F} \tag{1.47}$$

where $u(x)$ describes the deflection of the beam in the z direction at some position x, μ denotes the mass per unit length of the beam, E is Young's modulus and I is the area moment of inertia of the beam, taken as: $I = \int \int z^2 \, dy \, dz$. The quantity EI is collectively known as the *flexural rigidity*. The term on the right hand side of Eq. (1.47) is the distributed load, measured as the force per unit length, $\mathscr{F}$. For the case of a free vibration, i.e., where $\mathscr{F} - 0$, the general solution of this equation allows the natural frequencies of vibration to be evaluated and we can express these resonance frequencies as:

$$f_n = \frac{\omega_n}{2\pi} = \frac{\beta_n^2}{2\pi}\sqrt{\frac{EI}{\mu}} = \frac{\beta_n^2}{2\pi}\sqrt{\frac{EIl}{m}} \tag{1.48}$$

where β_n are constants, which depend on the vibrational mode, characterized by n, where $n = 1$ corresponds to the fundamental mode. As we mentioned earlier, one of the possible application of the cantilever in MEMS is its use as a detector for gases. Since the coating of the cantilever will imply a change of the effective mass of the cantilever, this will cause a shift in the resonance frequency, which is clearly a measurable quantity. From Eq. (1.48), we can determine the sensitivity of such as device. From the derivative of the frequency with respect to the mass, m, we find that the frequency change per mass change takes the form:

$$\frac{\Delta f_n}{\Delta m} = -\frac{\beta_n^2}{4\pi}\sqrt{EIlm} \tag{1.49}$$

We can now evaluate the mass change from the shift in the resonance frequencies before the mass change, $f_{n,(0)}$ and with the added mass, $f_{n,\Delta m}$, which gives:

$$\Delta m = \frac{\beta_n^4 EIl}{4\pi^2}\left(\frac{1}{f_{n,\Delta m}^2} - \frac{1}{f_{n,(0)}^2}\right) \tag{1.50}$$

We can understand the changes in the resonance frequency for the AFM tip due to the interaction between tip and surface in a similar manner. In this case, it is convenient to express the resonance frequency in terms of the spring constant of the cantilever, k. For the case of forced oscillations, and taking into account the damping inherent in the cantilever system, we can express the fundamental resonance frequency in the form (Heinrich and Dufour, 2015):

$$f = f_0\sqrt{1 - \zeta^2} = \frac{1}{2\pi}\sqrt{\frac{k}{m}(1 - \zeta^2)} \tag{1.51}$$

where $f_0 = (1/2\pi)\sqrt{k/m}$ is the undamped natural frequency, $\zeta = c/2\sqrt{km}$ and here c is a damping coefficient. So for a stiffening of the cantilever due to the tip-surface interaction, we can write the frequency shift as:

$$\delta f = f' - f = \frac{1}{2\pi}\sqrt{\frac{(1-\zeta^2)}{m}}(\sqrt{k'} - \sqrt{k}) \qquad (1.52)$$

Or alternatively, we can express the change in the spring constant in terms of the resonance frequencies as:

$$\delta k = k' - k = \frac{4\pi^2 m}{(1-\zeta^2)}(f'^2 - f^2) \qquad (1.53)$$

From the above, we can see that for increased tip-surface interaction, we expect that the cantilever should become stiffer, increasing the spring constant and therefore increasing the resonance frequency. Another consequence of the tip-surface interaction will be a modification of the mode shape. This will alter the displacement of the cantilever tip, an effect, which scales to the fourth power of the inverse wave number of the vibrational mode (Verbiest and Rost, 2016). Therefore, the higher the mode order, the less sensitive it is. The fundamental mode will be the most sensitive, having the largest displacement. This, therefore, indicates that the most reliable method for imaging in the tapping mode will be via this mode.

The load distribution on the cantilever can have a significant effect on the resonant modes and thus the frequencies. Loads will depend on the application of the cantilever. For example, in the case of AFM and SPM, a good approximation will be the end loading at the extremity of the cantilever. The expressions given above are made for this case. For other applications, such as molecular sensing, we can expect a more distributed load and hence modified expressions for the frequencies of resonance. These effects are considered by Zhang et al., (2017).

In addition to the operation of the cantilever in a vacuum or gaseous environment, it is possible also to perform measurements in a liquid. The dynamic mode operation in a liquid will be more difficult due to the larger damping forces at play due to the viscous medium. This will also severely affect the quality factor of the oscillation, making it more difficult to track changes in the resonance frequency. This will clearly deteriorate the measurement sensitivity. The increased damping will cause a shift in the measured resonance frequency to lower values with respect to the undamped case, see Figure 1.7. Also illustrated is the phase shift between driving force and cantilever tip, which is also strongly dependent on the damping factor, with

$$\phi = \arctan\left(\frac{-2\zeta r}{1 - r^2}\right) \tag{1.54}$$

where $r = f/f_0$, f_0 being the undamped resonance frequency.

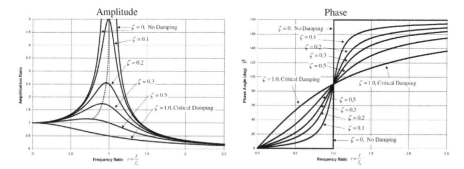

FIGURE 1.7 Forced vibration response for the amplitude of vibration and the phase shift as a function of frequency for different damping coefficients. As the damping increases, the peak height reduces and shifts to lower frequency values.

The quality factor for resonance is defined as:

$$Q = \frac{f_{res}}{\Delta f} \tag{1.55}$$

where Δf denotes the resonance linewidth at half-maximum. The quality factor is also related to the damping coefficient:

$$Q = \frac{1}{2\zeta} = \frac{\sqrt{km}}{c} \tag{1.56}$$

For well designed MEMS cantilever systems, the quality factor is typically in excess of 5.

1.3   Nanomechanical Properties of Carbon Nanotubes

Carbon nanotubes (CNTs) were introduced in Section 9.2 of Volume 1, where we outlined some details of their structural forms and general properties. Here, we will outline some of their mechanical properties and applications. As we noted earlier, CNTs offer quite unique physical properties and are easy to produce in large quantities under controlled environments. Their exceptional mechanical properties make them ideal candidates for the construction of NEMS structures and CNTs have also been used as strengthening additive components for composite materials.

The mechanical properties of CNTs are closely related to those found in graphite. The stiff sp^2-hybridized in-plane σ-bonds, which have a length of around 1.42 Å, provide strong links between the atoms and are the origin of the large values of Young's modulus found in these and related structures. The out-of-plane π-bonds are much weaker and are of van der Waals nature. These bonds are responsible for the links between different layers, such as found in multilayered graphene and MWNT systems. This is also the case for the bonding between SWNTs in the formation of bundles. Graphite is considered to be mechanically weak, and this is indeed due to the van der Waals forces between the hexagonal layers that are easily sheared. However, graphite has similar mechanical properties to graphene and CNTs along the planes. This is also true for the shearing properties that exist between the carbon nanotubes in the CNT bundle structures.

The electronic and thermal properties of CNTs and their related graphene structures are understood in terms of their structural arrangement and can be studied using band theory, which is derived from quantum physics. Despite this, the mechanical properties can be adequately expressed in terms of classical physics. The reason for this is because Hooke's law can be applied to a series expansion of any potential around the equilibrium position. This can be seen from the expression given in Eq. (1.10). The mechanical properties of SWNTs were given in Table 1.1. The Young's modulus, in this case, has been evaluated by many researchers, where a value of around 1 TPa is found. This value appears to be independent of the chirality and diameter (0.68–27 nm) of the CNT, (Lu, 1997; Hernández et al, 1998).

Early measurements of the mechanical properties of CNTs were performed using thermally induced vibrations in a TEM and measuring the amplitude of these vibrations (Treacy et al., 1996). In these measurements, one end of the CNT is fixed, while the other appears as a blurred image, due to the thermal excitation. Similar measurements were performed in a scanning electron microscope using an electric field to drive resonance (Dikin et al., 2003). The first two vibrational modes (fundamental and its second harmonic) are shown in an SEM image in Figure 1.8. The motion of the nanotube was modeled as a stochastically driven resonator, where Young's modulus is estimated from the Gaussian vibrational profile, whose standard deviation is expressed as:

$$\sigma^2 = \frac{16L^3 k_B T}{\pi E (D^4 - D_{int}^4)} \sum_n \beta_n^{-4} \tag{1.57}$$

where L denotes the length of the nanotube, D and D_{int} are in the outer and inner diameters. The factor β_n is a numerical constant, which depends on the vibrational mode number, n.

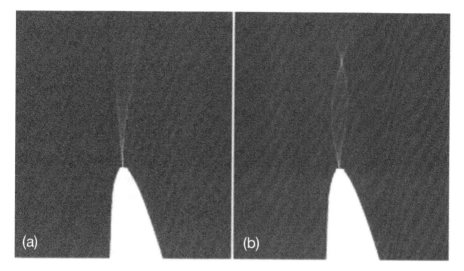

FIGURE 1.8 SEM images of the fundamental (a) and its second harmonic (b) vibrational modes due to an oscillating electric field in individual MWNTs. Reproduced from Rodney S. Ruoff, Dong Qian, & Wing Kam Liu, (2003). Mechanical properties of carbon nanotubes: theoretical predictions and experimental measurements, *C. R. Physique, 4*(9), 9931008. ©(2003) Elsevier Masson SAS. All rights reserved.

An alternative method was used by Poncharal et al. (1999), who studied MWNTs excited by an alternating electric field and Young's modulus related to the resonant frequencies of the first two-mode of oscillation. Here, the CNTs were fixed on a specially adapted TEM sample holder equipped with a piezoelectric stage. The application of an electric field is then converted to a mechanical motion, which drives resonance. Using the Bernoulli–Euler analysis, the frequencies are obtained as:

$$v_n = \frac{\beta_n^2}{8\pi L^2} \sqrt{\frac{(D^2 + D_{int}^2)E}{\rho}} \qquad (1.58)$$

where the symbols all have their usual meaning. For tubes of 10 nm diameter, values of Young's modulus of 1 TPa were obtained, while larger diameters exhibited reduced values down to 100 GPa.

The use of AFM methodologies was first applied by Wong et al. (1997), who measured MWNTs, which were deposited on smooth, low friction MoS_2 single crystal surfaces. Friction was further reduced by performing experiments in water. The nanotubes were fixed by depositing an array of square pads through a shadow mask, see Figure 1.9.

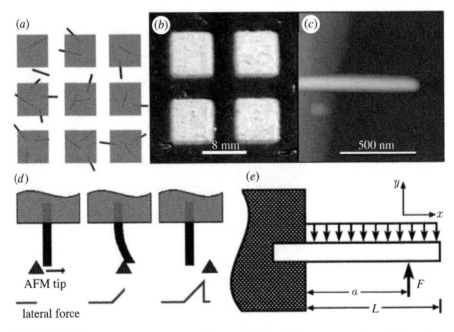

FIGURE 1.9 (a) Nanotubes are dispersed on a substrate and pinned down by SiO pads. (b) Optical micrograph of the sample. (c) AFM image of a SiC nanorod protruding from the pad. (d) The tip moves in the direction of the arrow. The lateral force is indicated at the bottom. During bending, the lateral force increases until the point at which the tip passes over the beam, which snaps back to its initial position. (e) The beam of length L is subjected to a point load F at $x = a$ and friction force f. From E. W. Wong, P. E. Sheehan, & C. M. Lieber, (1997). *Science, 277,* 1971. Reprinted with permission from AAAS.

The AFM was used to locate and measure the dimensions of the nanotubes as well as to apply a series of forces to the CNTs, which were laterally deformed and then allowed to snap back to their relaxed position. Force–distance curves were then measured as a function of position along with the structure. The lateral force can only be deduced as a proportional factor since the AFM cantilever was not calibrated. Uncertainty in the force could be eliminated by modeling the CNT as a cantilever and calculating the lateral force from the relation:

$$\frac{dF}{dy} \equiv k = \frac{3\pi D^4}{64x^3} E \qquad (1.59)$$

where k is the spring constant and x is the position along the nanotube. The experimental results are shown in Figure 1.10. The mean value of Young's modulus measured from the fit to Eq. (1.59) was obtained as $E = 1.3 \pm 0.6$ TPa.

By measuring the vertical deflections of nanotubes bridging the gaps in a porous membrane, Salvetat et al. (1999) were able to study the mechanical properties of SWNTs, SWNT bundles, and MWNTs. The nanotubes were deposited on porous alumina substrates and AFM measurements were performed as a function of an applied load. The deflection δ at the midpoint can be related to the applied normal force via the relation:

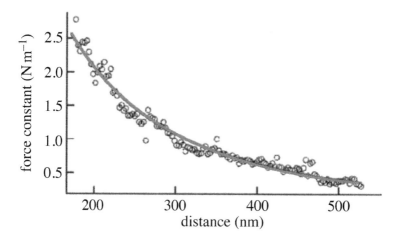

FIGURE 1.10 The lateral spring constant as a function of position on the beam. The curve is a fit to Eq. (1.59). From E. W. Wong, P. E. Sheehan, & C. M. Lieber, (1997). *Science, 277,* 1971. Reprinted with permission from AAAS.

$$\delta = \frac{FL^3}{192EI} \qquad (1.60)$$

where L here refers to the suspended length of the nanotube. In Figure 1.11(a), we show the lines of a CNT suspended over a pore. Once again, experimental values for Young's modulus of SWNTs are in agreement with previous measurements, with $E = 1$ TPa. The measurements on MWNTs showed little dependence on the diameter, with a mean Young's modulus of 870 GPa. This reduced value of Young's modulus is probably due to the high concentration of defects known to occur in MWNTs. Indeed, for catalytically prepared MWNTs, defect concentrations can be very high and measurements show that this leads to very low values of Young's modulus, down to around 12 GPa (Salvetat et al., 1999; Lukić et al., 2005). Also illustrated in Figure 1.11(b) is a plot of the bending (Young's) modulus versus the diameter for bundles of 12 SWNTs. In this case, it is necessary to further take into account the van der Waals forces between the CNTs in the bundles. These are seen to be strongly dependent on the CNT diameter in the bundle, where

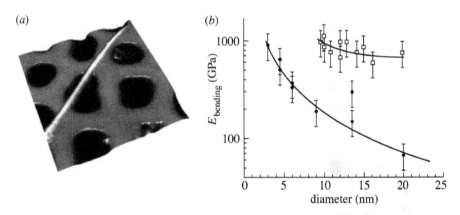

(a)

(b)

$E_{bending}$ (GPa)

diameter (nm)

FIGURE 1.11   (a) Three-dimensional rendering based on an AFM image of a 10 nm thick SWNT bundle on a porous substrate used by Salvetat et al. for their mechanical measurements (Salvetat et al., 1999a). (b) Values of the bending modulus for 12 SWNT bundles (filled circles) of different diameters. The measured $E_{bending}$ of thin bundles corresponds to Young's modulus, while for thick bundles, one obtains the value of the shear modulus G, (Salvetat et al., 1999c). MWNT (open squares) data are for arc-discharge grown tubes. Reprinted figure with permission from J.-P. Salvetat, G. Andrew D. Briggs, J.-M. Bonard, R. R. Bacsa, A. J. Kulik, T. Stöckli, N. A. Burnham, & L. Forró, (1999). *Phys. Rev. Lett. 82*, 944. ©(1999) by the American Physical Society.

the shearing forces are seen to be a weak link in the structure. This shows that in the rope, the CNTs behave as an assembly of individual tubes rather than a thick beam. The deflection, in this case, is expressed as:

$$\delta = \delta_b + \delta_s = \frac{FL^3}{192EI} + f_s\frac{FL}{GA} = \frac{FL^3}{192E_{\text{bending}}I} \quad (1.61)$$

where f_s denotes the shape factor (= 10/9 for a cylinder), G is the shear modulus and A the cross-sectional area of the beam. E_{bending} is the effective bending modulus, which is equal to Young's modulus, when the shear modulus can be neglected. For the thin ropes, the bending modulus approaches the value of Young's modulus for SWNT, while for thicker rope structures, the value tends towards the shear modulus, where shearing dominates the behavior.

The first direct measurement of the elastic properties of carbon nanotubes which were not based on bending was performed by Yu et al., (2000). In these studies, both SWNTs and MWNTs were measured under tensile strain, where the ends of the CNTs were attached to the ends of two AFM tips and observed inside a scanning electron microscope, see Figure 1.12(a) and (b). One of the tips was integrated with a rigid cantilever, of spring constant 20 Nm^{-1}, while the other was more compliant, having a spring constant of 0.1 Nm^{-1}. The rigid end was driven by a linear piezoelectric motor and the compliant end was bent under the applied tensile load. The measurement of the deflection of the compliant tip allowed the strain in the nanotube to be obtained under the applied force. Stress–strain curves were obtained, see Figure 1.12(c), from which Young's modulus was obtained, with values ranging

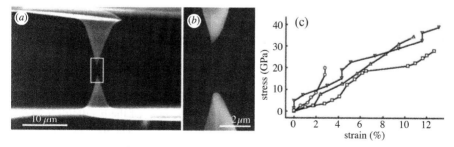

FIGURE 1.12 (a) A SEM image of MWNT mounted between two opposing AFM tips. (b) A close-up of the region indicated by a rectangle in (a). (c) A plot of stress versus strain curves for different individual MWNTs. Reprinted figure with permission from M.-F. Yu, B. S. Files, S. Arepalli, & R. S. Ruoff, (2000). *Phys. Rev. Lett., 84,* 5552. ©(2000) by the American Physical Society.

from 0.27–0.95 GPa. The examination of the broken ends of the nanotubes in a TEM revealed that the rupture of the nanotubes was via a *sword-in-sheath* mechanism, where only the outer layers of the MWNT carried the load.

It is interesting to note that while SWNTs are highly rigid in the axial direction, they can experience considerable structural instabilities in compression, torsion and bending in the radial direction. Buckling has been observed experimentally as well as being modeled using the Tersoff–Brenner potential (Yakobson et al., 1996). Their findings show that nanotubes subject to large deformations can reversibly switch into four different morphological patterns, each of which corresponds to an abrupt release of energy and a singularity in the strain energy vs. strain curve.

Under static or slow tensile conditions, quantum molecular dynamics simulations and dislocation theory predict a mechanical relaxation that occurs via a bond rotation known as a *Stone–Wales transformation*, (Stone and Wales, 1986). This is a defect that causes a local increase in the unit cell, as illustrated in Figure 1.13, and is the lowest energy defect, made up of two pentagons and two heptagons (5/7/7/5). This defect represents a dislocation

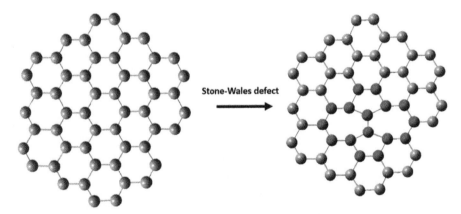

FIGURE 1.13   The Stone–Wales (5/7/7/5) defect in a graphene sheet.

dipole. These can be mobile at high temperatures, separating off in two 5,7 pairs that glide off leaving a plastically deformed wake in the nanotube. Under further tension, the Stone–Wales defect acts as a weak point in the structure allowing the formation of disordered cracks and large open rings, which will eventually lead to the complete rupture of the structure. In chiral structures, the passage of the Stone–Wales defects can lead to a small amount of ductility in the CNTs, which would normally fail in a brittle manner (Buongiorno Nardelli et al., 1998). It has been suggested that the Stone–Wales mechanism can also result from compression, leading to plastic deformation in SWNTs (Yakobson, 1998). However, Srivastava et al. (1999) have proposed another mechanism under compression, as modeled using a generalized tight-binding molecular dynamics method, which produces plastic deformation from collapsed graphitic sp^2 to a localized sp^3 diamond-like reconstruction.

Superplasticity has been observed in SWNTs by Huang et al. (2006), who studied tensile loading of nanotubes, exhibiting extraordinary elongations of up to 280%, with a 15-fold reduction in its diameters. The superplastic deformation occurred at high temperature under large bias voltages, assumed to be around 2000°C, and was thought to result from the nucleation and motion of kinks, possibly 5,7 pairs, as well as atom diffusion. In contrast, room temperature tensile strain measurements, with no voltage bias, showed much lower values of the failing tensile strain of less than 15%.

1.4 Applications for Micro- and Nano-Systems

In this section, we will provide a brief overview of some of the principal applications of MEMS and NEMS devices. This is a very broad area of research and development and also a fast-moving field of study. Given the volume of applications in a vast range of fields, we are unable to provide anything more than a brief introduction to this subject.

Frequently, the applications of micro- and nanomechanical systems require multi-functionality as a working principle. By this, we mean that the device design will be based on the coupling or multi-use of the physical properties of a system. For example, an actuator based on a micromotor may require an electrical signal to displace an object. Thus, it depends on both the mechanical and electrical properties of the device. Such transfer of physical signals is an essential concept for devices and MEMS and NEMS devices in particular.

Before we look at specific examples of MEMS and NEMS type devices, it is interesting to put into context the scaling laws that are applicable to these types of devices. For mechanical systems, we have discussed in some detail the use of cantilever systems. The cantilever is a mechanical device having spring-like qualities that are of great importance when studying the mechanical properties of materials. They are also at the root of many mechanical devices, including the tip for scanning probe microscopies. When the linear dimensions of the device are reduced, say by a factor f, the volume, and therefore the mass of the object, will be reduced by a factor of f^3. However, the mechanical flexure is scaled only by the linear factor, with the mechanical stiffness being expressed as:

$$k = \frac{wt^3 E}{4L^3} \qquad (1.62)$$

where w denotes the cantilever width, L its length and t its thickness. This explains, why the scaling is favorable for such a mechanical device; the mechanical strength reduces more slowly (f) than the inertial force it generates (f^3). This means that the MEMS structure can withstand enormous accelerations without breaking or even deforming its static structure. However, this benefit is slightly offset by the need for making a proof mass (in accelerometers, see below) to make them effective as a motion-detecting device.

In fluids, the behavior in reduced dimensions can be very different from that on the macroscale. For example, the Reynolds number, which is a measure of the turbulence in the fluid flow (where $Re < 2000$ represents laminar flow and $Re > 4000$ indicates turbulent flow), is a function of the scale of the fluidic system, as given by:

$$Re = \frac{\rho v D}{\eta} \qquad (1.63)$$

where ρ is the fluid density, v its characteristic velocity, D the characteristic length or diameter, and η its viscosity. Given these parameters, we can often find turbulent behavior being dominant at the macroscale. However, in microscopic systems, the flow is almost exclusively laminar in nature. This means that as the linear dimensions decrease, Re also decreases by the same linear factor. This can cause some difficulties in fluid mixing in microfluidic systems. Despite Eq. (1.63) defining the expected behavior of fluidic systems, this equation is not completely correct and corrections to this are required (Judy, 2001).

For electrical systems, the IC industry has illustrated that the behavior of components such as resistors, capacitors transistors, etc. can be scaled in a well-understood manner. However, for the case of more complex devices, such as electrostatic actuators, a more detailed analysis is required. A figure of merit for actuators is given by the density of the field energy, U, that can be stored in the gap between rotor and stator. In the case of the electrostatic actuator, this energy density can be expressed as:

$$U_{es} = \frac{1}{2}\varepsilon E^2 \qquad (1.64)$$

where ε represents the permittivity and E the electric field strength. This shows that the maximum energy density of electrostatic actuators is limited by the maximum electric field before the electrostatic breakdown. In a macroscopic system, the maximum field is constant and roughly 3 MV m^{-1}, which gives an energy density of only about J m^{-3}. In the case of magnetostatic actuators, the corresponding field energy density is given by:

$$U_{es} = \frac{1}{2}\frac{B^2}{\mu} \qquad (1.65)$$

where μ denotes the permeability and B is the magnetic flux density or magnetic induction. The maximum energy density for the magnetic actuator is typically limited by the saturation flux density B_{sat}, which is generally in the range of about 1 T. The corresponding energy density is now 400,000 J m^{-3}, which is four orders of magnitude larger than the electrostatic case. We conclude that magnetic actuators have the potential of storing much larger energy densities in the gaps between the rotor and stator. This relative situation is maintained even at lower length scales. However, as the air gap becomes smaller, fewer ionization collisions occur and a larger field can be effectively applied to the structure before cascade breakdown occurs. This situation continues until a large voltage is necessary before a breakdown occurs. On the other hand, for size reduction in the magnetic actuators, other practical considerations come into effect such as resistive power losses in windings and magnetic domain sizes, which are typically on the scale of micrometers for soft magnetic materials, such as permalloy. This means the macroscopic model used above is no longer valid. In reduced dimensions, for magnetic devices, the behavior needs to take into account the domain structure, which is also shape-dependent, and the fact that the micro- and nanoscale behavior will be dominated by single domain-like phenomena.

1.4.1 Micro-ElectroMechanical Systems: MEMS

Micro-ElectroMechanical Systems, or simply MEMS, have become important devices that can now be found in many everyday applications. The fabrication of these systems can be performed using a number of the processing technologies we discussed in Chapter 6 of Volume 1. Of particular importance are the deep lithography methods, such as LIGA in conjunction with the SU8 resist as well as micromachining and etching technologies. These will allow the definition and formation of the required structures, which can be free-standing and free to move. MEMS devices are designed for the purpose of sensing, actuating and controlling structures on the microscale, with effects generated at the macroscale.

The nature of MEMS means that it relies heavily on interdisciplinary fields for the design, engineering, and manufacture of devices, with expertise from a broad spectrum of technical areas, including materials science and engineering, mechanical engineering, optics, fluid engineering, integrated circuit technologies, chemical engineering, instrumentation, biosensing, to name a few. The complexity found in MEMS is reflected in the broad range of applications, and here again, we have a long list of products and markets such as consumer electronics, automobile, aeronautical, medical, communications and defense technologies. More specifically, we can name a number of devices, which are currently of interest for a large spectrum of applications; accelerometers, gyroscopes, blood and tire pressure sensors, microphones, projection display chips, optical switches, RF components, biosensors, and microfluidic lab-on-a-chip applications, etc.

The basic mechanics of MEMS devices can act as miniature mechanical systems such as pumps, valves, gears gauges as well as motors. It would, however, be a mistake to just see MEMS as a miniature mechanical component. MEMS is a manufacturing technology for creating integrated devices and systems using batch fabrication techniques that are more akin to integrated circuit technologies. It is also important to note, that while the technology exists for the fabrication of miniature devices, the costs of production for complex microscale devices can be prohibitive for the implementation of their production on a large scale. Some examples of such systems are illustrated in Figure 1.14.

MEMS has some distinct advantages for the manufacturing industry. The interdisciplinary nature of the subject coupled to the diversity of their applications means that MEMS can span a broad range of devices with synergies across fields in an unprecedented way. The batch production of devices allows products to be fabricated with increased reliability and performance as

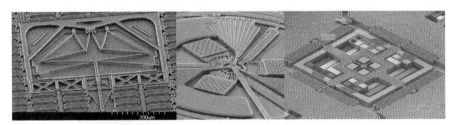

FIGURE 1.14   Some examples of MEMS structures.

well as taking advantage of size reduction, low volume, and weight as well as reduced costs per item. These factors are highly favorable for the future of the MEMS industry, with the MEMS market being expected to grow strongly. Despite this, MEMS is not expected to reach the same level of output as the IC industry, where the final product is the sum of many repeats of standard components such as transistors. Due to the complex nature of devices, MEMS are geared to more specific end-use applications and have structurally more complex procedures and this also makes them more expensive to produce. We can classify the extended range of devices for miniaturized systems as illustrated in Figure 1.15. As we can see, we can consider the interaction between different areas of study and their applications for microsystems, which include the mechanical, optical and electronic properties of materials. In this classification, MEMS is just one area in a larger area of device technologies. In this section, we include also the optical and electronic properties (MOEMS) of miniaturized systems. We will further include RF MEMS devices in this section to provide a more complete overview of the subject. MEMS has become a more or less universally applied name for all of these devices and is not really the most appropriate, as we can see. In the interests of clarity, we maintain this convention. Other terminologies refer to

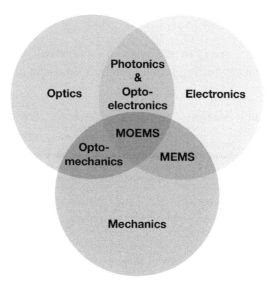

FIGURE 1.15 Classification of micromechanical systems in terms of physical properties and the interfaces between them.

a more generic name, such as microsystem technologies, or MST (Europe), or simply micro-machining (Japan).

One of the guiding principles for the operation of many MEMS type devices relies on a transduction mechanism, which is a physical effect that converts a signal from one form of energy to another form. This can include electrical, magnetic, mechanical, thermal, radiative and chemical forms of energy. There are many examples available, in which transductive signals have been employed in the operation of sensors and actuators. In Figure 1.16, we illustrate the various physical mechanisms that can be exploited for transduction conversion in sensor and actuator applications. It is common for more than one transduction mechanism to be used in series when performing sensing functions, with the final conversion being to an electrical signal. Conversely, for actuators, the opposite is true, where an electrical signal is applied and then converted to another form, and finally ending up with a mechanical signal to perform some motion operation. We will see some examples of these in the following, where we will provide a brief description of some of the more common MEMS devices and applications, where we also aim to give a brief description of the physical principles of operation of the devices.

1.4.1.1   Pressure Sensor

The pressure sensor is one of the earliest and simplest MEMS devices. These devices are used for measuring pressure in industrial and biomedical appli-

To / From	Electrical	Magnetic	Mechanical	Thermal	Chemical	Radiative
Electrical		Ampère's Law	Electrostatics, Electrophoresis	Resistive Heating	Electrolysis, Ionization	EM Transmission
Magnetic	Hall Effect, Mag. Resistance		Magnetostatics, Magnetostriction	Eddy Currents Hysteretic Loss	Magnetic Separation	Magneto-optics
Mechanical	Variable Cap. Piezoresistance Piezoelectricity	Magnetostriction		Friction	Phase Change	Tribo-luminescence
Thermal	Thermoelectric	Curie Point	Thermal Expansion		Reaction Rate Ignition	Thermal Radiation
Chemical	Electrochemical Potential	Chemomagnetic	Phase Change	Combustion		Chemo-luminescence
Radiative	Photoconductor, EM Receiving	Magneto-optics	Radiation Hardening	Photothermal	Photochemical	

FIGURE 1.16   Classification of transduction mechanisms that can be exploited for MEMS devices (Judy, 2001).

cations. The basic principle of operation is simple and typically based on piezoelectric, piezoresistive, capacitive, and resonant sensitive mechanisms. In either case, a membrane structure which covers a vacuum cavity is generally used. In the case of the piezoresistive detection, one common solution is to deposit four piezoresistive elements on the edges of the membrane, where variations in the ambient pressure (with respect to the enclosed cavity under the membrane) will cause the membrane to deform, see Figure 1.17. In this example, the device forms an on-chip integrated packaging-stress-suppressed suspension PS^3 structure. Stresses in the membrane are transferred to the

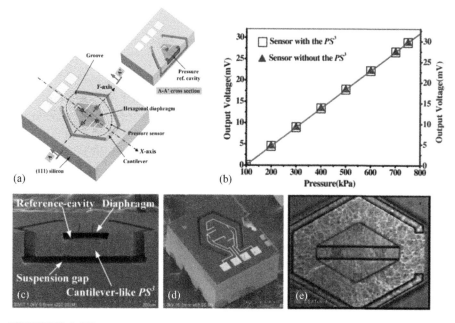

FIGURE 1.17 (a) 3D schematic of the pressure sensor integrated monolithically with the PS^3 structure. (b) Comparison of the sensor output between the sensor with the PS^3 structure and another sensor without the PS^3. Fabrication results of the pressure sensors with the PS^3 structure. (c) SEM image showing the cross-sectional cutting view of the sensor with the PS^3. (d) SEM of the fabricated pressure sensor chip with the PS^3. (e) Top-view infrared micrograph that confirms the formation of the hollowed vacuum reference-pressure cavity and the suspended PS^3 structure. Preprinted from Wang, J. & Li, X. (2013). *J. Micromech. Microeng., 23*, 045027. Reproduced with permission of the Institute of Physics Publishing in the format Book via Copyright Clearance Center.

piezoresistive elements causing them to change their electrical resistance. These sensing elements are connected in the form of a bridge with push-pull signals to increase sensitivity. The illustrated device can be made into the PS^3 configuration with no loss of functionality, exhibiting a sensitivity of 0.046 mV kPa^{-1} (3.3 V)$^{-1}$, as indicated in Figure 1.17(b).

Another scheme of detection for the MEMS pressure sensor is via a capacitive measurement between the silicon wafer and the diaphragm, which act as the plates of a capacitor on either side of the vacuum cavity. The pressure above the cavity will create a force on the diaphragm, as in the previous case, changing the effective separation between the plates of the cavity. The capacitance can be modeled as a parallel plate capacitor, which can be expressed as:

$$C_p = \frac{\varepsilon_r \varepsilon_0 A}{x} \tag{1.66}$$

where A represents the area of the diaphragm and x is the separation between the parallel plates. We note that for air or vacuum, the relative permittivity (dielectric constant) will be unity. The basic structure of the capacitive pressure sensor is shown in Figure 1.18(a). This device has a number of attractive features, such as its simplicity in design and fabrication, it is virtually temperature independent and consumes zero DC power. Furthermore, it can be easily incorporated into CMOS device circuits with readily interfaced signal processing. Despite this, the analysis of the mechanical response and the capacitance versus pressure is rather complex, but can be performed by computer simulations, as illustrated in Figure 1.18(c) and (d), (Miguel et al., 2018). In this device, the capacitance is modeled as:

$$C(p) = \int \int \frac{\varepsilon_r \varepsilon_0 \, dx \, dy}{t_g - w(r, \theta, p)} \tag{1.67}$$

where the integration is over the area of the diaphragm, t_g is the gap distance between the plates, w is the deflection of the plate at any position on the diaphragm, and p denotes the distributed load applied to the upper surface of the plate, which will depend on the pressure difference on either side of the diaphragm. This can be found from the classical analysis of a deflected diaphragm, see Miguel et al., (2018), and expressed as:

$$w(r, \theta, p) = \frac{pr^2}{8\pi D} \ln\left(\frac{a}{r}\right) + \frac{p}{16\pi D}(a^2 - r^2) \tag{1.68}$$

where a is the radius of the diaphragm and D is its flexural rigidity, expressed as: $D = Et^3/12(1 - v^2)$, where E, t and v are Young's modulus, thickness and Poisson's ratio of the diaphragm material, respectively. We note that in

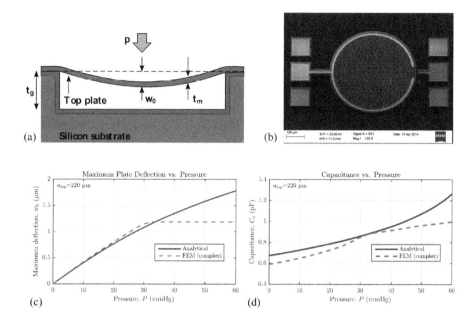

FIGURE 1.18 (a) Schematic illustration of the cross-sectional view of the capacitive pressure sensor. (b) SEM image of the device. Comparison between the analytical and FEM models for circular MEMS pressure sensor with top plate radius of 220 m. (c) Center deflection versus pressure response and (d) capacitance versus pressure response. Reprinted from Miguel,,J. A., Lechuga, Y., & Martinez, M. (2018). *Micromachines, 9*, 342, under Open Access License.

the model of the deflection and capacitance, a discontinuity occurs, where the upper and lower plates touch. This is not taken into account in the purely analytical model.

1.4.1.2 Accelerometer

The accelerometer is an example of an inertial sensor, which is used for detecting the variation of velocity and can be used for the deployment of airbags, detection of seismic activity or other safety applications, for example, in the automotive or aeronautical industries. As we noted earlier, the rigid mechanical properties at reduced dimensions mean that such device functioning is best achieved by the use of a proof mass, which is then attached to suspended compliant support and is anchored at strategic points. This will allow the proof mass to perform displacement and resonance motion when subject to accelerating forces. A common form of control and measurement is via a capacitive comb drive structure. A basic structure for a single-axis differential capacitive accelerometer is illustrated in Figure 1.19(a), where a movable mass is connected to two springs on each side to allow flexibility and stability to the system. Modification of the proof mass and support structure can allow double-axis detection to be performed, see for example Mohammed et al. (2018), which is schematically represented in Figure 1.20, for both single- and dual-axis devices (a) and (b). In Figure 1.20(c), we show the equivalent circuit for the single-axis accelerometer, from which we can base a model for the sensing of motion of the proof mass. We note that the device structure is based on two capacitances, C_1 and C_2, each having a capacitance based on their specific geometries, as expressed in Eq. (1.66).

We note that the MEMS accelerometer can be simplified as a second order damped spring-mass system, with a proof mass suspended by springs that are anchored to the substrate. The equation of motion for the assembly can be expressed as a classical damped oscillator in the form:

$$m\frac{d^2x}{dt^2} + b\frac{dx}{dt} + kx = F_{app} \tag{1.69}$$

The resonant frequency of such a system takes the form: $f_0 = \omega_0/2\pi = \sqrt{k/m}$, with a quality factor given by $Q = m\omega_0/b$. Here, b represents the damping of the system and k is the spring constant, which is determined, as we saw previously, by the Young's modulus of the materials used and the physical dimensions of the spring structure. For the motion off-resonance, $\omega \ll \omega_0$, the mechanical sensitivity is expressed as:

$$\frac{a}{m} \sim \frac{m}{k} = \frac{1}{\omega_0^2} \tag{1.70}$$

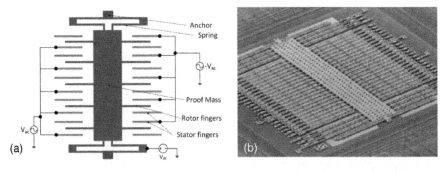

FIGURE 1.19   (a) Schematic illustration of a single-axis transverse comb accelerometer. (b) SEM image of a real single-axis MEMS accelerometer.

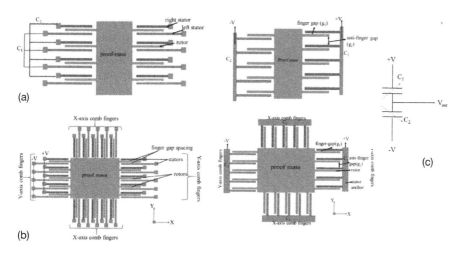

FIGURE 1.20 The arrangement of commonly used interdigitated fingers for a single-axis accelerometer (a) and a dual-axis accelerometer (b). The comb finger topology for (a) single-axis accelerometer and (b) dual-axis accelerometer is also illustrated. (c) The equivalent circuit of the single-axis accelerometer. Reprinted from Mohammed, Z., Dushaq, G., Chatterjee, A., & Rasras, M. (2018). *Mechatronics, 54*, 203, ©(2018), with permission from Elsevier.

This is inversely proportional to the square of the resonance frequency, so greater sensitivity can be achieved for lower values of the resonance frequency. Another factor of importance for the design of accelerometers is the Brownian noise, which limits the resolution of the device. This can be expressed as:

$$\sqrt{\frac{a_n^2}{\Delta f}} = \frac{\sqrt{4k_B T b}}{m} = \sqrt{\frac{4k_B T \omega}{mQ}} \qquad (1.71)$$

where a_n is the Brownian equivalent acceleration noise and Δf is the bandwidth. The noise is clearly dependent on the thermal energy $k_B T$ and can be reduced by increasing the proof mass or Q.

The sensing of the motion of the proof mass is made capacitively, since the separation between the interdigitated fingers of the two capacitances will change. In the case of the dual-axis accelerometer, the motion is detected by four capacitances, where changes in both separation and overlap will also affect the capacitance. By way of example, we shall consider the simpler case of the single-axis accelerometer with two capacitances. Considering the lateral displacement of the proof mass with respect to stators 1 and 2, the separation between the interdigitated electrodes will change by an amount, say δg, increasing for one while decreasing for the other by the same amount. We neglect the capacitance for the anti-gap spacings. For acceleration in one of the sensing directions, we can now write the capacitances for the two capacitors as:

$$C_1 = \frac{\varepsilon_r \varepsilon_0 NA}{g + \delta g} \tag{1.72}$$

and

$$C_2 = \frac{\varepsilon_r \varepsilon_0 NA}{g - \delta g} \tag{1.73}$$

In this case, the area A represents the area of overlap between the fingers, with $A = hL$, where h is the height and L the overlap length in the comb regions, and N denotes the number of pairs of sensing fingers. When the acceleration is non-zero, we can see that the two capacitances are unequal, with a difference of:

$$\Delta C = |C_1 - C_2| = 2\frac{\varepsilon_r \varepsilon_0 NA}{g^2}\delta g = 2C_0\frac{\delta g}{g} \tag{1.74}$$

where $C_0 = C_1 = C_2$, is the capacitance is without displacement ($\delta g = 0$). In the case of a longitudinal displacement, we can also evaluate the changes in the capacitances due to the variation of the overlap length. In this case, we can write:

$$C_1' = \frac{\varepsilon_r \varepsilon_0 Nh(L + \delta L)}{g} \tag{1.75}$$

and

$$C_2' = \frac{\varepsilon_r \varepsilon_0 Nh(L - \delta L)}{g} \tag{1.76}$$

which leads to a difference of:

$$\Delta C' = |C_1' - C_2'| = 2\frac{\varepsilon_r \varepsilon_0 h}{g}\delta L = 2C_0\frac{\delta L}{L} \tag{1.77}$$

Returning to the case of the lateral motion, we now consider the operation for the single-axis device. From Figure 1.20(c), we can express the operation with an applied voltage, which is modulated on the outer electrodes, as:

$$V_+ = V_0 \sin \omega t \quad \text{and} \quad V_- = -V_0 \sin \omega t \tag{1.78}$$

where V_0 is the modulation amplitude and ω the angular frequency of modulation. Since the charge is conserved in the system, the charge on both capacitors must be equal. This leads to the following relation:

$$C_1(\delta g)(V_+ - V_{out}) = C_2(\delta g)(V_{out} - V_-) \qquad (1.79)$$

This allows us to obtain the output voltage of the half-bridge as:

$$V_{out} = \frac{C_1 - C_2}{C_1 + C_2} V_0 \sin \omega t = \frac{\Delta C}{2C_0} V_0 \sin \omega t \qquad (1.80)$$

There are several parameters of importance in the operation of the accelerometer and the principal characteristics are sensitivity, resolution, operating range, bandwidth, cross-axis sensitivity, nonlinearity and offset. The cross-axis sensitivity refers to the output detected in a direction orthogonal to the sensing axis. For single-axis devices, this should be small and less than 1 %. For dual-axis operation, such cross-axis sensitivity is important and must be taken into account in the capacitance changes of the orthogonal capacitors, where relations (1.75) and (1.76) can be used. The variation of the differential capacitance, measured as the peak-to-peak output voltage, is seen to be linear with the applied acceleration, see (Mohammed et al., 2018) and (Kavitha et al., 2016). In addition to the capacitive measurement of the motion of the proof mass, the detection of acceleration can also be performed using piezoresistive elements with a good range of sensitivities (Wang et al., 2017).

1.4.1.3   Comb-Drive Actuator/Resonator

In addition to the comb-drive being used as a sensor as for the case of the accelerometer, this structure can also be adapted to actuator applications (Legtenberg et al., 1996). In this case, the application of a potential between the two sets of fingers in the interdigitated structure creates an electrostatic force, pulling or pushing the moveable part of the structure. The motion, due to the symmetry of the structure, will be principally along the direction of the finger, thus changing the overlap between the fingers. In this case, we generally consider just one set of interdigitated fingers, as illustrated schematically in Figure 1.21, with translation in a linear direction. In addition to the comb-drive being used as a sensor as for the case of the accelerometer, this structure can also be adapted to actuator applications (Legtenberg et al., 1996). In this case, the application of a potential between the two sets of fingers in the interdigitated structure creates an electrostatic force, pulling or pushing the moveable part of the structure. The motion, due to the symmetry of the structure, will be principally along the direction of the finger, thus changing the overlap between the fingers. In this case, we generally consider just one set of interdigitated fingers, as illustrated schematically in Figure 1.21, with translation in a linear direction. For the case of a positive force, this will

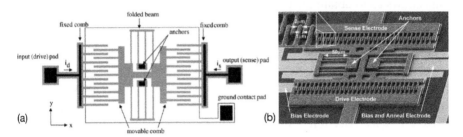

FIGURE 1.21   (a) Schematic illustration of a comb-drive actuator or resonator. (b) SEM image of the same structure.

increase the overlap and hence the capacitance between the fingers. This can be expressed as:

$$C = \frac{2\varepsilon_r\varepsilon_0 Nh(x + \delta x)}{g} \qquad (1.81)$$

with x being the initial comb finger overlap and δx its displacement due to an applied electric field. All symbols have the same meaning as in Eq. (1.75). The factor 2 arises due to the fact that the capacitor is made up of both sides of the fingers. For the case of voltage control, the lateral electrostatic force in the x direction will be equal to the negative derivative of the electrostatic energy $E_C = CV^2/2$ with respect to distance, from which we write:

$$F_{el} = -\frac{\partial E_C}{\partial x} = \frac{1}{2}\frac{\partial C}{\partial x}V^2 \qquad (1.82)$$

where V is the applied voltage between stator and rotor. Using Eq. (1.81), we thus obtain:

$$F_{el} = \frac{\varepsilon_r\varepsilon_0 Nh}{g}V^2 = \frac{C}{x}V^2 \qquad (1.83)$$

The force acting on the spring, to which the rotor is connected, results in the displacement by an amount equal to:

$$x = \frac{\varepsilon_r\varepsilon_0 Nh}{gk_x}V^2 \qquad (1.84)$$

As we noted earlier, some electrostatic forces along the direction transversal to the fingers can occur, which will reduce the transversal stator - rotor separation. This force in the axial direction takes the form:

$$
\begin{aligned}
F_{el} &= \frac{\varepsilon_r \varepsilon_0 N h(x + \delta x)}{2(g - y)^2} V^2 - \frac{c_r c_0 N h(x + \delta x)}{2(g + y)^2} V^2 \\
&= \frac{\varepsilon_r \varepsilon_0 N h(x + \delta x)}{2} V^2 \left[\frac{1}{(g - y)^2} - \frac{1}{(g + y)^2} \right]
\end{aligned}
\tag{1.85}
$$

In general, these forces should cancel out on either side of the rotor comb fingers. However, side instabilities can occur, which depend on the relative strength of the spring constant in the transversal direction k_y. The stability of the comb will be satisfied for the condition:

$$
k_y > \left(\frac{\partial F_{el}}{\partial y} \right)_{y \to 0} = 2 \frac{\varepsilon_r \varepsilon_0 N h(x + \delta x)}{g^3} V^2
\tag{1.86}
$$

When the driving voltage exceeds the stability condition limit, denoted as V_{SI}, the comb-drive will become unstable and transversal motion of the rotors can occur, which may result in the rotor sticking to the stator. We can express this side instability limit of the voltage using Eqs. (1.84) and (1.86) as:

$$
V_{SI}^2 = \frac{g}{2\varepsilon_r \varepsilon_0 N h} \frac{k_x x(x + \delta x) + k_y g^2}{x + \delta x}
\tag{1.87}
$$

We note that this expression contains the two spring constants for the motion of the moveable mass in the x-direction and of the comb fingers in the y-direction. This means that k_x refers to the spring constant of the folded beam structure, while k_y is the spring constant of the fingers themselves. Since we expect the spring constant for the y-direction to be much larger than that in the x-direction $(k_y \gg k_x)$, we can write:

$$
V_{SI}^2 \simeq \frac{g^3 k_y}{2\varepsilon_r \varepsilon_0 N h(x + \delta x)} = \frac{g^2 k_y}{C}
\tag{1.88}
$$

where we have made use of Eq. (1.81). Using the above expression in Eq. (1.84), we obtain the maximum deflection before the the plates touch can be expressed as:

$$
x_{SI} = \frac{g^2}{2(x + \delta x)}
\tag{1.89}
$$

From this equation and the previous relation, we see that the side instability will depend on the gap spacing. This shows that for comb-drive structures, small gap spacings are more susceptible to side instability. Such instabilities can be avoided as much as possible by the correct design of the comb structure and flexure. In Figure 1.22, we show the variation of the beam deflection as a function of the square of the driving voltage for two different gap spacings. From this measurement, it was possible to obtain an estima-

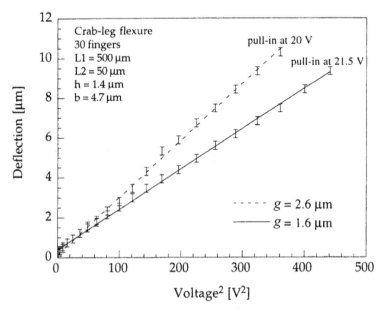

FIGURE 1.22 Measured deflection as a function of the driving voltage for two crab-leg flexure designs. The comb consisted of 30 fingers with a gap spacing g of 1.6 and 2.6 m. The crab-leg structure consisted of a thigh segment of length L_1 of 500 m and a shin of length L_2 of 50 m. Measured beam width b was 4.7 m and the beam thickness h was 1.4 m. Reprinted from R. Legtenberg, A. W. Groeneveld, & M. Elwenspoek, (1996). *J. Micromech. Microeng., 6,* 320. Reproduced with permission of Institute of Physics Publishing in the format Book via Copyright Clearance Center.

tion of Young's modulus of the finger structure in the comb-drive, which was fabricated from polysilicon, as 170 ± 20 GPa.

Returning to the construction of the device, as illustrated in Figure 1.21, we note that one set of comb-drive capacitors is used as the driver, while a second set is used as a sensing device, allowing direct measurement of the motion induced by the driving signal. This structure can also be used as a resonator, when the applied voltage is modulated at a given frequency, where resonance occurs for the general condition:

$$\omega_0 = \sqrt{\frac{k_x}{m_{eq}}} \tag{1.90}$$

This equation can be derived from Eq. (1.69), where $m = m_{eq}$ is the equivalent mass of the moveable assembly. In this case, the driving force will depend on the applied voltage at the input side of the assembly, which according to Eq. (1.83) takes the form:

$$F_{el} = \frac{\varepsilon_r \varepsilon_0 N h}{g} [V_P + v(t)]^2 \tag{1.91}$$

in which V_P is a dc polarization voltage and $v(t)$ the ac modulation voltage. The response as a function of the modulation frequency will take the form of a general mechanical resonance as illustrated in Figure 1.7.

1.4.1.4   MEMS Gyroscope

The MEMS gyroscope is a miniature inertial sensor, which can measure the rate of change of the angular position and can provide a low-cost solution for medium performance requirements. The automobile, aviation, and aerospace industries are large markets for such devices. In particular, the former demands low-cost gyroscopes for sensing yaw to compensate for braking and suspension systems for driving security and comfort. For example, in the automobile industry, accelerometers have seen widespread application in crash testing. Frontal accelerometers can process 35 g to 250 g inputs for triggering driver and passenger airbags. Side impacts are monitored using 250 g to 500 g sensors. Lower acceleration values in the 1 to 5 g range are employed for ride control purposes. The attraction of the MEMS type structures, while maybe not being as precise as spinning gyroscopes, is that they are robust and small devices with no bearings.

The main design approaches for MEMS are based on metallic ring structures or vibrating beam structures. Indeed, the latter is very similar in structure to the comb-drives and accelerometers of the previous sections. In this case, the moveable proof mass will have two sets of comb-drives placed in orthogonal directions, such as shown in Figure 1.20. One set, the linear drive, is used to drive the proof mass at its resonance frequency. The other direction is then used as the sensing direction. Ideally, a system is driven along one (drive) axis. Cross-axis sensing will occur if the system suffers angular rotation, which produces a Coriolis force in a direction orthogonal to both the sense drive and the angular axis of rotation, see Figure 1.23. This then allows

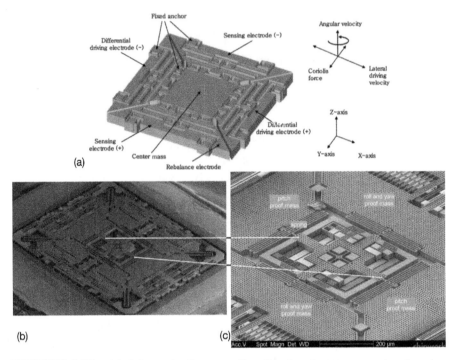

FIGURE 1.23   (a) Schematic diagram of a vibrational gyroscope, showing the main elements of the proof mass, differential driving electrodes, and sensing electrode. (b) SEM image of a MEMS gyroscope of the type used in an earlier version of the iPhone. (c) Close-up image of the central proof mass elements.

an energy transfer between the two vibrational modes of the structure, with the angular rotation about the z-axis. This device is a single-axis gyroscope and is only sensitive to the rotation about this axis. The device sensitivity, resolution and stability are the prime performance parameters.

We can briefly analyze the forces at play, starting with the Coriolis force, F_C. When an object moves in a rotating frame of reference, the structure will experience a Coriolis force given by:

$$\mathbf{F}_C = m\mathbf{a}_C \tag{1.92}$$

where m denotes the mass of the structure and $\mathbf{a}_C$ the Coriolis acceleration, which is proportional to the velocity, $\mathbf{v}$, of the moving structure and the angular rotational frequency, Ω, and can be expressed as:

$$\mathbf{a}_C = 2m\mathbf{v} \times \Omega \tag{1.93}$$

Using these expressions, we can deduce the displacement of the mass of the structure in the y-direction as:

$$y = \frac{F_C}{k} = \frac{m}{k} a_C = \frac{a_C}{\omega_0^2} \tag{1.94}$$

where k is the spring constant of the structure in the sense direction and $\omega_0 = \sqrt{k/m}$ is the resonance frequency of the structure. From this equation, we can see that this device is sensitive to acceleration, where the difference here is the fact that the acceleration is not a translational acceleration, as we saw above for the accelerometers. Here, the accelerometer must be driven at resonance, where the vibrational velocity must also be known and stable in order to evaluate the rotational rate.

We can make a simplified model of the system, with the drive mode in the x-direction and the sense mode in the y-direction and defining the spring-mass-damper system in these two orthogonal directions with stiffnesses k_d and k_s for the drive and sense directions, respectively, and corresponding resonant frequencies ω_d and ω_s. The suspended gyroscope is generally underdamped to allow for an appreciable displacement amplitude at resonance, which under the driven resonance can be expressed as a periodic function:

$$x(t) = Q_d \frac{F_d}{k_d} \sin \omega_d t \tag{1.95}$$

where F_m is the amplitude of the driving force and Q_d is the quality factor in the drive mode. Taking the derivative of this expression with respect to time yields the velocity in the drive sense at resonance as:

$$v_x(t) = Q_d \frac{F_d \omega_d}{k_d} \cos \omega_d t \tag{1.96}$$

The cross-axis transfer through the Coriolis force will alter the magnitude and phase as expressed in Eq. (1.94). Now, a mixing of the frequencies will be observed in the output due to the driving/resonance frequency, ω_d, and the frequency of angular rotation, ω_Ω. The sense mode displacement can be expressed as a function of time as:

$$y(t) = K_s Q_d \frac{F_d}{k_d} \frac{\omega_d \Omega_{zm}}{\omega_s^2} \{\cos[(\omega_d - \omega_\Omega)t + \phi_s] + \cos[(\omega_d + \omega_\Omega)t + \phi_s]\} \tag{1.97}$$

Here, K_s denotes the coupling amplitude between the two orientations of the drive and sense modes due to cross-axis transfer, Ω_{zm} is the amplitude of the rotation rate around the z-axis and ϕ_s is the phase of the sense signal with respect to the drive oscillations. Here, we see that the gyroscopic response is governed by the transfer function, K_s, which in ideal circumstances should be equivalent to the quality factor of the sense mode and the quality factor of the driven mode. The former condition arises for the case where the two resonance frequencies are equivalent (Xie and Fedder, 2003). Furthermore, the output signal is modulated, meaning that a demodulator is required to recover the strength of the rotation signal. Importantly, we note from Eq. (1.97) that the signal amplitude is directly proportional to the angular frequency or velocity to which the gyroscope is subjected.

A dual-axis functionality can be achieved for MEMS gyroscope devices, which can be used to detect two sense angular rotation in two lateral axes. In the design of Juneau et al., (1997), an inertial rotor is suspended by four beams, which are anchored to the substrate and provide torsional compliance about the three axes, see Figure 1.24. In order to generate Coriolis acceleration, the inertial rotor is driven into angular resonance about the z-axis, perpendicular to the plane of the substrate. In this condition, any rotation

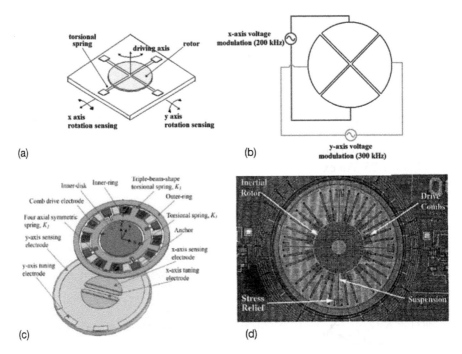

FIGURE 1.24 (a) Conceptual illustration of the dual-axis gyroscope. (b) Top view, showing four diffusion sense electrodes beneath inertial rotor: Note x- and y-axis have different modulation frequencies, so two rotation rates may be independently resolved. (c) Detailed schematic illustration of the dual-axis gyroscope. (d) SEM image of the fabricated device.

of the substrate about the *x*-axis will induce a Coriolis angular acceleration about the *y*-axis, which will induce a tilting oscillation of the rotor about the *y*-axis. The circular symmetry of the device means that the same is true for rotational motion about the *y*-axis. These dynamics are demonstrated in the Coriolis acceleration on each axis, as expressed in the following:

$$a_{Cor,x} = 2I_x \Omega_y \omega_z \cos \omega_z t \qquad (1.98)$$

$$a_{Cor,y} = 2I_y \Omega_x \omega_z \cos \omega_z t \qquad (1.99)$$

where $I_{x,y}$ denote the *x* and *y* inertial moments, ω_z is the resonant drive frequency and $\Omega_{x,y}$ are the rotation rates with respect to the in-plane axes. Tilting oscillations about these axes will be at the same frequency as that of the drive oscillation. These tilting oscillations are amplitude modulated at different frequencies, as shown in Figure 1.24(b), so the rotation rates can be easily demodulated at the correct frequency to distinguish the two different rotation rates. Cross-axis sensitivity is an important issue and needs to be correctly understood. The main source of cross-axis coupling is mechanical in nature and can arise from irregularities from imperfect fabrication. This can result in inertial, viscous and elastic cross-coupling.

Another design for the detection of rotation is the vibrating ring gyroscope, where a ring structure is driven into resonance in the plane of the substrate and anchored at its center. The vibration of interest is a fundamental mode, which forms an elliptical pattern. In the absence of external rotation, four fixed nodal points at ±45° from the drive axis are observed, see Figure 1.25. An external rotation about the *z*-axis will generate a Coriolis

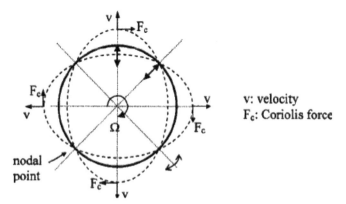

FIGURE 1.25   Illustration of the vibrational mode a vibrating ring gyroscope.

force, which will perturb the vibrational motion along the 45° axis, meaning that the nodal point is no longer stationary. The motion is detected capacitively be a series of electrodes placed around the perimeter of the ring, see Figure 1.26(a). The output of the device, shown in Figure 1.26(b), shows good linearity for the angular rate as a function of the detected signal. The sensitivity, rate range, and bandwidth are programmable. Typical bandwidths selected are between 15 and 50 Hz. Nonlinearity is less than 0.2%. The noise is less than 0.8°/s root mean square (RMS). The device can operate between −40°C and 85°C. The most important device parameter related to chip-scale packaging in vacuum for this sensor is the quality factor, Q, of the device. The Q value of a resonant peak was determined by dividing the peak frequency by the 3 dB peak width. In the open air, the Q values of the ring structure range from 100 to 200. In a vacuum, the Q values can go up to 2,000 (Sparks et al., 1999).

1.4.2   MOEMS–Optical MEMS

As we have seen from the above regarding MEMS devices, there is a large range of mechanical manipulations that can be performed and measured. Such technologies allow for the precise mechanical control of rotation and linear positioning with a good degree of reliability. This further allows the enabling of technologies for microelectromechanical optical devices with applications ranging from visual information displays, fiber optic communications, and bar-code reading. MEMS technologies offer a wealth of possibilities for the miniaturization of optical applications, which promise to not

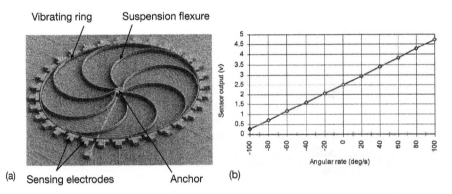

FIGURE 1.26   (a) SEM image of the vibrating ring gyroscope. (b) Angular rate sensor output. Reprinted from D. Sparks, D. Slaughter, R. Beni, L. Jordan, M. Chia, D. Rich, J. Johnson, & T. Vas, (1999). *Sens. Mater., 11*, 197. Open Access License.

only reduce the size and bulkiness of devices, but reduce costs, power consumption while improving efficiency and reliability. Here, we will outline just a few examples of promising devices to illustrate some of the potential developments.

One of the early devices to be developed was in display applications, where Texas Instruments initiated work in the digital mirror device (DMD) in the 1980s, that was commercialized in the 1990s. Essentially, the DMD consists of a large array of micromirrors, which are mounted on hinged supports, enabling them to be tilted towards or away from a light source. This essentially means that each mirror can act as a pixel in the formation of an image projected on a surface. The design and construction of the device are illustrated in Figure 1.27. Each mirror element is 16 m by 16 m square and can rotate to $\pm 12°$, corresponding to the "on" or "off" states. The motion of each mirror is controlled by an electrostatic actuator placed on its underside at two diametrically opposed corners, as illustrated in Figure 1.27(a). The assembly is integrated onto the CMOS chip with the electronics for the control of the array of devices. Gimballed mirror arrays have also been produced with two-axis beam steering capabilities, see Figure 1.28. The structure is positioned using a self-assembly technique during the final stages of the release of the mirror processing sequence.

The grating light valve (GLV) is a unique MEMS device that acts as a dynamically tunable optical grating, whose purpose is to vary the amount of light, usually from a laser source, that is diffracted or reflected. The structure of the GLV consists of a series of metallic bands, that can be moved up or down over small distances by controlling the strength of electrostatic potentials between the bands and the substrate. Each of the reflecting ribbons can be addressed by the appropriate signal to move into a position, where it will either reflect or diffract light. The device, also referred to in some texts as a diffractive spatial light modulator, shown in Figure 1.29(a) and (b), can modulate light intensities via diffraction rather than by reflection. The two states of the system, shown in Figure 1.29(c), illustrate the principle of operation. The GLV can perform high-speed modulation, fine gray-scale attenuation, and scalability to small pixel dimensions. Each pixel consists of six ribbons with a reflective coating, which also acts as an electrode for switching between states. The on–off switching can be as fast as 20 ns and is faster than the switching speed to the mirror displays discussed previously (Kim et al., 2004). The GLV can be made into two-dimensional arrays for projection display applications.

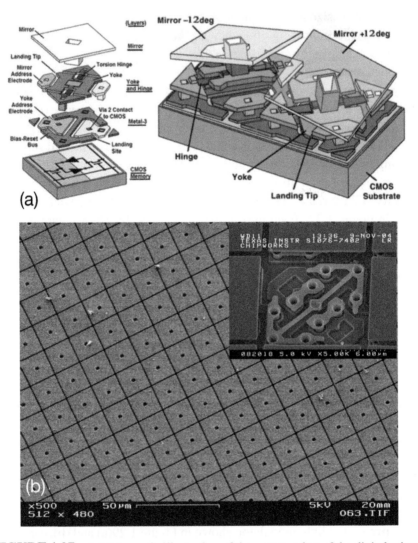

FIGURE 1.27 (a) Schematic illustration of the construction of the digital mirror device. (b) Micrograph images of the DMD array. The inset shows the structure under the mirror.

A somewhat different approach to image formation is used in the retinal scanning display (RSD). Here, a low power light source is optimized to create a single pixel, which is then scanned via a single mirror to project an image directly into the viewer's retina. This method can reduce the excellent spatial and color resolution (Wu et al., 2012). RSD systems usually employ two uniaxial scanners or one biaxial scanner. The combination of two actuation

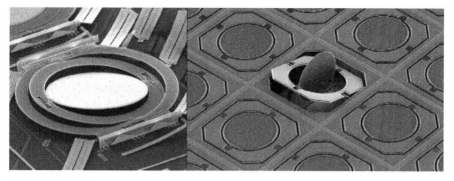

FIGURE 1.28 SEM micrographs of a surface machined two-axis micromirrors.

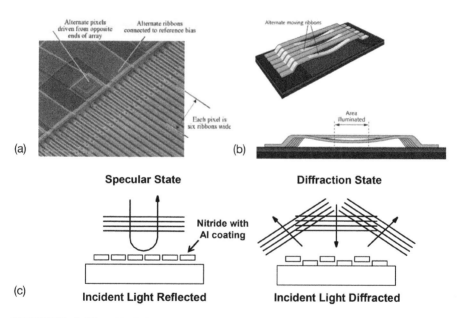

FIGURE 1.29 (a) SEM micrographs of a grating light valve structure. (b) Schematic illustration of the device. (c) Cross-section illustrating the specular and diffractive states of the GLV device (Bloom, 1997).

mechanisms, electrostatic (for fast response times) and electromagnetic (for larger actuation forces) can be used to provide more versatile operation.

Today, the use of data storage and transfer has reached global proportions with the rise of the internet and all that we have come to rely on in our daily lives. Internet banking, messaging, social networks, etc. all require huge quantities of data transfer and storage. Systems for dealing with such issues need to be fast, robust, reliable and affordable. Optical communications systems have emerged as a practical means to address some of the complex issues of data traffic. Systems for dealing with the transfer and switching of data traffic require many fast and efficient components, which include optical sources, cross-connectors, routers, variable optical attenuators, optical switching units, and tuning filters. The devices we have considered so far in this section can be classified as *free-space optics*, since the moving structures (mirrors and grating beams) interact with light in free space propagation. However, for many communications applications, we require the optics to be integrated, which frequently means using optical waveguides or optical fibers, which need to be laid out on the device. Since the direct propagation of free propagating light is incompatible with many functions, alternatives can be found using guided light and evanescent coupling with dielectric structures such as bridges or cantilevers suspended over the waveguide. Such structures are essentially phased modulators and function by altering the effective refractive index of the waveguide and hence the group velocity of the light.

Nano-positioning is critical functionality for applications such as tunable laser cavities, where the cavity length will define the lasing wavelength. Such precision can be achieved using MEMS capabilities in conjunction with vertical-cavity surface-emitting laser (VCSEL) technologies. VCSEL was first introduced in the late 1980s (Kinoshita et al., 1988; Jewell et al., 1989) to produce solid-state lasers, which emit light in a vertical direction. This can be useful in many practical applications. For the MEMS implementation of this device, the reflector, which can be a DBR (distributed Bragg reflector) or a HCG (high contrast grating), is mounted on a moveable structure, either a cantilever or a membrane, see Figure 1.30. This forms one end of the tunable cavity, which is adjusted by the resonance wavelength over a range of values for the device. For the structure illustrated, the SC half-VCSEL consists of an InGaAlAs-based active region, two heat- and current-spreader layers, a buried tunnel junction (BTJ) and a hybrid bottom DBR mirror. In order to enhance high-speed operation, the electrical, optical, and thermal designs of the SC half-VCSEL are optimized for the wavelength range around 1550 nm. The bulk micromachined, directly modulated MEMS VCSEL can be tuned over 64 nm of wavelength, centered at 1550 nm, by electrothermal actuation of the MEMS DBR (Zogal et al., 2015).

The tunable gratings of the GLV type can be described as digital, since their actuation results in a change of grating period and modulation profile by a discrete amount. This is set by the dimensions of the structures and their

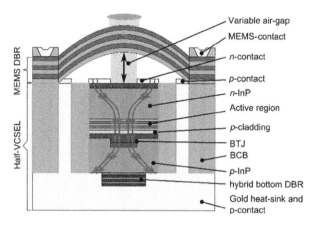

FIGURE 1.30   Cross-section of a MEMS tunable VCSEL (Zogal et al., 2015).

period. In such a case, reconfigurability is limited by the smallest feature dimensions attainable. Improvement of this reconfigurability can be achieved using analog tunable gratings. These devices are constructed using the transverse actuation of a flexure grating structure via piezoelectric actuators with electrostatic comb-drives. To do this, the grating grooves are mounted on a floating deformable membrane. When the membrane is actuated by the thin film piezoelectric, the membrane is physically stretched. The induced strain can cause a 0.3% extension, resulting in a 0.3% change in the grating separation (Wong et al., 2003).

The two-dimensional MEMS optical switch is essentially a cross-bar switch, with a $N \times N$ array of micromirrors that can be selectively activated to reflect a series of optical beams to orthogonal ports or if not activated, to pass to the next mirror or to pass straight through the switching array. Such a system is referred to as a 2D switch, since the optical beams are switched in a two-dimensional plane. A 3D switch adds a third dimension to the switching capabilities. An example of the 2D switch is illustrated in Figure 1.31. We can see that by activating the switches in an appropriate manner, we can re-order the different beams in any sequence we desire. The optical beams can, of course, be of the same color. The beams will be collimated to reduce diffraction losses. The micromirrors are actuated using a linear drive to pop-up out of the substrate plane. The OFF state then consists of the mirror lying in the plane and in the ON state, the mirror sits vertical to the substrate plane. The transition between states will see the mirror change angle via the motion of the actuator, This mechanical operation must be reproducible over many cycles. The first such devices were produced in 1996 by Toshiyoshi

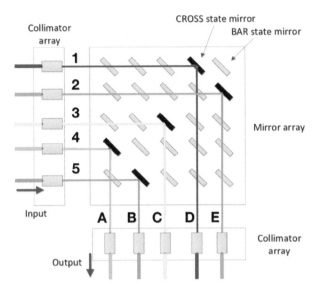

FIGURE 1.31   Schematic illustration of the operation of a 2D optical MEMS switch.

and Fujita (1996). In Figure 1.32, we show a micro-hinged pivot, which can flip the mirror between the ON and OFF states in a switching time of 700 s (Lin et al., 1999). Mechanical stoppers assist the switching process, allowing repeatability of better than 0.1° for the mirror positioning. The drawback of this method is the issue of wear and tear of the free rotating hinges and actuators and can limit the reliability of the switch.

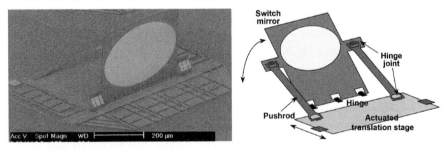

FIGURE 1.32   Schematic and SEM of the surface-micromachined free-rotating hinged mirrors reported by AT&T.

More sophisticated optical switching networks with a large scaling potential can be implemented using three-dimensional switching architectures. The network consists of arrays of two-axis mirrors to guide the reflected optical beams from input fibers to the output fibers. This requires high precision position control of the mirror along the two rotational axes of the gimbal mount, such as illustrated in Figure 1.28.

A MEMS tunable dielectric lens has recently been developed with varifocal capabilities and can have a number of applications in imaging and optical beam scanning (Arbabi et al., 2018).

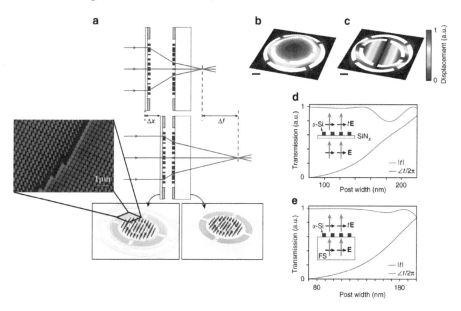

FIGURE 1.33 Schematic illustration of the tunable doublet and design graphs. (a) Schematic illustration of the proposed tunable lens, comprised of a stationary lens on a substrate, and a moving lens on a membrane. With the correct design, a small change in the distance between the two lenses ($\Delta x \sim 1$ m) results in a large change in the focal distance ($\Delta f \sim 35$ m). (Insets: schematics of the moving and stationary lenses showing the electrostatic actuation contacts.) Also illustrated is a SEM micrograph of the lens structure on the support membrane. (b) The first and (c) second mechanical resonances of the membrane at frequencies of ~2.6 and ~5.6 kHz, respectively. The scale bars are 100 m. (d) Simulated transmission amplitude and phase for a uniform array of α-Si nano-posts on a ~213 nm thick SiN$_x$ membrane versus the nano-post width. The nano-posts are 530 nm tall and are placed on the vertices of a square lattice with a lattice constant of 320 nm. (e) Simulated transmission amplitude and phase for a uniform array of α-Si nano-posts on a glass substrate versus the nano-post width. The nano-posts are 615 nm tall and are placed on the vertices of a square lattice with a lattice constant of 320 nm. FS: Fused silica. Reprinted from E. Arbabi, A. Arbabi, S. M. Kamali, Y. Horie, M. S. Faraji-Dana, & A. Faraon, (2018). *Nat. Comms., 9*, 812, under the Creative Commons Attribution 4.0 International License.

The lens is constructed of a metasurface doublet composed of a converging and a diverging metasurface lens with an electrically tunable focal distance. The large and opposite-sign optical powers of the two elements, as well as their very close proximity, make it possible to achieve large tuning of the optical power ($\sim$60 diopters, corresponding to about 4%) with small movements of one element ($\sim$1 micron). This device has demonstrated to have an effective focal length, which is variable over 60 m, with tuning between 565 to 629 m. It is possible to fabricate arrays of these devices on the same chip to allow multiple lenses with different focal lengths, scanning different depths with frequencies of up to several kHz.

1.4.3   RF-MEMS

The increasing demands of wireless communications and the ballooning number of devices that require remote control are feeding a growing industry, which is ever hungry for new and faster solutions. Not only is the number of mobile phones increasing as well as their added functionalities, but cordless telephony, paging, global positioning systems, remote devices, and internet-controlled applications are all adding to these demands. In response, RF-MEMS has seen some important progress, much in line with developments in other areas of MEMS technologies, since their component structures are of a similar nature. Among the range of RF-MEMS components, there are two principal categories that can be identified; lumped components, such as tunable reactive elements (capacitors and inductances) and microswitches, and secondly, complex networks such as reconfigurable filters and phase shifters. These latter are based on the suitable combination of the basic RF-MEMS components in the first category.

MEMS variable capacitors are an essential component of many RF applications, particularly in bandpass and bandwidth filters, matching networks and telecommunications systems in general. The most important characteristics that lumped capacitors exhibit are tunability range and quality factor, both of which should be as high as possible. The former will enhance reconfigurability, while the latter ensures high selectivity for passive filters and better performance characteristics. Variable capacitors, or varactors, are widely used in the semiconductor industry and can be realized with reversed biased p-n junctions. Improvements are found, however, using MEMS devices. Such components in MEMS are fabricated with structures that have at least one of the capacitor plates movable with respect to the other. This is used to modify the separation of the capacitor plates and hence vary their electrical capacitance, which can then be modified to optimize the circuit function. The displacement of the capacitor plate can be performed in a number of ways, with electrostatic, piezoelectric, thermal and magnetic actuation being the most common. In Figure 1.34(a), we show a typical construction for a MEMS variable capacitor with two parallel plates, where the lower is fixed to the substrate and the upper one is formed by a metallic plate supported by a flexible beam structure. The upper plate can be moved towards the lower one by the suitable application of a DC bias between the two plates due to electrostatic attraction.

When no bias is applied to the electrodes, the MEMS capacitor is in its rest position, Figure 1.34(b). The plates will be at their furthest from each other and the capacitance will have its minimum value. By applying a small DC bias between the electrodes, the electrostatic forces will cause them to be attracted and the suspended plate will move towards the fixed plate, Figure 1.34(c). As the distance between the electrodes reduces, the value of the capacitance will increase, as seen from Eq. (1.66). The range of capacitances that can be achieved is limited by the instability in the structures once the distance between the plates reaches around one-third of the rest distance. This is because of the critical balance of the attractive electrostatic force and the mechanical restoring force. Once this limit has been reached, the phenomenon of *pull-in* occurs, and the upper plate abruptly snaps down onto the fixed

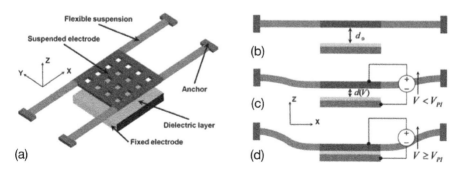

FIGURE 1.34   (a) An electrostatically controlled radio-frequency (RF) MEMS-based variable capacitor based on an electrode kept suspended by deformable slender beams over a fixed electrode. Cross-section of the variable capacitor in (a), when (b) the switch is in the rest position, (c) the applied bias is smaller than the pull-in voltage $(V < V_{PI})$, and (d) the applied bias is larger than the pull-in voltage $(V \geq V_{PI})$. Reprinted from *Practical Guide to RF-MEMS* by Iannacci, Jacopo. Reproduced with permission of Wiley in the format Book via Copyright Clearance Center.

plate. The bias level of this point is referred to as the pull-in voltage, V_{PI}. This situation is illustrated in Figure 1.34(d). To prevent shorting, a thin dielectric layer is deposited on the lower capacitor plate. At this point, the capacitance will jump to its maximum value. The variation of the displacement of the upper plate as a function of the applied bias potential is shown in Figure 1.35(a), where we can clearly see the range of tunability for the device and the pull-in voltage. The corresponding variation of the capacitance is shown in Figure 1.35(b).

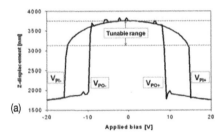
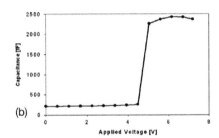

FIGURE 1.35   (a) Pull-in/pull-out characteristic of a MEMS switch. The range of continuous tunability is highlighted in the plot. (b) Experimental C-V characteristic of a RF-MEMS variable capacitor. Reprinted from *Practical Guide to RF-MEMS* by Iannacci, Jacopo. Reproduced with permission of Wiley in the format Book via Copyright Clearance Center.

In Figure 1.36, we show a scanning electron microscope image of a capacitive RF MEMS switch. The capacitor is placed above the signal line of a coplanar waveguide with the two ground planes on either side.

The construction of inductances has also been successfully performed with a number of possible design strategies. The most common form is a spiral coil, which can be formed in a planar geometry above the substrate using a sacrificial layer and suspension supports. As with the capacitor structures, the inductances can be readily incorporated within a coplanar waveguide structure compatible with the experimental characterization on a probe station. The spiral geometry of the inductance requires a connection to the center of the coil via an overpass or underpass, meaning that it can cross the windings of the structure to bring the RF signal to the output. As with the capacitor, the quality factor is an important factor in any RF circuit, and the Q factor for the MEMS inductor can be improved by choosing a low-loss substrate as well as by reducing the coupling of the windings with the substrate with the use of a good insulating layer between them. This is best achieved by using a suspended structure rather than just a dielectric layer. Examples of

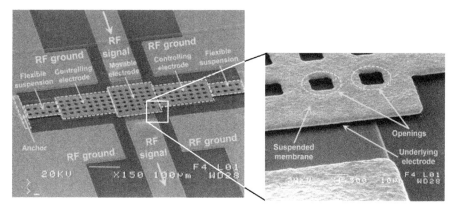

FIGURE 1.36   SEM photograph of a RF-MEMS variable capacitor (or capacitive switch). The suspended capacitor plate is connected to two controlling electrodes and to flexible suspensions. On the right, we see a close-up of a portion of the central suspended metal membrane. The underlying RF line and the openings on the suspended metal layer are visible. Reprinted from *Practical Guide to RF-MEMS* by Iannacci, Jacopo. Reproduced with permission of Wiley in the format Book via Copyright Clearance Center.

these types of structures are shown in the SEM images of Figure 1.37. Other

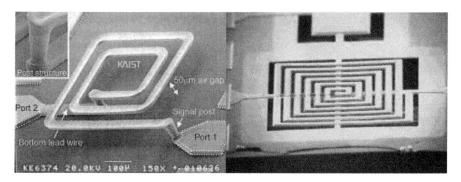

FIGURE 1.37   SEM micrograph of two different RF-MEMS inductors with planar geometry suspended above the substrate.

design architectures include toroidal structures and self-assembly techniques that can be tuned by thermal stressing. In the examples illustrated in Figure 1.38, the solenoid and toroid structures have an air core to improve the quality factor and are rigid and self-supporting. The inductors can be fabricated with 20 and 25 turns and 280–350 m heights having an inductance from 34.2 to 44.6 nH and a quality factor from 10 to 13 at frequencies ranging from 30 to 72 MHz (LÃ^a et al., 2017). Also illustrated in Figure 1.38(b) is a MEMS transformer, which consists of two intertwined toroidal solenoids.

The switching function in RF MEMS devices relies on similar principles to those found in the variable capacitor. Micromachined switches offer superior performance and lower insertion losses than off-chip solid-state devices, providing switching between the receiver and transmitter signal paths. These components are essential for contracting phase shifters, tunable antennae, and filters. Two different RF-MEMS switches have been developed; a series inline switch and a shunt capacitive switch. For the capacitive switch, the device consists of a conductive membrane, typically Al or Au alloy, suspended above a coplanar contact by an air gap of the order of a few microns. For RF or microwave applications, direct metal-to-metal contacts are not necessary, what is required is a step-change in the plate-to-plate distance and hence ca-

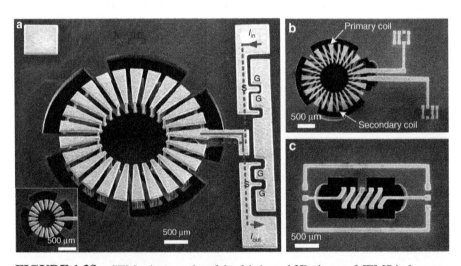

FIGURE 1.38   SEM micrographs of the fabricated 3D air-core MEMS inductors, (a) toroidal inductors with 16 mm^2 (1.5 mm outer radius, 0.75 mm inner radius, and 25 turns) and 4 mm^2 footprint (inset). Presented by the lines and arrows, the current flows from the top wire bonding pad, through the TSV interconnects, then passes through the windings and exits at the lower pad. The measurement pads are designed in a ground-signal-ground configuration at both terminals for wafer-level probing. Four 800 m by 800 m pads at the corners are for flip-chip bonding. (b) 1:1 toroidal transformer. The primary coil has larger conductors than that of the secondary coil. (c) Solenoid inductor. Reprinted from H. T. Le, I. Mizushima, Y. Nour, P. T. Tang, A. Knott, Z. Ouyang, F. Jensen, & A. Han, (2017). *Microsystems and Nanoengineering*, *3*, 17082, under the Creative Commons Attribution 4.0 International License.

pacitance to perform the switching function. The operation is illustrated in Figure 1.39. When the switch is in the ON state, the membrane is in the high position, resulting in a small plate-to-plate capacitance with a minimum high-frequency signal coupling. The OFF state is when the membrane is at the pull-in position, with a large capacitance in the device, resulting in strong high-frequency coupling (low insertion loss). The device consumes very little power and is attractive for low power portable applications, and the switching cycles can number in the millions (Young et al., 2007; Jaafar et al., 2014).

The series or metal-to-metal switch has important applications in interfacing large bandwidth signals including DC. This device is usually constructed from a cantilever beam or a clamped bridge with a metallic contact pad positioned below the beam/bridge. Actuation is again via an electrostatic potential between the suspended structure and the contact pad on the substrate. Such switches typically exhibit actuation voltages of tens of volts, with response times in the order of 20 s and can withstand 10^9 actuations.

1.4.4 *Microfluidics*

Microfluidics refers to the area of MEMS technologies dedicated to the control of fluidic systems on a miniature scale. There are a host of devices and applications that stem from MEMS fabrication techniques and technologies and are of particular interest in biological and medical applications. This area is often termed BioMEMS and is closely related to the *lab-on-a-chip* concept

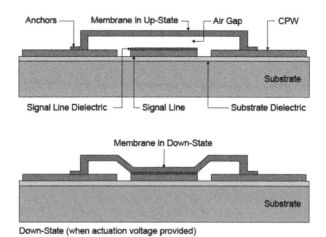

FIGURE 1.39 Cross-section illustration of the RF MEMS capacitive shunt switch.

as well as micro total analysis systems or TAS, where most biological analyzes are performed in a liquid environment and involve the transport of fluids as well as the analysis of reactions. These devices offer the integration of multiple components and processes that can be achieved on a single device. As such, microfluidic devices are very well-suited to point-of-care (POC) diagnostics. Since the amount of fluids required in these applications is small, the reactions are quick and low-cost. In microfluidic devices, which are applied to biological processes, liquid handling is of critical importance, as its quality can significantly affect the end results. According to the way a small amount of liquid is handled and manipulated, the field of microfluidics is further classified as continuous-flow microfluidics and digital (droplet-based) microfluidics (Nguyen et al., 2017).

By way of demonstration, we can note that a single nanoliter (1 nl) of fluid is contained in a cube with a side of 100 m. It is smaller than a grain of salt. Microfluidic devices and systems handle fluids in these quantities for a range of applications. Flow within microfluidic devices, as we noted earlier, is almost always laminar as opposed to turbulent, meaning that mixing generally occurs by molecular diffusion. While mixing based on diffusion could take days in conventional flask-based systems, the small distances within microfluidic channels enable complete mixing within seconds or minutes (Tarn and Pamme, 2013). Microfluidic devices and systems need to be able to handle sample fluids for a range of functions such as blood analysis, biochemical detection, chemical synthesis, drug screen and delivery, protein analysis, DNA sequencing, inkjet printing, etc. To perform the various tasks presented in these various applications, microfluidic systems will generally consist of a platform, in which various components are required for fluid sampling, control, monitoring, mixing, pumping, transporting, etc. These challenges are met with devices such as micropumps, mixers, microvalves, microfilters, etc.

The principal techniques used for the fabrication of microfluidic devices include micromachining, soft lithographies, embossing, injection molding, and laser ablation. While the first methods used were micromachines of silicon, this material is not the most suitable for microfluidic applications, since it is opaque, costly and poses difficulties in component integration. The most common chip material employed nowadays is the flexible elastomer, poly(dimethylsiloxane) or PDMS. This material lends itself to soft lithography, molding, embossing, and micromachining. These methods are relatively fast and less expensive than glass and silicon micromachining. PDMS is also transparent, making it suitable for the observation of microfluidic mixing and monitoring in general.

As we noted earlier, the Reynolds number of fluid flow describes the flow regime as either turbulent or laminar. The former is chaotic and unpredictable in the sense that we cannot determine the position of a particle in the fluid as a function of time. The Reynolds number can be expressed as:

$$Re = \frac{\rho v D}{\eta} \tag{1.100}$$

where ρ is the fluid's density, v the characteristic velocity of flow, D the characteristic length or hydraulic diameter, depending on the cross-section of the channel, in which the fluid is flowing, and η the viscosity of the fluid. Since the geometries for microchannels are very small, the value of the characteristic diameter will be correspondingly small, which means that for most practical cases, the Reynolds number will be smaller than the critical value of 2300, where the flow enters the turbulent regime, and flow will be laminar. In this case, the flow is more predictable and can be more easily modeled.

The resistance in microchannels depends on their geometry and will affect the flow rate according to the relation:

$$Q_f = \frac{\Delta P}{R} \tag{1.101}$$

where ΔP denotes the pressure drop across the channel and R is the resistance of the channel. The most common form of geometry is the circular cross-section tube, in which the resistance can be expressed as:

$$R = \frac{8\eta L}{\pi r^4} \tag{1.102}$$

where L is the length of the channel and r is the channel's radius. Pressure-driven flow, or hydrodynamic pumping, is commonly encountered in microfluidic systems. In this type of flow, the fluid is pumped through the device by means of positive displacement pumps. One of the important assumptions in fluid dynamics for such pressure-driven flow is that the flow velocity at the walls of the channel is zero, this is referred to as no-slip boundary conditions. In this case, the velocity of the fluid has only a resultant component in the direction of flow. The velocity profile can be obtained from the solution of the Navier–Stokes equation, which yields:

$$v_z(r) = \frac{1}{4\eta} \frac{\partial P}{\partial z} (r^2 - h^2) \tag{1.103}$$

This velocity profile is illustrated in Figure 1.40(a). This form of pumping requires components with moving parts, such as micropumps and microvalves, which can have complex designs and can develop mechanical failure. An alternative to this form of flowing is electro-osmotic flow (EOF). This form of pumping occurs when a potential difference is applied across a microchannel that has charged surfaces. An electrode double-layer is formed at the channel surface and the application of a voltage drives the solution through the channel. In this case, the pumping requires no mechanical components and the flow (fluid velocity) profile is almost completely flat, as illustrated in Figure 1.40(b).

Micropumps are the driving force for many operations. These come in many different forms and can be categorized according to their operating principles. The main categories are displacement pumps and dynamic pumps. In the former, a periodic force is applied to effectively push the liquid through the systems. Rotary and peristaltic pumping are examples of this form of action. In Figure 1.41(a) we show the schematic workings of a peristaltic three-phase pump. The pump consists of three pumping chambers connected in series. The sequential actuation of each of the actuation chambers ensures the movement of the fluid in the positive direction, as indicated by the arrows, Figure 1.41(b). Each actuation chamber consists of a pair of electrodes, which undergo pull-in when an appropriate electrostatic potential is applied between the upper and lower electrodes. In Figure 1.41(c), we show a micrograph of the structure, illustrating the three chambers as well as the channels for the flow of the fluid and the venting system. Figs. 1.41(d) and (e) illustrate the sequential actuation of the pumping chambers.

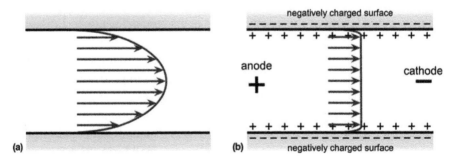

FIGURE 1.40   (a) Parabolic flow profile of a fluid as it is passed through a microchannel via hydrodynamic pumping. (b) Flat profile generated by electro-osmotic flow (EOF) through a microchannel.

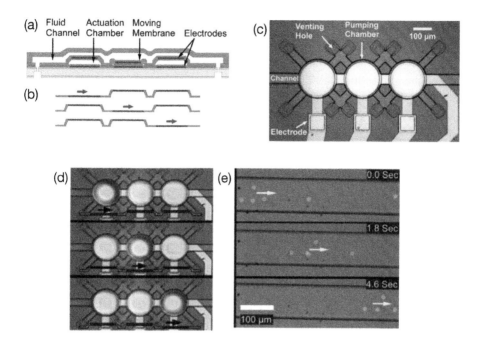

FIGURE 1.41   Schematic of the micropump design and operation. (a) Micropump design. (As an example, the middle moving membrane is pulled down by electrostatic force). (b) Three-phase peristaltic sequence. Arrows indicate pumping direction. (c) Micrograph of a fabricated device. (Each pumping chamber is 200 mm in diameter. Venting holes are connected to the actuation chambers and separated from the fluid channel.) (d) Snapshots from three-phase peristaltic actuation. Schematic view of the cross-section was added to indicate the phase sequence. (e) Tracking of beads (highlighted for clarity) for flow rate measurement. J. Xie, J. Shih, Q. Lin, B. Yang, & Y.-C. Tai, (2004). *Lab Chip, 4*, 495. Reproduced with permission of the Royal Society of Chemistry in the format Book via Copyright Clearance Center.

Non-mechanical pumps have also been developed, which impart momentum to the fluid by directly converting non-mechanical energy into kinetic energy with no mechanical movement. Since there are no moving parts in this form of valve, structures can be more easily produced in miniature form. Since viscous resistance in microchannels increases to second order with miniaturization, mechanical pumps cannot deliver the necessary power for high fluidic impedance on the microscope. Driving forces in non-mechanical systems can be based on electric, magnetic, thermal, chemical or surface tension forces. This class of pump is usually designated as one of the following: capillary pumps, electro-hydrodynamic (EHD) pumps, electrokinetic pumps, magneto-hydrodynamic (MHD) pumps and phase difference pumps.

We saw previously, that the flow of a fluid can be produced by the application of potential at the walls of a microfluidic channel, see Figure 1.40(b). This will create a thin ion layer of liquid at the wall surface, which is oppositely charged to the wall charge. This is called a Stern layer and these ions are fixed. The interior of the channel is referred to as the diffuse layer and consists of the fluid, which is free to move under the application of potential along the channel. This process is called electro-osmotic flow and forms the basis of an EHD pump. Another form of EHD arises due to the simple acceleration of charged molecules in the fluid by an applied potential. These forces can be counterbalanced by frictional forces. The net motion is characterized by the mobility, $\mu = v/E$, where v is the velocity of the ions and E the applied electric field. The mobility is independent of the particle size for uniform fields. This form of motion is called electrophoresis. One form of MHD pump is illustrated in Figure 1.42. This pump uses a ferrofluid as the rotor for the pumping action, where the pump works by moving a plug of ferrofluid around the fluid-filled circle with a rotating magnet. Since the ferrofluid plug is immiscible with the fluid in the circle, fluid is pumped as the plug moves. The plug merges with the stationary plug as the arm swings past the stationary magnet. A new plug is created as the arm moves past the stationary plug and begins a new cycle.

The control of fluid motion can also be achieved in conjunction with the use of microvalves. These come in two varieties, the active and the passive valve. The type of valve used will principally depend on the degree of control that is required for a particular application. Active valves require actuation via some form of an external signal. The pull-in method can be used to con-

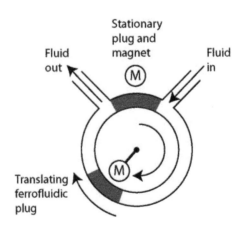

FIGURE 1.42   Schematic diagram for a ferrofluidic pump.

tract and stop flow in a microchannel, for example. Another form of active valve requires chemical stimulation and eliminates the need for an external power supply. In this case, a hydrogel valve can be used, which undergoes a volume change in response to the acidity of a fluid. The valve is actuated via the introduction of a basic the solution, which will close the valve, and flushing with an acidic solution will deactivate the valve as it returns to its normal dimensions. This is demonstrated in Figure 1.43.

Passive valves can be used to limit fluid flow in a certain direction, to remove air or to temporarily block the flow of a fluid. Hydrophobic surfaces, which can consist of a nanopatterning of a surface, have also been used as a method to create valve action. Such a valve will initially cause a build-up of pressure and eventually breaks down, allowing the flow of the fluid. This can be used as a pressure switch. Different surface patterns will create different surface energies, as is illustrated in Figure 1.44. The surface energy controls the angle of curvature and hence the maximum pressure that a wall can sustain.

An important component in microfluidics is a micromixer. As the name suggests, this device is used as a tool for mixing fluids and plays an important

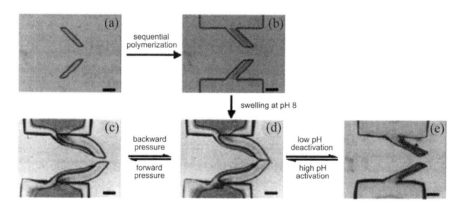

FIGURE 1.43   Fabrication and operation of the bistrip hydrogel valve. The valve was fabricated by simultaneous photopolymerization of the pH-sensitive strips (a), followed by photo-polymerization of pH-insensitive strips and anchors (b). When exposed to pH=8 phosphate buffer, the hydrogel changes its shape and size to form a closed check valve (d); when exposed to pH=3 buffer, the valve is deactivated due to shrinking (e). The activated valve allows forward fluid flow when forward pressure reaches a threshold value (c) while resisting backflow (d). Scale bars are 500 m. Reprinted from D. T. Eddington & D. J.Beebe, (2004). *Advanced Drug Delivery Review, 56,* "Flow control with hydrogels," 199–210, Copyright (2004), with permission from Elsevier.

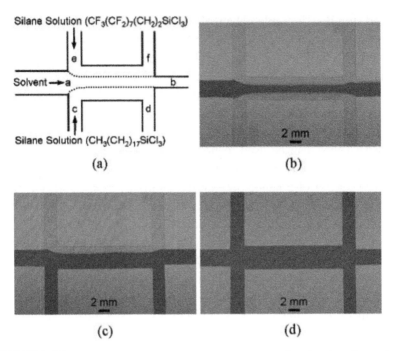

FIGURE 1.44   Demonstration of a pressure switch. (a) The laminar flow scheme for patterning two different surface energies in a microchannel. Images of Rhodamine B solution at various water-column heights: (b) 10 mm, (c) 26 mm, and (d) 39 mm.

role in microfluidic applications. Since the flow of the fluid in microfluidic systems is laminar, the only mechanism for mixing is diffusion. This can be visualized as shown in Figure 1.45, where two different fluids meet at an interface and gradually intermix. The degree of intermixing will depend on the diffusion coefficient. Therefore, for efficient intermixing at the output, the length and width of the microchannels must be appropriately dimensioned. In the 1D case, the distance a particle will move in time t can be modeled as $d = \sqrt{2Dt}$, where D is the diffusion coefficient.

In microfluidics, there are two principal methods used to achieve mixing, which is via passive and active mixing. In the former case, this can be performed by methods such as a distributive mixer, static mixer, a T-mixer and a vortex mixer. We have already illustrated the principle of diffusive mixing using a T-mixer, as shown in Figure 1.45. Two further examples of mixers are illustrated in Figure 1.46. In the first case, Figure 1.46(a) and (b), exploitation the Coanda effect allows two fluids to be split and recombined. In the second example, Figure 1.46(c) and (d), chaotic advection is induced in a mi-

Input: stream X

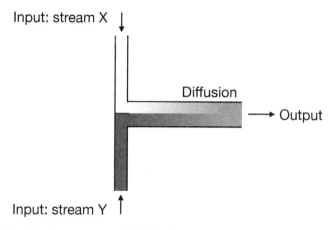

Diffusion

Output

Input: stream Y

FIGURE 1.45 Two streams of fluids in contact will gradually intermix due to the diffusion gradient across the interface, creating a mixed fluid at the output.

crochannel, which has a 3D serpentine structure. Active mixing requires an external supply to drive mixing, which in general will be more efficient than passive mixing. Once again, a number of possibilities exist including piezo-electric devices, electrokinetic mixers, chaotic advection, and magnetically driven devices. The choice of the mixer will depend on the size limitations and applications involved.

In addition to the mixing function, certain applications require the separation and analysis of mixtures. The separation of particles and cells has become an important function in biomedical applications, for example. Surface-functionalized particles can bind target analyses in a sample, allowing the separation of the target from a solution. Cell sorting and counting are particularly important for biomedical analyses such as the separation of blood cells from plasma. A number of physical forces can be used as a mechanism for separation such as magnetism, acoustophoresis, surface acoustic waves (SAWs), optical tweezers, electrophoresis, etc. (Tarn and Pamme, 2013; Nguyen et al, 2017). These can be applied to the trapping of particles/cells within a channel for separation. Continuous flow separations involve pumping a sample through a microchamber and deflecting the particles via a lateral force. In this way, the particles/cells can be separated into different channels for analysis. Figure 1.47 illustrates the principles of trapping and separation.

So far, we have essentially discussed the case for continuous flow microfluidic devices. Droplet and digital microfluidics (DMF) use the injection and control of small quantities or droplets of a fluid, which typically size in the order L or less. In this case, it is necessary to separate discrete volumes of a dispersed phase in an immiscible continuous phase. This separation of phases can be produced by a T-junction or with a flow-focusing method, as illustrated in Figure 1.48. By modifying the surface properties of the microchannels and the flow rates of the immiscible solutions, it is possible to produce water-in-oil droplets as well as droplets within droplets, which can then be transported through the system in discrete quantities. It is also possible to create microbubbles of a gas in a similar fashion. Droplets of liquid or gas can be produced at elevated rates, of thousands of droplets per second in a highly regular reproducible manner. Multiple reagents can be introduced into the droplet and mixed rapidly, within milliseconds.

Microfluidic systems have become popular in both academic research and in the industry for a wide range of applications over the past few decades. One of the most important applications is the Lab-on-a-chip concept, which incorporates a number of devices working on a specific application on a single block or chip. The architectures vary depending on the specific requirements and the control of fluids, which can be mixed, transported and analyzed all on the same chip structure. Two particularly important developments have

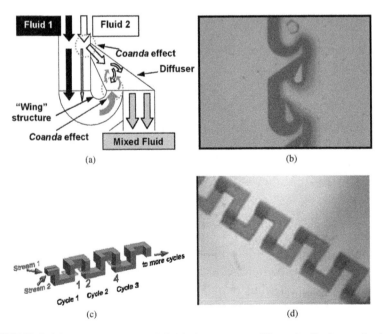

(a)

(b)

(c)

(d)

FIGURE 1.46   Two streams of fluids in contact will gradually intermix due to the diffusion gradient across the interface, creating a mixed fluid at the output.

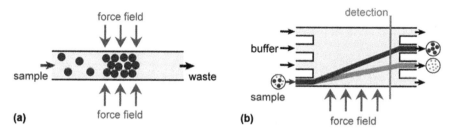

FIGURE 1.47 Particle and cell processing in microfluidic devices. (a) Trapping of particles, allowing separation or reactions. (b) Continuous deflection of particles, enabling separations or even reactions if the particles are deflected through laminar streams containing reagents. In both cases, physical barriers can also be employed rather than applied forces. Reprinted from N. Pamme, (2007). *Lab Chip, 7*, 1644. Reproduced with permission of the Royal Society of Chemistry in the format Book via Copyright Clearance Center.

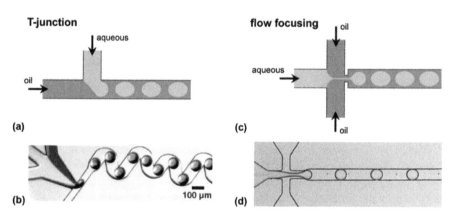

FIGURE 1.48 Common chip designs for droplet generation: (a) T-junction generation, with (b) a photograph showing formation of the droplets in a fabricated device (Song et al., 2003). (c) Flow focusing generation, with (d) a photograph of droplet production. Reprinted from W.-A. C. Bauer, M. Fischlechner, C. Abell, & W. T. S. Huck, (2010). *Lab Chip, 10*, 1814. Reproduced with permission of the Royal Society of Chemistry in the format Book via Copyright Clearance Center.

been soft lithography in PDMS (polydimethylsiloxane) and the incorporation of pneumatically activated valves, mixers, and pumps based on the soft lithographic procedures. Some examples of microfluidic chemostats and lab-on-a-chip systems are illustrated in Figure 1.49. The range of applications includes micro reactions, drug delivery, biosensors, cell and DNA analysis, ink-jet printers, etc. Microfluidics has reached a level of maturity, where individual and multiple integrated processes can be performed for very specific

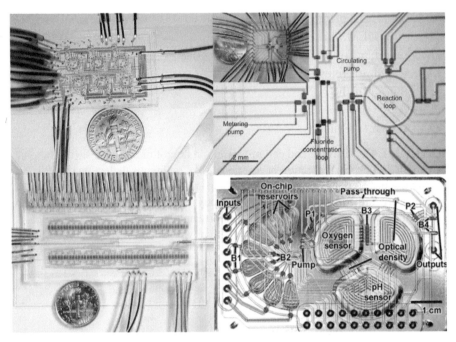

FIGURE 1.49   Microfluidic devices: microfluidic chemostats are used to study the growth of microbial populations routinely incorporate intricate plumbing. These devices include a high density of pneumatic valves, pumps, channels and sensors to perform specific tasks. The colors are dyes introduced to trace the channels. Reprinted by permission from Nature: G. M. Whitesides, (2006). *Nature, 442*, 368, ©(2006); Reprinted from K. S. Lee, P. Boccazzi, A. J. Sinskeyb, & R. J. Ram, (2011). *Lab. Chip, 10*, 1730. Reproduced with permission of the Royal Society of Chemistry in the format Book via Copyright Clearance Center.

tasks. Future development will concern the commercialization of devices that can be operated by the non-specialist and will require a standardization of components and real-world interfacing. This will allow intuitive handling and operation by unskilled end-users. The research will continue to furnish specialized applications and developments, much of which will be dedicated to developments on the biological and chemical front.

1.4.5   *Nano-ElectroMechanical Systems: NEMS*

While MEMS technologies have advanced significantly in the last few decades with marketable products and have reached a reasonable level of maturity, the corresponding nano-electromechanical systems (NEMS) are still at an early stage. This is in large part due to the higher level of sophistication and costs involved in fabrication. It is, for the moment, the main research area, while MEMS has a reasonably steady market, though research is obviously important for future developments. NEMS is characterized by dimensions, which are relevant for the functioning of the device and going well beyond those of the corresponding MEMS. Critical feature sizes can range from a few to hundreds of nm. This means that due to scaling laws, the relevant resonant frequencies of NEMS resonators lie in the microwave range with values of up to 100 GHz. Here, mechanical quality factors can reach values of the order of 10^5, which means a low dissipation of energy. Also, due to the reduced sizes and volumes, the effective masses involved can be in the order of femtograms with mass sensitivities in the attogram range. Further consequences of the size reduction will include low power consumption, with an estimated range of attowatts, and extremely high integration levels, approaching 10^{12} elements per square centimeter (Ke and Espinosa, 2005).

The principle devices for NEMS applications are based around the mechanical properties of cantilever systems or doubly clamped beams, which are of manometric dimensions. The materials that are used for the active components are typically based on Si, SiC, carbon nanotubes as well as Au and Pt. Since Si is one go the basic materials used in the IC industry as well as a well-established material for MEMS devices, this has also been studied for an extension to NEMS devices. In Figure 1.50, we show an example of SEM images of arrays of suspended silicon nanowires defined by ion beam implantation, which are used to study the mechanical coupling and combination with CMOS technology. Despite their compatibility with CMOS devices, silicon

 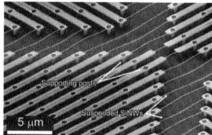

FIGURE 1.50 SEM micrographs of suspended Si nanowires structures with a cross-section of about 26 nm. Reprinted from J. Llobet, G. Rius, A. Chuquitarqui, X. Borrisé, R. Koops, M. van Veghel, & F. Perez-Murano, (2018). *Nanotechnology, 29,* 155303. Nanotechnology by Institute of Physics (Great Britain); American Institute of Physics Reproduced with permission of IOP Publishing in the format Book via Copyright Clearance Center.

suffers from a number of effects such as surface oxidation and reconstructions as well as thermoelastic damping, which limits the strength and flexibility of the material producing inferior performance and low-quality factors. Carbon nanotubes, on the other hand, provide a number of advantages. They naturally have large aspect ratios and, as we have seen, they are characterized by excellent mechanical properties making them ideal for resonant structures. Additionally, carbon nanotubes have good electrical properties, which again provides a good basis for their application in NEMS devices.

As with MEMS systems, the operation of devices will require either the induction or detection of motion. While such challenges have been adequately met for MEMS devices, the shrinking of the device dimensions means that not all methods can easily be transposed or adapted for NEMS applications. For example, the optical methods used in the detection of the motion of scanning probe devices are not suitable for NEMS, since the light will suffer from diffraction effects due to their dimensions.

As in the case of MEMS, electrostatic actuation of nanostructured objects is a viable method using an applied electric field such as is the case for nanotweezers (Kim and Lieber, 1999). Magnetic fields can also be used via a Lorentz force induced in a conductive beam such as one carrying an alternating current, where a large magnetic field can be applied transversally. In this case, we require a doubly clamped beam so as to be able to establish an electrical current in the beam (Husain et al., 2003; Badzey et al., 2004). Other actuation methods include piezoelectric induction and thermal actuation using a bilayer system with different thermal expansion coefficients. Motion detection can be achieved by a number of methods, the most straightforward of which is by direct observation in an electron microscope (Treacy et al., 1996; Kim and Lieber, 1999; Poncharal et al., 1999), which has the relevant spatial resolution, where motion is best detected in the plane perpendicular to the incident electron beam. Clearly, this method is not suitable for device operation. Another method that can be employed for the measurement of motion is via the electron tunneling effect and is very sensitive to small changes in distance. Detection can be easily made in the sub-nanometer range due to the exponential dependence of the tunnel current on the separation between tunnel electrodes. The capacitive sensing of motion has been widely used in MEMS devices and is capable of being used in NEMS, with a resolution of a few nanometers (Zhu et al., 2005) and can potentially be improved to the sub-nanometer range.

A significant proportion of NEMS devices are resonant structures that are stimulated via a transducer to perform some form of mechanical vibration at a specific frequency, which depends on the geometry of the structure. A second transducer can be coupled to the system to modify and control the mechanical vibration and acts as a perturbation to the normal modes of vibration. The mechanical response, i.e., the displacement of the nanostructure due to the stimuli, is transduced back into an electrical signal for measurement purposes. The perturbation of the mechanical vibration can be considered as a form of control, since it will modify the vibration characteristics of the structure such as its resonance frequency and the quality factor of resonance. In this aspect, such a NEMS device will be a three-terminal device, providing input, output, and control.

Doubly clamped beam structures with a length, l, and thickness, t have a frequency that varies linearly with the geometric factor t/l^2 according to the relation (Ekinci and Roukes, 2005):

$$f = 1.05\sqrt{\frac{E}{\rho}}\frac{t}{l^2} \tag{1.104}$$

where E denotes Young's modulus of the beam and ρ the mass density of the material. In Figure 1.51, we show the plots for the frequency versus the geometric factor t/l^2 for SiC, Si and GaAs nanostructured beams. We note that the variation of the experimental result is in good agreement with the predicted behavior expressed in Eq. (1.104). It is worth pointing out that the frequencies of resonance, for structures of the same dimensions, are greatest for the SiC devices, being three times those for the GaAs structures. This increase is a reflection of the increased phase velocity, as expressed by $\sqrt{E/\rho}$, for stiffer materials. From this, we note that SiC has the largest value of $\sqrt{E/\rho}$ of the three materials tested. A comparison of the values of the velocity of sound, $v = \sqrt{E/\rho}$, in these materials from the measurements and the literature show good agreement. For example, from the measurement, $v_{SiC} = 1.5 \times 10^4 \text{ m s}^{-1}$, which compares to $v_{SiC} = 1.2 \times 10^4 \text{ m s}^{-1}$, similarly good agreements are found for Si and GaAs (Yang et al., 2001).

The quality factors obtained for semiconductor NEMS of this type lie typically in the range $10^3 - 10^5$, which is far superior to those of electrical oscillators. Such high values of the Q factor mean that the operating power levels are very low and the resonators are highly sensitive. This also translates into low insertion losses. While this does not necessarily imply reduced

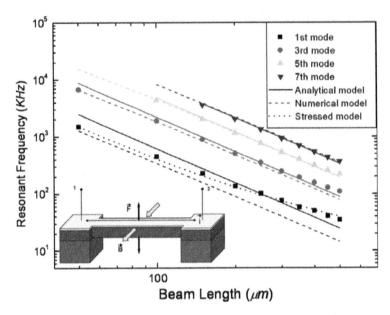

FIGURE 1.51 Resonant frequency of double-clamped beam resonators with an analytical model (solid line) and numerical model (dashed line) fits. In the inset, the actuation principle has been schematized. E. Sillero, O. A. Williams, V. Lebedev, V. Cimalla, C.-C. Rohlig, C. E. Nebel, & F. Calle, (2009). *J. Micromech. Microeng.*, *19*, 115016. ©IOP Publishing. Reproduced with permission. All rights reserved.

bandwidths, resonators operating at 1 GHz with a high Q of around 10^5 and bandwidths of 10 kHz are attainable.

To understand the power levels required for the operation of resonant NEMS devices, we can consider the resonator as a lossy energy storage device. The energy transmitted to the system will be dissipated in a time interval $\tau \sim Q/\omega_0$, where ω_0 denotes the resonance frequency of the device and τ is referred to as the ring-up or ring-down time for the resonator. The minimum operating energy for the system is defined as the energy, which will drive the system to amplitudes comparable to those due to thermal fluctuations. Given that the energy due to thermal fluctuations can be expressed as $k_B T$, the minimum input power can be estimated as:

$$P_{min} \sim \frac{k_B T \omega_0}{Q} \qquad (1.105)$$

For NEMS resonators, this means that power levels can be in the region of 10 aW (10^{-17} W). Even if we multiply this value by a 10^6, to attain a robust signal-to-noise ratio, and consider a million devices working in tandem to realize some signal processing operation, the total power required will still only be in the W regime. This compares very well to the power dissipated in current systems working on similar levels of complexity based on digital devices working in the electronics domain.

Opto-mechanical systems have also generated much interest. The coupling between the mechanical and optical degrees of freedom can, for example, be enhanced by using sub-wavelength optical mode profiles. The photon–phonon coupling produces a radiation pressure the can enable actuation, tuning, damping, and amplification of a resonator. In a cavity-based optomechanical system with an optical resonance frequency ω_c, that depends on the resonator position, both the sensitivity of the displacement and the magnitude of the radiation pressure forces are governed by the coupling strength between the acoustic and optical degrees of freedom, as expressed by the resonant frequency ω_c and the natural linewidth of the cavity, expressed as κ (Leijssen and Verhagen, 2015). The strength of the coupling, g_0, is a fundamentally important factor and will control the modification of the resonant frequency and damping through dynamical back-action, and scales as g_0^2/κ. It is important to improve this ratio to increase sensitivity and to obtain the best optical control of the mechanical resonator. While a reduction of κ can be useful, it can also have some drawbacks, since narrow linewidths mean that the demands on excitation sources are more stringent as will be the tolerances for fabrication, increasing the structural limitations on the device. The phonon-photon coupling rate can be expressed as $g_0 = G x_{zpf}$, where $G = \partial \omega_c / \partial x$ expresses the frequency shift (control) per unit displacement, with $x_{zpf} = \sqrt{\hbar/2m_{\mathrm{eff}}\Omega_m}$ denoting the zero-point fluctuations of a resonator with effective mass m_{eff} and frequency Ω_m. The strength of g_0 is maximized in suitably engineered NEMS systems, since G and x_{zpf} have small cavity dimensions and small resonators mass, respectively.

A displacement induced frequency shift of an optical mode depends on the fraction of energy density that is located near its moving dielectric boundaries. Therefore, to obtain significant photon-phonon coupling rates, the cavity mode should be located at positions, where it is most sensitive to the motion of the mechanical resonator. It is crucially important to optimize the optical confinement along with the directions in which the mechanical mode is also strongly localized. An example of an optomechanical resonator is illustrated in Figure 1.52.

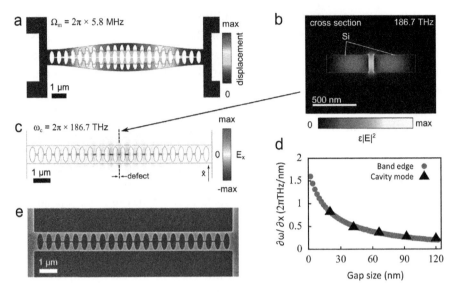

FIGURE 1.52 Resonances and geometry of the sliced nanobeam. (a) Simulated displacement profile of the fundamental (in-plane) mechanical resonance of the structure. (b) The cross-section in the center of the sliced nanobeam (indicated by the dashed line in c), showing the simulated energy density distribution of the fundamental optical cavity mode of the structure. (c) Simulated transverse electric field profile of the fundamental optical cavity mode of the structure. (d) Simulated frequency shift as a result of an outward displacement of 1 nm. The cavity mode shift was determined by simulating the full nanobeam and introducing a uniform displacement along the beam. (e) Electron micrograph of a fabricated device. The thickness of the structure is 200 nm, both in the simulations and in the fabricated device. Reprinted from R. Leijssen & E. Verhagen, (2015). *Sci. Rep.*, *5*, 15974, under the Creative Commons Attribution 4.0 International License.

Here, the doubly clamped beam is sliced along its length so as to resemble a pair of doubly clamped beams.

The small width of 80 nm of the narrowest parts of the beam reduces the effective mass ($\simeq$ 2.4 pg) and the small spring constant leads to large values of the zero-point fluctuations, x_{zpf}. The motion of the nanobeam will modify the local optical properties, which are strongly localized at the surface perpendicular to the motion, but extend over several microns in either direction from the position of maximum deflection. The coupling will be strongest when the light is concentrated in the sub-wavelength gap between the two halves of the beam structure. The periodic pattern enables confinement along the beam length and acts as a photonic crystal with a quasi-band gap for transverse electric (TE) polarized modes guided by the beam. For example, in Figure 1.52(b), the waveguide mode is illustrated for the confined mode, which appears at the lower edge of the bandgap and is physically confined in the nanoscale gap between the half-beams. This mode is illustrated in Figure 1.52(c). The simulation of the frequency response, Figure 1.52(d), for this defect cavity mode, shows the dependence on the gap size between the two half-beams, where an enhancement of the frequency is observed for decreasing separations.

Another form of coupling for a photomechanical transducer can be achieved using a tuning fork nanocavity, which exhibits high-frequency (f) and quality factor (Q) products, fQ. Nanoscale systems naturally have much higher resonant frequencies than micrometer-scale structures due to scaling effects. However, it is frequently observed that there is a trade-off between the device frequency and its quality factor due to increased losses associated with the clamping. Using a microdisk optical resonator with an intrinsic optical quality factor of $Q > 6 \times 10^5$, coupled via the near optical field to a doubly clamped tuning fork (mechanical) resonator, it has been possible to attain both high-frequency and high-quality factors, for devices with $f \simeq 29$ MHz and $Q \simeq 2.2 \times 10^5$, providing a fQ product of around 6.4×10^{12} Hz, (Zhang et al., 2015).

The mechanical manipulation of nanoscale objects is not a simple matter. One solution to this issue was provided by the invention of nanotweezers, which was first performed by Kim and Lieber (1999) and later by Akita and co-workers in 2001, (Akita et al., 2001). Both systems were based on the use of MWNTs, which were used as the arms of the tweezers and are actuated using an electric field. The construction of the tweezers is performed by depositing electrically independent freestanding electrodes onto tapered glass micropipettes, which have end diameters of around 100 nm. The arms of the tweezers, consisting of MWNT or SWNT bundles with diameters in the range of 25–50 nm, were then attached to the electrodes. The electromechanical response of the nanotweezers was then investigated by applying different bias voltages to the electrodes while simultaneously imaging the nanotubes under dark-field illumination in an optical microscope, as shown in Figure 1.53. As the potential is increased, the separation between the tips of the tweezers is observed to reduce gradually and then abruptly close at an applied potential of around 8.5 V, see Figure 1.53(F). The experiment was successfully repeated more than ten times with identical results. This shows that mechanical behavior is elastic with no apparent inelastic deformation of the carbon nanotubes. The approach by Akita et al. (2001) involved depositing three Al

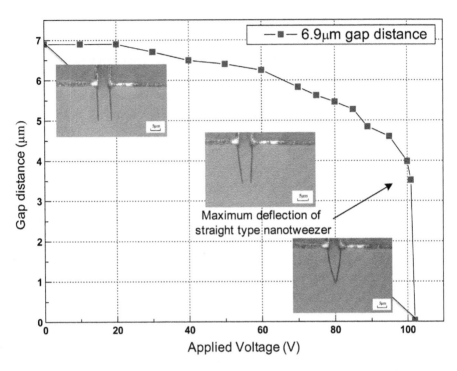

FIGURE 1.53 Straight type tweezers show a relatively large actuation range which is 3.4 μm from the initial gap distance of 6.9 μm. The applied voltage at snap-down was 102 V. A travel length about 1/2 of initial gap distance is measured. Insets show the actuation of straight type tweezers of 300 nm in diameter and 20 μm in length, from no voltage applied to the case of pull-in. Reproduced from J. Chang, B.-K. Min, J. Kim, S.-J. Lee, & L. Lin, (2009). *Smart Mater. Struct., 18*, 065017. ©IOP Publishing. Reproduced with permission. All rights reserved.

interconnects onto a Ti/Pt coated Si AFM tip, which was then split into two parts using focused ion beam etching. The two halves were independently connected to the Al interconnects. Nanotubes were then attached to the two sides and as with the Kim and Lieber device, the actuation is provided by the application of a DC potential between the two arms of the tweezers. Again, repetition of the actuation was performed, showing that there was no permanent deformation of the CNT arms.

In 2003, a carbon nanotube-based rotational motor was developed by Fennimore et al. (2003). The conceptual illustration is shown in Figure 1.54(a) and the SEM image is given in Figure 1.54(b). The rotational element (R) is a solid metal plate, which acts as a rotor, having a length dimension of about 300 nm, and is supported on a suspended shaft, which makes electrical contact with the conducting anchors (A1, A2). The rotor plate assembly is surrounded by three fixed stator electrodes: two of which are in the same plane (S1, S2) and are horizontally opposite one another, while the third (S3) is buried beneath the surface. The electrodes (A1, A2, S1, and S2) are all supported on silicon oxide (SiO_2) surface. The device is actuated via four independent (DC and/or an appropriately phased ac) voltage signals; one to the rotor plate and three to the different stators. These are applied to control the position, speed and direction of the rotation of the rotor plate. The key to the operation of the NEMS motor is a single MWNT, which serves simultaneously as the rotor plate support shaft and as the electrical feedthrough to the rotor plate itself. Most importantly, it allows the rotational freedom required for the motion of the device.

To provide the free movement required by a motor, a simple but effective method was employed. This involved the modification of the MWNT, which was mechanically fatigued to shear the outer nanotube shell by the successive application of very large stator voltages of the order of 80 V DC. This is sufficient to torque the outer shell past the elastic limit, eventually leading to the partial or complete failure of the outer shells. This results in

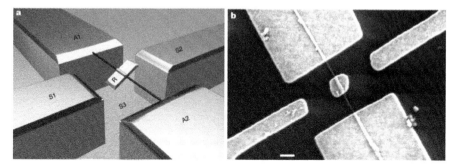

FIGURE 1.54 Integrated synthetic NEMS actuator. (a) Conceptual drawing of nanoactuator. A metal plate rotor (R) is attached to a multi-walled carbon nanotube (MWNT), which acts as a support shaft and is the source of rotational freedom. Electrical contact to the rotor plate is made via the MWNT and its anchor pads (A1, A2). Three stator electrodes, two on the SiO_2 surface (S1, S2) and one buried beneath the surface (S3), provide additional voltage control elements. The SiO_2 surface has been etched down to provide full rotational freedom for the rotor plate. The entire actuator assembly is integrated on a Si chip. (b) Scanning electron microscope (SEM) image of nanoactuator just prior to HF etching. The actuator components can be identified by comparing this image to (a). Scale bar: 300 nm. Reprinted by permission from A. M. Fennimore, T. D. Yuzvinsky, Wei-Qiang Han, M. S. Fuhrer, J. Cumings, & A. Zettl, (2003). *Nature, 424*, 408, ©(2003) Nature.

a dramatic increase in the rotational freedom of the rotor plate. In this state, the rotor plate was still held by the undamaged inner nanotubes and could be azimuthally positioned by applying the appropriate combination of stator signals. The intrinsic low friction bearing means that complete rotation and control of the rotor plate can be achieved. In Figure 1.55, a sequence of SEM images shows that the device actuation, in this case via a set of quasi-static DC stator voltages, allows the device to be set to any rotational position.

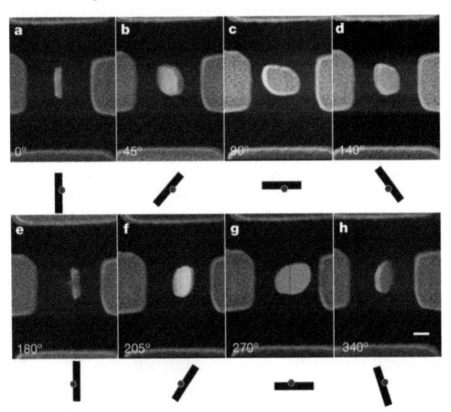

FIGURE 1.55   Sequence of SEM images showing the actuator rotor plate at different angular displacements. The MWNT, barely visible, runs vertically through the middle of each frame. The schematic diagrams located beneath each SEM image illustrate a cross-sectional view of the position of the nanotube/rotor-plate assembly. Scale bar: 300 nm. Reprinted by permission from A. M. Fennimore, T. D. Yuzvinsky, Wei-Qiang Han, M. S. Fuhrer, J. Cumings, & A. Zettl, *Nature, 424,* 408 (2003), ©(2003) Nature.

Finite frequency operation was also possible, where a variety of suitably phased AC and DC voltage signals to the three stators was applied. For example, in one simple operation, an out-of-phase common frequency sinusoidal voltage signal was applied to stators S1 and S2 with a frequency-doubled signal to S3 and a DC offset voltage to the rotor plate R; $V_{S1} = V_0 \sin(\omega t); V_{S2} = V_0 \sin(\omega t - \pi); V_{S3} = V_0 \sin(2\omega t + \pi/2); V_R = -V_0$. While the spatial and temporal drive forces are quite complex, the sequence allowed the rotor plate to be sequentially electrostatically attracted to the next available stator. In principle, the high-frequency operation should be possible, though measurements were limited by the image capture rate in the SEM.

The operation of nanocantilevers has also been demonstrated using feedback control (Ke and Espinosa, 2004). The device was made from a MWNT, which was placed on a microfabricated step. A bottom electrode allows the actuation via a potential applied between it and the carbon nanotube. With a voltage of $U < V_{PI}$, where V_{PI} denotes the pull-in voltage, the electrostatic force is balanced by the elastic force from the deflection of the nanotube cantilever. In this case, the nanocantilever remains in the "upper" equilibrium position. When the applied voltage exceeds the pull-in voltage, the electrostatic forces overcome the elastic force and the nanotube will accelerate towards the lower electrode. When the tip of the nanotube is in close proximity with the lower electrode (i.e., for a gap of around 0.7 nm), there will be a significant tunnel current that will pass between the tip and the electrode. Instabilities can occur in the system due to the resistance in the control circuit. If there are damping forces, this will allow the system to stabilize to a "lower" equilibrium position. If the applied potential is decreased, the tip of the nanotube will retract. Reducing to a certain value will allow the system to pull-out to the upper equilibrium point. In this manner, a pull-in and the pull-out process is observed to follow a hysteretic loop for tip position, Δ, as a function of the applied voltage, U. In Figure 1.56 we show the variation of the cantilever deflection as a function of the applied voltage. Such systems offer a number of potential applications such as the testing of materials and structures, gap sensing, NEMS switches, logic devices, and memory elements.

Carbon nanotube-based resonators and oscillators have been reported by a number of groups. Typically, such systems are fabricated from a doubly clamped nanotube. The resonance frequency of such systems can be widely tuned and devices can be employed to transduce very small forces. In a typical example, a single or few multi-walled nanotubes, with diameters in the range of 1–4 nm, have been suspended over a micron-sized trench, between two Au/Cr electrodes, (Sazonova et al., 2004). Actuation and detection are performed using electrostatic forces using a gate electrode beneath the nanotube. A gate voltage, V_G, will induce an additional charge o the nanotube, which is expressed as $q = C_G V_G$, where C_G denotes the capacitance between the nanotube and the gate electrode. The electrostatic attraction between the charge q on the nanotube and the opposite charge on the electrode will cause an electrostatic force on the nanotube towards the gate electrode. If we denote the derivative of the gate capacitance with respect to the distance between the nanotube and the gate as $C'_G = dC_G/dz$, the total electrostatic force on the nanotube can be expressed as:

$$F_{el} = \frac{1}{2}C'_G V_G^2 \simeq \frac{1}{2}C'_G V_G^{DC}(V_G^{DC} + 2\delta V_G) \tag{1.106}$$

This expression assumes that the gate voltage has both a static (DC) component and a small time-varying (AC) component. The DC component of the gate voltage, V_G^{DC} produces a static force on the nanotube that can be used to control its tension. The AC component, δV_G, will induce a periodic electric

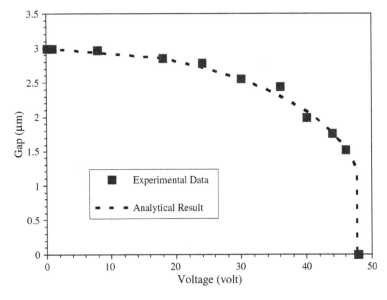

FIGURE 1.56 Comparison of the cantilever deflection as a function of applied voltage for experimental data and theoretical prediction in the finite kinematics regime. The analytical model includes finite kinematics, the van der Waals force and charge concentration at the free end of the nanotube cantilever. Reprinted from C.-H. Ke, N. Pugno, B. Peng, & H. D. Espinosa, (2005). Experiments and modeling of carbon nanotube-based NEMS devices, *Journal of the Mechanics and Physics of Solids, 53,* 1314–1333, Copyright (2005), with permission from Elsevier.

force, which can be used to set the nanotube in motion. As with any driven resonance, as the driving frequency, ω, approaches the natural resonance frequency, ω_0, of the nanotube, the displacement will reach a maximum.

The detection of the vibrational motion of the nanotube relies on the semiconducting properties of the nanotube with its small bandage. The transiting behavior is used to detect the conductance changes, which are proportional to the change in the induced charge:

$$\delta q = \delta(C_G V_G) = C_G \delta V_G + V_G \delta C_G \tag{1.107}$$

The first term is the transistor gating effect, the modulation of conductance due to the variation of the gate at the driving frequency. The second term is non-zero only if the nanotube moves (when the driving frequency approaches the resonance); the distance to the gate changes, resulting in a variation δC_G in its capacitance. Indeed, the current through the nanotube as a function of the driving frequency can be used to detect the vibrational motion. This is shown in Figure 1.57, where the shift of the resonance frequency illustrates the importance of the gate voltage and its effect on the resonance frequency.

The ultimate limit of the force sensitivity attainable with this device is limited by the thermal vibrations of the nanotube, which can be expressed as:

$$\delta F_{min} = \sqrt{\frac{4kk_B T}{\omega_0 Q}} \tag{1.108}$$

which for typical values can be evaluated as 20 aN $Hz^{1/2}$. Sensitivities were observed to be 50 times lower in reality, which is linked to the low values of transconductance for the measured nanotubes, with sensitivity increasing as the temperature is lowered. Estimates give a value of the ultimate sensitivity in the range of about 5 aN at low temperatures. The combination of high sensitivity, tenability, and high-frequency operation makes this device promising for a number of scientific and technological applications.

1.5 Summary

The mechanical properties of solids have been studied for centuries and determine many of the physical applications of materials. One of the most important of the physical properties is that which concerns the relation between the force applied to the solid and its relative deformation. This can be neatly summarised by Young's modulus, which is defined as the ratio of the stress (force per unit cross-sectional area) and the strain (extension divided by the original length). When considering the response of a solid object to an applied force we can plot these two physical quantities. For the linear region, the solid is said to be elastic, and the value of Young's modulus is taken as the gradient of this region. Further loading of the material will lead to per-

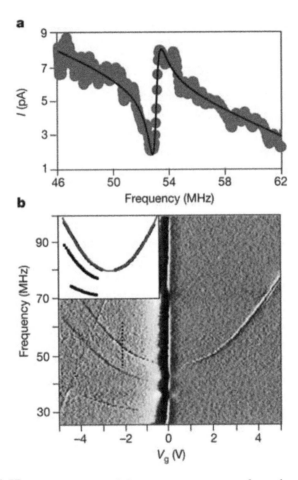

FIGURE 1.57   Measurement of the resonant response of a carbon nanotube oscillator. (a) Detected current as a function of the driving frequency taken at $V_G = 2.2$ V, $\delta V_G = 7$ mV. The solid black line is a Lorentzian fit to the data with an appropriate phase shift between the driving voltage and the oscillation of the tube. The fit yields the resonance frequency of 55 MHz and quality factor $Q = 80$. (b) Detected current (plotted as a derivative in color scale) as a function of gate voltage. The dashed black line shows the trace taken for the curve in (a). Reprinted by permission from V. Sazonova, Y. Yaish, H. Üstúnel, D. Roundy, T. A. Arias, & P. L. McEuen, (2004). *Nature, 431*, 284,©(2004) Nature.

manent or plastic deformation of the solid. Another form of characterization of a solid is via the bulk modulus. This quantity is defined by the variation of an applied pressure with the volume strain of a body. The volume strain indicates the normalized volume change of the solid under the applied pressure. Its inverse is referred to as the compressibility of the solid.

We can further characterize the mechanical properties of the solid using the shear modulus or modulus of rigidity. This is defined as the ratio of the shear stress with the shear strain. A final quantity frequently used in the characterization of the mechanical properties of a solid is the Poisson ratio. This ratio concerns the deformations along the direction of an applied force to that along is the perpendicular axis, i.e., the ratio of the contractile and tensile strains.

The study of micro and nano-objects presents specific problems in the measurement and testing of these properties since it can be rather difficult to physically manipulate the samples. Test structures can be made though much of the test of the mechanical properties are performed using cantilever and doubly clamped beam structures. The cantilever is a well-known structure and its mechanical properties have been studied in much detail. The deflection of the beam can be related to its physical dimensions as well as Young's modulus. This, therefore, provides a standard test method for MEMS and NEMS objects. Indeed, the cantilever has many applications for these technologies. Furthermore, the natural resonances of the cantilever as a function of the frequency of the applied force are also well-known and can also provide an accurate method for mechanical testing.

Micro-electromechanical systems or MEMS are important technological devices, which have found many applications. These are typically objects with dimensions of a few to a few hundred microns and can be used as miniaturized mechanical systems, which can either be controlled electronically or can supply electrical information concerning a mechanical force. These are used in sensing and actuating applications such as pressure sensors, accelerometers, resonators, and gyroscopes. Such devices are used in many everyday applications, such as in smartphones and airbag deployment systems. MEMS structures have also found many uses with optical systems and are referred to as MOEMS or optical MEMS. High definition projection systems have been produced using large arrays of MEMS mirrors, each acting as a single pixel. Fast optical switching applications can also be performed by MOEMS. Radiofrequency applications in the GHz frequency range are also of great interest since these are the natural resonance frequencies of MEMS type structures. They can also be incorporated as high-frequency electronic components, such as tunable capacitors, inductors, and switches.

Another class of applications for micron-sized devices is in the area of microfluidics. Such systems are concerned with the control of fluidic systems with important uses, particularly for biological and biomedical applications. These are the basis for lab-on-a-chip, micro-total analysis systems as well as point-of-care diagnostics. Microfluidic devices include pumps, valves, switches as well as systems for mixing fluids and detection systems, such as micros chemostats. Devices are typically made using soft lithographies with PDMS and have the advantage of being transparent, which means the fluids can often be visualized.

Further miniaturizing the MEMS objects has proven to be a complex task and technologies are not yet ripe enough to be able to make the same type of MEMS structures on the nanoscale. Despite this, there has been huge progress in NEMS. Given the reduced dimensions, device operations can be much faster, with resonators functioning in the microwave range with frequencies up to 100 GHz. The most common form of device is based on the cantilever or the doubly clamped beam. Structures are typically fabricated from Si and can be compatible with CMOS technologies. The coupling of mechanical and optical properties has also progressed, where periodic beam structures can be used with specific photonic properties coupled to the mechanical response of the system. Carbon nanotubes of both the SWNT and MWNT variety also lend themselves to NEMS structures and can be used in cantilever like resonators for example. They have also been used to make electromechanical tweezers.

One particularly interesting application is the use of a multi-walled carbon nanotube as a shaft for a nano-actuator. The MWNT has a metal rotor plate attached to the outer wall of the nanotube, which is anchored to a fixed structure. The rotor is driven until it is sheared from the rest of the outer CNT and is free to rotate on the remaining intact interior portion of the MWNT. Actuation is achieved via a number of electrodes placed in close proximity to the rotor.

1.6   Problems

(1) Show that gravitational forces scale as S^4.
(2) Evaluate the suitability of using a radioactive source to propel a micro-rocket. Make the necessary assumptions to justify your answer.

(3) A MEMS cantilever is made of poly-silicon with the following dimensions: $L = 3$ m; $w = 0.8$ m, $t = 0.1$ m. Evaluate:

  (a) The spring constant for the cantilever.
  (b) The force necessary to produce a deflection of 0.2 m.
  (c) The resonance frequency of the cantilever.

  Poly-silicon has the following physical properties: $E = 160$ GPa; $v = 0.22$; $\rho = 2.33 \times 10^3$ kg m^{-3}.

(4) The figure below shows a comb drive unit used in a MEMS actuator. Calculate the maximum force, maximum displacement, and resonance frequency that you expect to achieve with this design.

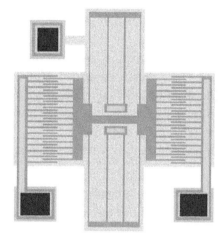

  Use the following data: Width of each finger: 2 m; Thickness of finger: 3.5 m; Spacing between fingers: 2 m; Total number of fingers: 62; Length of fingers: 85 m; Finger overlap: 30 m; Maximum applied voltage: 30 V; Mass: 5×10^{-11} kg; spring constant: 8.3 N m^{-1}.

(5) Why is downscaling for an amperometric device more advantageous than the miniaturization of a potentiometric device? Use equations to prove your point.

(6) Derive an expression for the sensitivity ($\delta R/R$) of a strain gauge, which is based on the resistance change of electrical wire in terms of its dimensions and the Poisson ratio of the wire itself.

(7) For pressure sensing diaphragms of the order of 100 m^2, why is it impractical to use capacitance to measure deflection?

(8) Consider a MEMS-based condenser acoustic transducer as shown in the figure below. The diaphragm is made of silicon nitride and the backplate

is perforated. A stress gauge was used to measure the stress in the silicon nitride and was found to be 1.5×10^8 N m^{-2}.

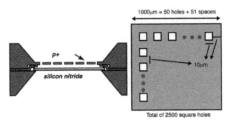

Thickness of air gap = 5 m; Thickness of diaphragm = 5 m; Density of $Si_3N_4 = 3 \times 10^3$ kg/m^3.

(a) Write the differential equation governing the dynamic behavior of the movable diaphragm.

(b) Solve the differential equation under the three cases of overdamped, underdamped, and critically damped conditions.

(c) Calculate the mass of the diaphragm, the small-signal capacitance, and the air-streaming resistance as per the layout shown if the device was to be operated in air and water.

(d) Calculate the cut-off frequency for the overdamped case in air and in water.

(9) A certain process forms Al contacts to n$^+$ Si through a 10 m×10 m contact window resulting in contact resistance of 0.5 Ω.

(a) What is the specific contact resistivity for this contact?

(b) What will the contact resistance be if the contact windows are reduced to 1 m×1 m?

(10) Describe Flexural Resonators. What are their main types? What are their typical frequencies and quality factors? How is possible to control their parameters? How are flexural resonators coupled to external electronic circuitry?

(11) How does the frequency of doubly-clamped resonator depend on its mass, length, thickness and Young's modulus? How can it be actuated and sensed?

(12) Describe the geometry and the principle of work of tunable carbon nanotube electromechanical oscillators. At what frequencies do they resonate? How can actuation and detection of nanotube motion be achieved? What are the different regimes of oscillations and their oscillation modes? How could the motion of nanotube be detected?

References and Further Reading

Akita, S., Nakayama, Y., Mizooka, S., Takano, Y., Okawa, T., Miyatake, Y., Yamanaka, S., Tsuji, M., & Nosaka, T., (2001). *Appl. Phys. Lett., 79,* 1691.

Arbabi, E., Arbabi, A., Kamali, S. M., Horie, Y., Faraji-Dana, M. S., & Faraon, A., (2018). *Nat. Comms., 9,* 812.

Badzey, R. L., Zolfagharkhani, G., Gaidarzhy, A., & Mohanty, P., (2004). *Appl. Phys. Lett., 85,* 3587.

Bauer, W.-A. C., Fischlechner, M., Abell, C., & Huck, W. T. S., (2010). *Lab Chip, 10,* 1814.

Beebe, D. J., Mensing, G. A., & Walker, G. M., (2002). *Annu. Rev. Biomed. Eng., 4,* 261.

Bhushan, B., (2007).*Springer Handbook of Nanotechnology*, B. Bhushan (Ed.), pp. 1305–1336, Springer Science+Business Media Inc., Berlin, Heidelberg, New York.

Bloom, D. M., (1997). *Proc. Int. Soc. Opt. Eng. (SPIE), 3013,* 167.

Buongiorno Nardelli, M., Yakobson, B. I., & Bernholc, J., (1998). *Phys. Rev. Lett., 81,* 4656.

Chang, J., Min, B.-K., Kim, J., Lee, S.-J., & Lin, L., (2009). *Smart Mater. Struct., 18,* 065017.

Dikin, D. A., Chen, X., Ding, W., Wagner, G. J., & Ruoff, R. S., (2003). *J. Appl. Phys., 93,* 226.

Eddington, D. T., & Beebe, D. J., (2004). *Adv Drug Del. Rev., 56,* 199.

Ekinci, K. L., & Roukes, M. L., (2005). *Rev. Sci. Instrum., 76,* 061101.

Fennimore, A. M., Yuzvinsky, T. D., Wei-Qiang Han, Fuhrer, M. S., Cumings, J., & Zettl, A., (2003). *Nature, 424,* 408.

Field, J. E., (2012). *Rep. Prog. Phys., 75,* 126505.

Fraga, M. A., Pessoa, R. S., Barbosa, D. C., & Trava Airoldi, V. J. (2017). *Sensors and Materials, 29,* 39.

Heinrich, S. M., & Dufour, I., (2015). *Resonant MEMS–Fundamentals, Implementation and Applications*, First Edition, Brand, O., Dufour, I., Heinrich, S. M., & F. Josse (Eds.), Wiley–VCH, Verlag GmbH & Co.

Hernández, E., Goze, C., Bernier, P., & Rubio, A., (1998). *Phys. Rev. Lett., 80,* 4502.

Huang, J. Y., Chen, S., Z.Wang, Q., Kempa, K., Wang, Y. M., S.Jo, H., Chen, G., M.Dresselhaus, S., & Z.Ren, F., (2006). *Nature, 439,* 281.

Husain, A., Hone, J., H. W. Ch. Postma, Huang, X. M. H., Drake, T., Barbic, M., Scherer, A., & Roukes, M. L., (2003). *Apps. Phys. Lett., 83*, 1240.

Iannacci, J., (2013). *Practical Guide to RF-MEMS*, Wiley-VCH, Verlag GmbH & Co, Weinheim, Germany.

Jaafar, H., Beh, K. S., Yunus, N. A. M., Hasan, W. Z. W., Shafie, S., & Sidek, O., (2014). *Microsyst. Technol., 20*, 2109.

Jewell, J., Scherer, A., McCall, S., Lee, Y., Walker, S., Harbison, J., & Florez, L., (1989). *Electr. Lett., 25*, 1123.

Judy, J. W., (2001). *Smart Mater. Struct., 10*, 1115.

Juneau, T., Pisano, A. P., & Smith, J., (1997). *Proc., IEEE 1997 Int. Conf. on Solid-State Sensors and Actuators* (Transducers – 97), 883–886.

Kavitha, S., Joseph Daniel, R., & Sumangala, K., (2016). *Measurement, 93*, 327.

Ke, C.-H., Pugno, N., Peng, B., & H.Espinosa, D., (2005). *J. Mech. Phys. Sol., 53*, 1314–1333.

Ke, C., & Espinosa, H. D., (2005). *Handbook of Theoretical and Computational Nanotechnology*, Rieth, M., & W. Schommers (Eds.), vol. 1, pp. 1–38, American Scientific Publishers.

Ke, C., & Espinosa, H. D., (2004). *Appl. Phys. Lett., 85*, 681.

Kim, P., & Lieber, C. M., (1999). *Science, 126*, 2148.

Kim, S., Barbastathis, G., & Tuller, H. L., (2004). *J. Electroceramics, 12*, 133.

Kinoshita, S., Morito, K., Koyama, F., & Iga, K., (1988). *Electr. Lett., 24*, 699.

Kis, A., & Zettl, A., (2008). *Phil. Trans Roy. Soc. A, 366*, 1591.

Kulik, A. J., Kis, A., Gremaud, G., Hengsberger, S., Luengo, G. S., Zysset, P. K., & Forró, L. (2007).*Springer Handbook of Nanotechnology*, B. Bhushan (Ed.), pp1107–1136, Springer Science+Business Media Inc., Berlin, Heidelberg, New York.

Lang, H. P., Hegner, M., & Gerber, C., (2007). *Springer Handbook of Nanotechnology*, B. Bhushan (Ed.), pp. 443–461, Springer Science+Business Media Inc., Berlin, Heidelberg, New York.

Le, H. T., Mizushima, I., Nour, Y., Tang, P. T., Knott, A., Ouyang, Z., Jensen, F., & Han, A., (2017). *Microsystems and Nanoengineering, 3*, 17082.

Lee, C.-C., et al., (2005). *Science, 310*, 1793.

Lee, K. S., Boccazzi, P., Sinskeyb, A. J., & Ram, R. J., (2011). *Lab. Chip, 10*, 1730.

Legtenberg, R., Groeneveld, A. W., & Elwenspoek, M., (1996). *J. Micromech. Microeng., 6*, 320.

Leijssen, R., & Verhagen, E., (2015). *Sci. Rep., 5*, 15974.

Li, S., & Wang, G., (2008). *Introduction to Micromechanics and Nanomechanics*, World Scientific Publishing Co. Pte. Ltd.

Lin, L. Y., Goldstein, E. L., & Tkach, R. W., (1999). *IEEE Photonics Technology Letters, 11*, 1253.

Liu, W. K., Karpov, E. G., & Park, H. S., (2006). *Nano Mechanics and Materials: Theory, Multiscale Methods, and Applications*, 1st Edition, J. Wiley and Sons, Chichester, UK.

Llobet, J., Rius, G., Chuquitarqui, A., X. Borrisé, Koops, R., M. van Veghel, & Perez-Murano, F., (2018). *Nanotechnology, 29*, 155303.

Lu, J. P., (1997). *Phys. Rev. Lett., 79*, 1297.

Lukić, B., et al., (2005).*Appl. Phys. A, 80*, 695.

Madou, M., (2002). *Fundamentals of Microfabrication: The Science of Miniaturization*, CRC Press, Boca Raton.

Miguel, J. A., Lechuga, Y., & Martinez, M., (2018). *Micromachines, 9*, 342.

Mohammed, Z., Dushaq, G., Chatterjee, A., & Rasras, M., (2018). *Mechatronics, 54*, 203.

Nanot, S., Thompson, N. A., J.-Kim, H., Wang, X., Rice, W. D., Hárose, E. H., Ganesan, Y., Pint, C. L., & Kono, J., (2013). *Springer Handbook of Nanomaterials*, R. Vajtai (Ed.), pp. 105–146, Springer Science+Business Media Inc., Berlin, Heidelberg, New York.

Nelson, B. J., & Dong, L., (2007). *Springer Handbook of Nanotechnology*, B. Bhushan (Ed.), pp. 1545–1571, Springer Science+Business Media Inc., Berlin, Heidelberg, New York.

Nguyen, N.-T., Hejazian, M., C. Hong Ooi, & Kashaninejad, N., (2017). *Micromachines, 8*, 186.

Pamme, N., (2007). *Lab Chip, 7*, 1644.

Poncharal, P., Wang, Z. L., Ugarte, D., & de Heer, W. A. (1999). *Science, 283*, 1513.

Ruoff, R. S., Qian, D., & Liu, W. K., (2003). *C. R. Physique, 4*, 993.

Salvetat, J.-P., Kulik, A. J., Bonard, J. M., Briggs, G. A. D., T. Stókli, K. Méténier, Bonnamy, S., Béguin, F., Burnham, N. A., & Forró, L. (1999). *Adv. Mater., 11*, 161.

Salvetat, J.-P., Bonard, J. M., Thomson, N. H., Kulik, A. J., L. Forró, Benoit, W., & Zuppiroli, L., (1999). *Appl. Phys. A, 69*, 255.

Salvetat, J.-P., Andrew, G., Briggs, D., Bonard, J.-M., Bacsa, R. R., Kulik, A. J., Stöckli, T., Burnham, N. A., & Forró, L. (1999). *Phys. Rev. Lett. 82*, 944.

Sazonova, V., Yaish, Y., Üstúnel, H., Roundy, D., Arias, T. A., & McEuen, P. L., (2004). *Nature, 431*, 284.

Sillero, E., Williams, O. A., Lebedev, V., Cimalla, V., C.-Rohlig, C., Nebel, C. E., & Calle, F., (2009). *J. Micromech. Microeng., 19*, 115016.

Sinha, S., Shakya, S., Mukhiya, R., Gopal, R., & B.Pant, D., (2014). ISSS International Conference on Smart Materials, Structures, and Systems, Bangalore, India, July 08–11.

Song, H., Bringer, M. R., Tice, J. D. Gerdts, C. J., & Ismagilov, R. F., (2003). *Appl. Phys. Lett., 83*, 4664.

Sparks, D., Slaughter, D., Beni, R., Jordan, L., Chia, M., Rich, D., Johnson, J., & Vas, T., Sens. Mater., *11*, 197 (1999).

Srivastava, D., Menon, M., & Cho, K., (1999). *Phys. Rev. Lett. 83*, 2973.

Stone, A. J., Wales, D. J., (1986). *Chem. Phys. Lett., 128*, 501.

Tabor, D., (1979). *Gases, Liquids and Solids*, Cambridge University Press, Cambridge, UK.

Tarn, M. D., & Pamme, N., (2013). *Microfluidics*, Reference Module in Chemistry, Molecular Sciences and Chemical Engineering, Elsevier.

Toshiyoshi, H., & Fujita, H., (1996). *IEEE J. Microelectromechanical Systems, 5*, p. 231.

Treacy, M. M. J., Ebbesen, T. W., & Gibson, J. M., (1996). *Nature, 381*, 678.

Walton, A. J., (1987). *Three Phases of Matter*, Oxford University Press, Oxford, UK.

Wang, J., & Li, X., (2013). *J. Micromech. Microeng., 23*, 045027.

Wang, Y., Ding, H., Le, X., Wang, W., & Xie, J., (2017). *Sensors and Actuators A, 254*, 126.

Whitesides, G. M., (2006). *Nature, 442*, 368.

Wong, E. W., Sheehan, P. E., & Lieber, C. M., (1997). *Science, 277*, 1971.

Wong, C.-W., Jeon, Y., Barbastathis, G., & S.-Kim, G., (2003). *Appl. Opt., 42*, 621.

Wu, M. C., Tsai, J.-C., Piyawattanametha, W., & Patterson, P. R., (2012).*Microsystems and Nanotechnology*, Zhou, Z., Wang, Z., & L. Lin (Eds.), Springer Science+Business Media Inc., Berlin, Heidelberg, New York.

Xia, D., Kong, L., & Gao, H., (2015). *Sensors, 15*, 28979.

Xie, H., & Fedder, G. K., (2003). *J. Aerospace Engin., 16*, 65.

Xie, J., Shih, J., Lin, Q., Yang, B., & Tai, Y.-C., (2004). *Lab Chip, 4*, 495.

Yakobson, B. I., Brabec, C. J., & Bernholc, J., (1996). *J. Comput.-Aided Mater. Des., 3*, 173.

Yakobson, B. I., (1998). *Appl. Phys. Lett. 72*, 918–920.

Yang, Y. T., Ekinci, K. L., Huang, X. M. H., Schiavone, L. M., Roukes, M. L., Zorman, C. A., & Mehregany, M., (2001). *Apple. Phys. Lett, 78*, 162.

Young, D. J., Zorman, C. A., & Mehregany, M., (2007). *Springer Handbook of Nanotechnology*, B. Bhushan (Ed.), pp. 415–442, Springer Science+Business Media Inc., Berlin, Heidelberg, New York.

Yu, M.-F., Lourie, O., Dyer, M. J., Moloni, K., Kelly, T. F., & Ruoff, R. S., (2000). *Science, 287*, 637.

Yu, M.-F., Files, B. S., Arepalli, S., & Ruoff, R. S., (2000). *Phys. Rev. Lett., 84*, 5552.

Zhang, R., Ti, C., Davanço, M. I., Ren, Y., Aksyuk, V., Liu, Y., & Srinivasan, K., (2015). *Appl. Phys. Lett., 107*, 131110.

Zhao, B., Moore, J., & Beebe, D., (2001). *Science, 291*, 1023.

Zhu, Y., Moldovan, N., & Espinosa, H. D., (2005). *Appl. Phys. Lett., 86*, 013506.

Zogal, K., Paul, S., Gierl, C., Meissner, P., & KŸppers, F., (2015). Conference: Optical Communication (ECOC), European Conference, Valencia, Spain.

Chapter 2

Electronic Properties at the Nanoscale

2.1 Introduction and Overview of Electronic Properties of Solids

The electrical properties of solids are among the most sought after of the physical characteristics for device applications. This is also true for nanodevices, especially since on this scale quantum effects can be exploited even at room temperature. While metals, with their large number of charge carriers, are an obvious choice of material, semiconductors are extremely important in both electrical and optical devices, as will be discussed in Chapter 3. Indeed, the optical and electronic properties of solids are very closely related, as will become clear from this and the following chapter.

It is informative to provide a short review of the electronic properties of solids as a precursor to the study of the electrical properties of nanomaterials. In this section, we will provide a brief overview of some of the more important basic concepts in electrical transport.

The basic importance to the electronic properties of nano- and mesoscopic systems is the deviation of behavior that arises from a limitation of the dimensions, in which the electrons or charge carriers are free to move. Importantly, this gives rise to a number of quantum effects in the discussion of transport properties of such low-dimensional systems. The fundamental interest is the deviation from the classical behavior as described by Ohm's law; $I = V/R$. This basically states that the current I passing through a sample is a linear function of the applied voltage V. The coefficient of proportionality being related to the resistance R of the material. In this representation, the resistance appears as an extrinsic parameter. The intrinsic property of the material, given as the resistivity ρ, is expressed in the form:

$$R = \frac{\rho L}{A} = \frac{L}{\sigma A} \tag{2.1}$$

where L is the length and A the cross-sectional area of the wire. The second form is expressed in terms of the conductivity, $\sigma = 1/\rho$. Eliminating all dimensional factors, we can express Ohm's law in terms of the current density, $\mathbf{j}$, and the applied electric field, $\mathbf{E}$, as: $\mathbf{j} = \sigma \mathbf{E} = \mathbf{E}/\rho$, we note that the bold symbols refer to the vectorial nature of these quantities. The classical theory of electrical conductivity is generally expressed in terms of the Drude model, dating from 1900. Since the electric force acting on a charge is given by $\mathbf{F}_e = q\mathbf{E}$, the product of the charge of the particle q and the electric field, $\mathbf{E}$, or for electrons $\mathbf{F}_e = -e\mathbf{E}$, we can express the drift velocity of the electrons as:

$$\mathbf{v}_d = -\frac{e\mathbf{E}\tau}{m} \tag{2.2}$$

where τ is the average time the electron travels between collisions and m is the mass of the electron. Given that the current density can be expressed in terms of the density of charge carriers, $\rho_c = n_c e$, where n_c is the concentration of carriers, we can write, $\mathbf{j} = -nce\mathbf{v}_d$, which with Eq. (2.2), gives:

$$\mathbf{j} = \frac{n_c e^2 \tau}{m} \mathbf{E} \tag{2.3}$$

Comparing this with the previous expression for the current density, we find that the conductivity can be expressed in terms of the material properties, n_c and τ, as:

$$\sigma = \frac{n_c e^2 \tau}{m} \tag{2.4}$$

This expression is referred to as the Drude conductivity. We note that the electron density in a metal can be expressed in terms of the Fermi velocity and Fermi density of states at the Fermi level, E_F, as, $n_c \simeq N_F v_F^2 m/2 = N_F E_F$. Often, the mobility of charge carriers is used to characterize the transport properties of a material. This quantity is defined as: $\mu_m = e\tau_m/m$. The quantity τ_m is the collision relaxation time, given by:

$$\tau_m = \frac{\lambda_{mfp}}{v_{th}} \tag{2.5}$$

with the ratio of the mean free path of the electrons, λ_{mfp}, and the thermal velocity of the electrons, v_{th}. The conductivity can be given in terms of the mobility as:

$$\sigma = n_c e \mu_m \tag{2.6}$$

We can relate the thermal energy via kinetic theory to the thermal energy of the charge carriers as: $v_{th} = v_{rms} = \sqrt{3k_B T/m}$. This yields:

$$\sigma = \frac{n_c e^2 \tau}{m} = \frac{n_c e^2 \lambda_{mfp}}{m v_{th}} = \frac{n_c e^2 \lambda_{mfp}}{\sqrt{3mk_B T}} \tag{2.7}$$

Typical values of the mean free path of electrons in metals is of the order of 100 nm, where the relaxation time is of the order of 10^{-14} s. With charge densities of the order of 10^{28} m^{-3}, the conductivities of bulk metals is of the order of $\sim 10^7$ Ω^{-1}m^{-1}.

In low-dimensional systems, the above description is insufficient to understand the transport of charge carriers. There are a number of effects which are encountered, these include, tunneling, energy quantization, single-electron effects and ballistic transport behavior. In this chapter, we will address these main issues and a few others besides. We will also look at some applications and devices to illustrate then how progress has been made in recent decades.

2.2 Length Scales for Transport Processes

The idea of the *length scale* to define physical phenomena is useful in determining at which effective size of object certain effects become noticeable. This is analogous to the mean free path concept that was discussed in Chapter 2 of Volume 1 to define viscous and molecular flow in vacua. In this example, we saw that once the mean free path for molecules in a gas becomes comparable with the dimensions of the vacuum system, molecular flow becomes the dominant flow regime. In the following, we define specific length and time scales for electrons or charge carriers in a solid materials.

2.2.1 *The Fermi Wavelength*

The Fermi wavelength is related to the Fermi energy, E_F, of a material by the equation:

TABLE 2.1   Dimensionality and Fermi Wavelength

d	λ_F
3	$2^{3/2}(\pi/3n_3)^{1/3}$
2	$\sqrt{2\pi/n_2}$
1	$4/n_1$

$$\lambda_F = \frac{h}{\sqrt{2m^*E_F}} \tag{2.8}$$

where E_F denotes the Fermi energy and m^*, the effective mass of the charge carriers in the solid. Incidentally, we can also define the Fermi wave vector and Fermi velocity in terms of the Fermi energy as follows:

$$k_F = \frac{\sqrt{2m^*E_F}}{\hbar} \tag{2.9}$$

and

$$v_F = \frac{\hbar k_F}{m^*} = \frac{h}{m^*\lambda_F} \tag{2.10}$$

which is consistent with the relation: $\lambda_F = 2\pi/k_F$. Size quantization effects occur on length scales of $l \sim \lambda_f$, though quantization can also be observed for larger sized objects. The Fermi wavelength decreases with electron density, n_d, where the d denotes the dimensionality of the system, as illustrated in Table 2.1.

2.2.2 Elastic Scattering Length and Time

The average time between scattering events, τ_q, is related to the quantum scattering length (or quantum mean free path), l_q, via the relation:

$$l_q = v_F\tau_q \tag{2.11}$$

where v_F is the Fermi length as given in Eq. (2.10). This equation indicates the average distance traveled by electrons at the Fermi energy between elastic collisions. It should be noted that this does not determine the resistivity. The electrical resistivity, in the Drude model, for example depends on the average scattering time. This can be simply derived from Newton's second law applied to the drift velocity of the electron, $\langle \mathbf{v} \rangle$, supposing the mean time between collisions to be $\langle\tau\rangle$:

$$m^* \frac{\langle\mathbf{v}\rangle}{\langle\tau\rangle} = -e\mathbf{E} \tag{2.12}$$

From the relation of the current density; $\mathbf{J} = \sigma\mathbf{E} = -ne\langle\mathbf{v}\rangle$, we obtain:

$$\sigma = \frac{ne^2\langle\tau\rangle}{m^*} \tag{2.13}$$

The momentum relaxation, which leads to the momentum relaxation time, τ, should also take into account the scattering angle, ϕ, is weighted by a factor $(1 - \cos\phi)$. The elastic mean free path, l_e is defined as: $l_e = v_F\tau$, and is the average distance covered by an electron between subsequent strong scattering events. For example, in a 2DEG HEMT, $l_e = (\hbar/e)\mu\sqrt{2\pi n_2}$, where $\mu = e\tau/m^*$ denotes the mobility of the n_2 electrons of the two-dimensional electron gas (2 DEG) in the high-electron mobility transistor (HEMT); $l_e = 8$ m.

2.2.3 The Diffusion Constant

The diffusion constant originates from the diffusion equation and relates the gradients in the electron density with the movement of charge carriers. In the 1D model for diffusion, this equation can be expressed as:

$$\frac{dn}{dt} = D\frac{d^2n}{dx^2} \tag{2.14}$$

where D is the diffusion constant.

Brownian motion of the electrons gives rise to a Brownian force, denoted as $b(t)$, which averages to zero for large time intervals. This can be taken into account in the equation of motion by writing $F(t) \to F(t) + b(t)$, giving rise to the Langevin equation for electrons:

$$m^*\left(\frac{dv}{dt} + \frac{v}{\tau}\right) = F(t) + b(t) \tag{2.15}$$

For zero external force, a statistical analysis allows the evaluation of the diffusion coefficient from the autocorrelation function, $C_v(t)$, of the electron velocity:

$$D = \int_{t=0}^{\infty} C_v(t)dt \tag{2.16}$$

where,

$$C_v(t) = \int_{t=0}^{\infty} v(t')v(t+t')dt' \qquad (2.17)$$

Taking the derivative of $C_v(t)$, and replacing dv/dt with the Langevin equation, we obtain:

$$\frac{dC_v(t)}{dt} = \frac{C_v(t)}{m^*} - \frac{C_v(t)}{\tau} \qquad (2.18)$$

The first term vanishes, since there is no correlation between the velocity at t' and the Brownian force at $t+t'$. The solution of Eq. 2.18 leads to:

$$C_v(t) = C_v(0)e^{-t/\tau} \qquad (2.19)$$

and

$$D = C_v(0)\tau \qquad (2.20)$$

Since $C_v(0)$ represents the averaged square of the electron velocity, we can use the equipartition theorem to establish the relation:

$$\frac{1}{2}m^* C_v(0) = \frac{1}{2}k_B\Theta \qquad (2.21)$$

for which we can write the Einstein relation:

$$D = \frac{k_B\Theta\tau}{m^*} = \frac{k_B\Theta}{e}\mu \qquad (2.22)$$

The diffusion constant may also be obtained from the relation: $D = (1/2)v_F^2\tau$. From Eq. (3.107), the conductivity can written as:

$$\sigma = \frac{ne^2}{k_B\Theta}\int_{t=0}^{\infty} C_v(t)dt \qquad (2.23)$$

This is known as the Kubo formula.

2.2.4   *Phase Coherence Length and Dephasing Time*

As we saw above, elastic scattering events will determine the electron mobility via the scattering or relaxation time, τ. At low temperatures, no dephasing will occur, since the phase shift experienced by scattered electrons is reproducible and no fluctuations in the scattering are experienced. Dephasing can occur for various types of processes, such as spin-flip scattering or electron-phonon scattering events. Electron–electron scattering can also lead to dephasing, since energy is transferred between scatterers. However, this latter does not cause resistance, since the total momentum change of the electron

TABLE 2.2 Size and Regimes in Electronic Devices

Macroscopic	Mesoscopic
$L \gg l_q$ – Diffusive	$L \leq l_q$ – Ballistic
$L \gg l_\phi$ – Incoherent	$L \leq l_\phi$ – Coherent
$L \gg \lambda_F$ – No size quantization effects	$L \leq \lambda_F$ – Size quantization effects
$e^2/C < k_B\Theta$ – No single electron charging	$e^2/C \geq k_B\Theta$ – Single electron charging effects

system remains unchanged. Transport effects, which rely on the interference of electronic wave functions, can be used to determine the dephasing time, τ_ϕ. The phase coherence length is related to the dephasing time and refers to the distance an electron travels before its phase becomes randomized. For $\tau_\phi < \tau$, $l_\phi = v_F \tau_\phi$, while for $\tau_\phi > \tau$, electrons are scattered elastically within the phase coherence time and the distance that they can travel becomes reduced. For $\tau_\phi \gg \tau$, usually at low temperatures, $l_\phi = \sqrt{D\tau_\phi}$. In mesoscopic media, $\tau_\phi \sim 1$ ps.

The existence of size effects can be summarized as illustrated in Table 2.2.

2.3 Electrons and Wave Packets

The construction of a wave packet is well-known in quantum mechanics and can be used to describe the quantum mechanical representation of the electron as a wave. The general form of the wave function can be expressed in the form:

$$\psi_{\mathbf{k}}(\mathbf{r},t) = \frac{1}{\sqrt{V}} e^{i[\mathbf{k}\cdot\mathbf{r} - E(\mathbf{k})t/\hbar]} \tag{2.24}$$

Here, $E(\mathbf{k}) = \hbar^2 k^2/2m^*$ is the electron energy. In Eq. (2.24), the electron is considered as occupying a state, which is spread out in space and has an effective volume V. The quantity $|\psi_{\mathbf{k}}|^2$, the probability density function, is independent of position coordinates. In most nanostructures, there will be many electrons and the spin 1/2 fermions will be governed by the Pauli exclusion principle, which ensures that no two electrons can occupy the same quantum state, as defined by the four quantum numbers (n, l, m_l, m_s). If we consider a cube in reciprocal space, centered at position $\mathbf{k}$, with sides $dk_x, dk_y, dk_z \ll |\mathbf{k}|$, the number of states available inside the cube will be $2V dk_x dk_y dk_z/(2\pi)^3$. The fraction of states that are occupied inside the cube is referred to as the filling factor, $f(\mathbf{k})$. The particle density, n, the energy density, ε and the current density j can be expressed as:

$$\begin{bmatrix} n \\ \varepsilon \\ j \end{bmatrix} = \int 2\frac{d^3 k}{(2\pi)^3} \begin{bmatrix} 1 \\ E(\mathbf{k}) \\ e\mathbf{v}(\mathbf{k}) \end{bmatrix} f(\mathbf{k}) \tag{2.25}$$

$\mathbf{v}(\mathbf{k}) = \hbar\mathbf{k}/m^*$ is the electron velocity. The form of $f(\mathbf{k})$ at equilibrium for a given electrochemical potential μ and temperature T can be expressed using the Fermi–Dirac function:

$$f_{eq}(\mathbf{k}) = f_{FD}[E(\mathbf{k}) - \mu] \equiv \frac{1}{1 + e^{[E(\mathbf{k}) - \mu]/k_B T}} \tag{2.26}$$

Here, note that at $T = 0$ K, $\mu = E_F$.

For electrons, which are subject to the electrostatic potential, $U(\mathbf{r},t)/e$, the wave function $\psi(\mathbf{r},t)$ is no longer a plane wave and will obey the time-dependent Schrödinger equation:

$$i\hbar\frac{\partial \psi(\mathbf{r},t)}{\partial t} = \hat{H}\psi(\mathbf{r},t) = \left[\frac{-\hbar^2}{2m^*}\nabla^2 + U(\mathbf{r},t)\right]\psi(\mathbf{r},t) \tag{2.27}$$

For a stationary potential, $U(\mathbf{r},t) = U(\mathbf{r})$, the wave function becomes a stationary function with a temporal dependence:

$$\psi(\mathbf{r},t) = e^{-iEt/\hbar}\psi_E(\mathbf{r}) \tag{2.28}$$

for which we have:

$$\psi_E(\mathbf{r}) = \hat{H}\psi_E(\mathbf{r}) = \left[\frac{-\hbar^2}{2m^*}\nabla^2 + U(\mathbf{r})\right]\psi_E(\mathbf{r}) \tag{2.29}$$

For the case of a 1D potential, where the electrons are confined in a waveguide configuration of rectangular cross-section and are free to move in the x direction, we can set the potential to zero for $|y| < a/2; |z| < b/2$ and ∞ elsewhere. Thus, impenetrable walls are formed in the y and z directions. The electrons will be fully reflected from such walls, thus changing the sign of the wave vectors; $k_y \rightarrow -k_y; k_z \rightarrow -k_z$. The solution of the Schrödinger equation is a superposition of incident and reflected waves:

$$\psi(x,y,z) = e^{ik_x x} \sum_{S_y S_z = -,+} C_{S_y S_z} e^{S_y i k_y y} e^{S_z i k_z z} \tag{2.30}$$

From this, we must have; $\psi(x, y = \pm a/2, z) = \psi(x, y, z = \pm b/2) = 0$. This can be simply stated as the walls of the waveguide must be nodes of the standing wave in both y and z directions. Therefore, the specific values of the wave vectors in these directions must be quantized such that:

$$k_y^{(n)} = \frac{\pi n_y}{a}, \tag{2.31}$$

$$k_z^{(n)} = \frac{\pi n_z}{b}. \tag{2.32}$$

where $n_{y,z} > 0$ are integers. Using $n = (n_y, n_z)$, the wave function can now be expressed as:

$$\psi_{k_x, n}(x, y, z) = \psi_{k_x}(x) \phi_n(y, z) \tag{2.33}$$

with

$$\psi_{k_x}(x) = e^{ik_x x} \tag{2.34}$$

and

$$\phi_n(y, z) = \frac{2}{\sqrt{ab}} \sin[k_y^{(n)}(y - a/2)] \sin[k_z^{(n)}(z - b/2)] \tag{2.35}$$

Equation (2.33) shows that the transverse motion of the electrons is quantized. Thus, the electron in state n (where n indicates the mode number in wave theory or transport channels in nano-device physics) has only one degree of freedom corresponding to the 1D motion along the x direction. The energy spectrum will consist of 1D branches, which are shifted by a channel dependent energy, E_n, as given by:

$$E_n(k_x) = \frac{(\hbar k_x)^2}{2m^*} + E_n; \tag{2.36}$$

where

$$E_n = \frac{\pi^2 \hbar^2}{2m^*} \left(\frac{n_y^2}{a^2} + \frac{n_z^2}{b^2} \right) \tag{2.37}$$

If we place a small potential barrier along the waveguide of the form:

$$U(x) = \begin{cases} U_0; & 0 < x < d \\ 0; & \text{elsewhere} \end{cases} \tag{2.38}$$

The solutions outside the barrier for a given n and energy will have the form of $\psi_{k_x}(x)$ given above. We note that there are two possible solutions, which

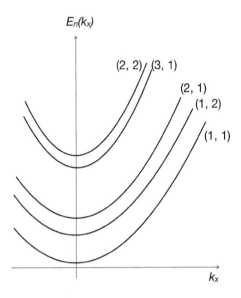

FIGURE 2.1 Energy spectrum for 1D confinement (quantum wire), showing the different branches for the various quantum channels in the system.

have $k_x = \pm k = \pm\sqrt{2m^*(E - E_n)}\hbar$, corresponding to waves propagating to the right (+) or left (-). For a wave arriving, from the left, at the barrier, the wave function can be expressed in terms of the incident (i), reflected (r) and transmitted (t) components as:

$$\psi(x) = \begin{cases} e^{ikx} + re^{-ikx}; & x < 0 \\ Be^{iKx} + Ce^{-iKx}; & 0 < x < d \\ te^{ikx}; & x > 0 \end{cases} \tag{2.39}$$

where $K = \sqrt{2m^*(E - E_n - U_0)}/\hbar = \sqrt{k^2 - 2m^*U_0}/\hbar$. We have normalized the incident intensity ($i = 1$). The wave function and its x derivative must be continuous at $x = 0$ and $x = d$. These four conditions allow the evaluation of the constants r, B, C and t. Of principal interest is the evaluation of r and t, which are related to the reflection and transmission coefficients:

$$T(E) = |t|^2; \quad R(E) = |r|^2 = 1 - T(E) \tag{2.40}$$

It is a relatively simple (though a little lengthy) matter to show:

$$T(E) = \frac{4k^2K^2}{(k^2 - K^2)^2 \sin^2(Kd) + 4k^2K^2} \tag{2.41}$$

In the classical analogue, for $E < U_0$, all particles would be reflected; i.e., $T(E) = 0$, while all particles with $E > U_0$, all would be transmitted, $T(E) = 1$. In the quantum mechanical calculation this is not the case, and an oscillatory solution is observed, as shown in Figure 2.2. For the case

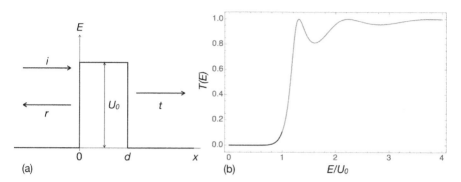

(a)   (b) E/U_0

FIGURE 2.2   (a) Schematic illustration of a potential barrier with incident (i), reflected (r) and transmitted components. (b) Variation of the transmission coefficient as a function of the incident energy E.

where $E < U_0$, K will be imaginary and $T(E)$ is finite with a small possibility of transmission; $T(E) \propto e^{-2d\sqrt{2m^*(U_0+E_n-E)}/\hbar} \ll 1$. This referred to as *tunneling*.

2.4   Quantum Confinement Effects

In the previous section, we have briefly described confinement effects for the case of a waveguide like structure where the walls are constructed of infinite barriers and the superposition of the electron waves create a standing wave for specific values of the electron energy (wavelength) and thus produce discrete allowed energy states of stationary values. The dimensions of the confined system will therefore have a direct bearing on the energy quantization of the system in a very specific manner. This is referred to as *quantum confinement*. We can consider the implications of the Heisenberg uncertainty principle on such a situation in which the electron is confined to a specific region of space. For the 1D case, the uncertainty relation can be expressed in terms of the confinement length Δx and particle momentum as:

$$\Delta p_x = \frac{\hbar}{\Delta x} \tag{2.42}$$

Here we note that if the particle is only confined to the x direction, but free to move in the y and z directions, the x-confinement increases a kinetic energy of the particle (electron) by an amount equal to:

$$E_{\Delta x} = \frac{(\Delta p_x)^2}{2m^*} = \frac{\hbar^2}{2m^*(\Delta x)^2} \tag{2.43}$$

This confinement energy will be significant if it is comparable to or greater than the kinetic energy of the electrode to its thermal motion; i.e., for the condition:

$$E_{\Delta x} = \frac{\hbar^2}{2m^*(\Delta x)^2} \gtrsim \frac{k_B T}{2} \tag{2.44}$$

which leads to the quantum confinement condition:

$$\Delta x \leq \frac{\hbar}{\sqrt{m^* k_B T}} \tag{2.45}$$

For room temperature and an effective mass of $m^* = 0.1 m_e$, significant quantum effects will be observed for $\Delta x \sim 5$ nm. For lower temperatures, Δx will increase.

2.5   Quantum Wells/Wires/Dots

Size quantization can occur in one, two or three dimensions of space. For the confinement in one direction, we define the structure as a quantum well, for confinement in two dimensions, we refer to a quantum wire, while three-dimensional confinement leads to a quantum dot structure.

2.5.1   *The Quantum Well*

The confinement of an electron to a single layer leads to the discretization of the energy of the electrons in the system. Such a structure can be realized using a thin film structure where the electrons can move freely in two dimensions and frequently this is referred to as a two-dimensional electron gas or 2DEG. In the simplest case, we can consider that the electric potential which confines the electron gas is zero within the well and infinity outside. This will make the walls of the well impenetrable to the electrons and allows a simple treatment of the system. The infinite *square-well* potential is a well-known problem in quantum mechanics and we will illustrate this below. The potential can be expressed as:

$$U(x) = \begin{cases} 0; & -a < x < a \\ \infty; & |x| > a \end{cases} \tag{2.46}$$

In the regions of infinite potential, there will be zero probability of finding the electron and as such we can set the wave functions to zero for $|x| > a$; $\psi(x) = 0$. In addition to this, we need to consider the boundary conditions of the wave function and its first derivative at the limits of the quantum well; $x = \pm a$. Since the electron cannot exist in the region of infinite potential, we must have $\psi(\pm a) = 0$, which means that $d\psi(\pm a)/dx$ must be discontinuous at the boundaries, where the potential makes an infinite jump. Such a discontinuity in the derivative is only permitted when the potential makes an infinite jump.

Within the potential well itself, where the potential is defined as zero, the solution of the one-dimensional time-independent Schrödinger equation, expressed as:

$$E\psi(x) = \frac{-\hbar^2}{2m^*} \frac{d^2\psi(x)}{dx^2}; \quad -a < x < a \tag{2.47}$$

Real solutions of this equation can be expressed in a form that will also take into account the boundary conditions mentioned above, as:

$$\psi(x) = A\cos(kx) + B\sin(kx) \tag{2.48}$$

where $k = \sqrt{2m^*E/\hbar^2}$, and $A\cos(ka) = 0; B\sin(ka) = 0$. To avoid the trivial case where $\psi(x) = 0$, we class the solutions as even and odd. In the case of odd solutions we can write:

$$\psi_n(x) = A_n\cos(k_n x) \tag{2.49}$$

with $k_n = n\pi/2a; n = 1,3,5,...$ Normalizing Eq. (2.49), we can write:

$$\psi_n(x) = \frac{1}{\sqrt{a}}\cos\left[\left(\frac{n\pi}{2a}\right)x\right]; \quad n = 1,3,5,... \tag{2.50}$$

In a similar fashion, we obtain the even class solutions with:

$$\psi_n(x) = \frac{1}{\sqrt{a}}\sin\left[\left(\frac{n\pi}{2a}\right)x\right]; \quad n = 2,4,6,... \tag{2.51}$$

The de Broglie wavelengths are expressed as:

$$\lambda_n = \frac{2\pi}{k_n} = \frac{4a}{n} \tag{2.52}$$

The energy of the discrete states are obtained as:

$$E_n = \frac{\hbar^2 k_n^2}{2m^*} = \frac{\hbar^2 n^2 \pi^2}{8m^* a^2}; \qquad n = 1, 2, 3, \ldots \qquad (2.53)$$

The energy spectrum for the electrons in the infinite quantum well will therefore consist of an infinite number of discrete energy states, where each level is described by an individual wave function. In Figure 2.3, we show the energy levels for the different modes, illustrated as $\psi_n(x)$. These are just the standing wave solutions for a wave confined between the fixed boundaries, such as a string. Also shown in Figure 2.3 are the probability density functions; $P_n(x) = |\psi_n(x)|^2$ for the first five modes. If we consider the case for the zero point energy, i.e., the lowest energy state possible for the system, this does not correspond to zero energy. This will be obtained for $n = 1$, where we can write: $E_1 = \hbar^2 \pi^2 / 8m^* a^2 \simeq \hbar^2 / m^* a^2$, and is not very different from the energy we obtained in Eq. (2.44) for $\Delta x = a$.

The full energy relation for this case can be expressed as:

$$E_n(k_y, k_z) = \frac{\hbar^2 (k_y^2 + k_z^2)}{2m^*} + \frac{\hbar^2 n^2 \pi^2}{8m^* a^2}; \qquad n = 1, 2, 3, \ldots \qquad (2.54)$$

where we take into account the three dimensional aspects of the well potential and the fact that the electrons are free to move in the y and z directions. The energy dispersion curve is represented in Figure 2.4. For each energy state E_n, the energy curve will have a parabolic shape in the y and z k-space directions, as indicated in Eq. (2.54). This is similar to the case we illustrated previously for the waveguide system, c.f. Eq. (2.37).

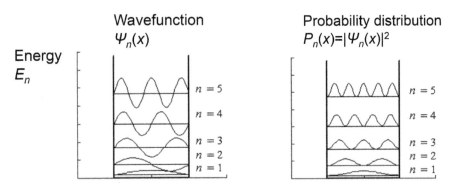

FIGURE 2.3   Illustration of the wave function and probability density function for the first five modes for an infinite square well potential.

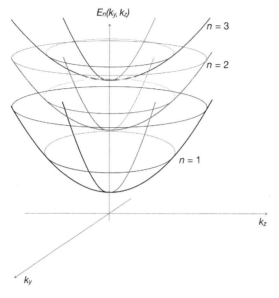

FIGURE 2.4 Energy curves for the first three states, $n = 1, 2$ and 3. Each energy band has a parabolic shape due to the k_y^2 and k_z^2 dependence, see Eq. (2.54).

A more complex situation arises when we consider the case of a finite square well potential, since now the wave functions can penetrate into the regions, which are classically forbidden, i.e., outside the well, even when the electron energies are lower than the potential. The potential in this can be expressed in the form:

$$U(x) = \begin{cases} -U_0; & |x| < a \\ 0; & |x| > a \end{cases} \tag{2.55}$$

This is illustrated in Figure 2.5. As with the infinite square well potential, we have chosen to make the potential symmetric, which allows us to look for even and odd solutions. Contrary to the infinite potential, where all states are bound states, in the finite square potential we need to distinguish between the case where the electrons are of an energy greater than the well height. If not, then they will form bound states within the well, otherwise, they will be free states, though when the particle passes between regions of different potentials, then it may suffer partial reflections. We will consider the bound states first:

i) Bound states: For the potential that we have expressed in Eq. (2.55), the bound states will have $E < 0$. In this situation, we need to express

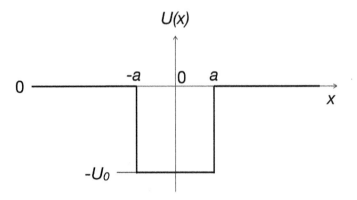

FIGURE 2.5   Illustration of the finite square well potential.

the Schrödinger equation both within the well ($|x| < a$) and outside the well ($|x| > a$):

$$\frac{d^2\psi(x)}{dx^2} + k_1\psi(x); \qquad |x| < a \tag{2.56}$$

and

$$\frac{d^2\psi(x)}{dx^2} - k_2\psi(x); \qquad |x| > a \tag{2.57}$$

where

$$k_1 = \sqrt{\frac{2m^*}{\hbar^2}(U_0 + E)} = \sqrt{\frac{2m^*}{\hbar^2}(U_0 - |E|)} \tag{2.58}$$

and

$$k_2 = \sqrt{-\frac{2m^*}{\hbar^2}E} = \sqrt{\frac{2m^*}{\hbar^2}|E|} \tag{2.59}$$

where $|E| = -E$ represents the binding energy of the electron.

As we stated before, the choice of a symmetric potential allows us to represent the solutions in terms of even or odd parity. For the even parity case, we can consider only the positive values of x, where for $x > 0$, we can write:

$$\psi(x) = A\cos(k_1 x); \qquad 0 < x < a \tag{2.60}$$

and

$$\psi(x) = Ce^{-k_2 x}; \qquad x > a \tag{2.61}$$

In Eq. (2.61), we have taken into account that the wave function cannot have a term of the form $De^{k_2 x}$, which would increase to infinity as $x \to \infty$, so we set $D = 0$. The boundary conditions for the case of the finite square well potential will be that $\psi(x)$ and $d\psi(x)/dx$ are continuous at $x = a$. From this, we find:

$$A\cos(k_1 a) = Ce^{-k_2 a} \tag{2.62}$$

and

$$-k_1 A\sin(k_1 a) = -k_2 Ce^{-k_2 a} \tag{2.63}$$

which yields:

$$k_1 \tan(k_1 a) = k_2 \tag{2.64}$$

Repeating this procedure for the odd parity solutions, we have:

$$\psi(x) = B\sin(k_1 x); \qquad 0 < x < a \tag{2.65}$$

and

$$\psi(x) = Ce^{-k_2 x}; \qquad x > a \tag{2.66}$$

Applying the boundary conditions, we find:

$$B\sin(k_1 a) = Ce^{-k_2 a} \tag{2.67}$$

and

$$k_1 B\cos(k_1 a) = -k_2 Ce^{-k_2 a} \tag{2.68}$$

which yields:

$$k_1 \cot(k_1 a) = -k_2 \tag{2.69}$$

By setting $\xi = k_1 a$ and $\eta = k_2 a$, we obtain:

$$\xi \tan \xi = \eta; \qquad \text{even modes} \tag{2.70}$$

$$\xi \cot \xi = -\eta; \qquad \text{odd modes} \tag{2.71}$$

Both ξ and η are positive and can be expressed as:

$$\xi^2 + \eta^2 = \gamma^2 \tag{2.72}$$

and

$$\gamma = \sqrt{\frac{2m^* a^2}{\hbar^2} U_0} \tag{2.73}$$

The simultaneous solution of Eq. (3.80) with Eq. (3.76) for the even modes and with Eq. (3.77) for the odd modes allows us to obtain the energy levels for the system. These are usually computed and will depend explicitly on the depth of the well (U_0) and its width $(2a)$. The number of bound states will also directly depend on these two parameters. In all of the above, we have maintained the effective mass of the electron, m^*, which will also have an influence on the allowed energies and the number of bound states in the quantum well structure. In Figure 2.5(a), we illustrate the case for three bound states. It is important to notice that the wave functions now extend into the region outside the well, which is classically forbidden. The extension will depend on the energy difference between the state E_n and the top of the potential. In Figure 2.5(b), we compare the cases of infinite and finite square well potentials for a well width of 0.4 nm and the finite well of depth 75 eV.

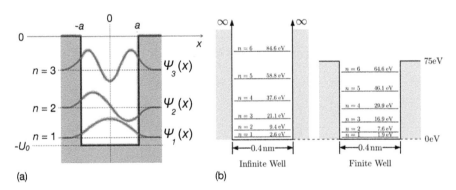

FIGURE 2.6   (a) Illustration of the wave functions for the first three bound states for a finite square well potential. (b) Comparison of the energy levels for the infinite square well potential and finite well (with $U_0 = 75$ eV) and a well width of 0.4 nm.

(ii) Unbound states: We now consider the case, where the electron energy is above the well, which in our case corresponds to $E > 0$. The solution of the Schrödinger equation for the regions with $U(x) = 0$ can be expressed as:

$$\psi(x) = \begin{cases} Ae^{ik_1 x} + Be^{-ik_1 x}; & x < -a \\ Ce^{ik_1 x}; & x > a \end{cases} \tag{2.74}$$

where $k_1 = \sqrt{2m^*E/\hbar^2}$ and A, B and C are constants. This solution takes into account that we have considered the electron to be incident from the left (A) and there is a reflected component (B), for $x < -a$, while in the region $x > a$, only the particle travelling in the $+x$ direction (C) is permitted. In the region of the well itself, $-a < x < a$, where $U(x) = U_0$, we have:

$$\psi(x) = Fe^{ik_2x} + Ge^{-ik_2x}; \quad -a < x < a \tag{2.75}$$

where $k_2 = \sqrt{2m^*(E+U_0)/\hbar^2}$. We note from Eqs. (2.74) and (2.75) that the solutions are of the form of plane waves moving along the x direction. These solutions only consider the x dependence, since we are concerned with the confinement direction. The full solutions will also retain the plane wave behavior in the y and z directions. Comparing the above with the potential barrier, we see that the situation is indeed similar. Any wave incident upon a change in the potential can be either reflected or transmitted, whether that change be a decrease or an increase in potential. The transmission and reflection probabilities will reflect these differences. These are obtained by taking into account the boundary conditions. For the current situation, we find the transmission, $T = |C/A|^2$, and reflection coefficients, $R = |B/A|^2$, to be of the form:

$$T = \frac{4k_1^2k_2^2}{4k_1^2k_2^2 + (k_1^2 - k_2^2)^2 \sin^2(2k_2a)} = \frac{4E(U_0+E)}{4E(U_0+E) + U_0^2 \sin^2(2k_2a)} \tag{2.76}$$

$$R = \frac{(k_1^2 - k_2^2)^2 \sin^2(2k_2a)}{4k_1^2k_2^2 + (k_1^2 - k_2^2)^2 \sin^2(2k_2a)} = \frac{U_0^2 \sin^2(2k_2a)}{4E(U_0+E) + U_0^2 \sin^2(2k_2a)} \tag{2.77}$$

The sum of these being unity; $T + R = 1$, as expected. We further note that the potential barrier can be obtained by setting $U_0 = -U_0$. The transmission coefficient will be zero for $E = 0$, and will increase as E increases. However, due to the sine function in Eq. (2.76), the variation of $T(E)$ will be oscillatory and will tend towards unity as E becomes very large. The maxima in T, also known as *resonant transmission*, occur for the condition: $2k_2a = n\pi$, while the minima occur for $2k_2a = (2n+1)\pi/2$. The resonant transmission condition can be used to find:

$$E_{res,n} = \frac{\hbar^2 n^2 \pi^2}{8m^*a^2} \tag{2.78}$$

c.f. Eq. (2.53). In Figure 2.7, we illustrate the variation of the transmission coefficient with the particle energy for two different values of the well depth as given by γ from Eq. (3.77).

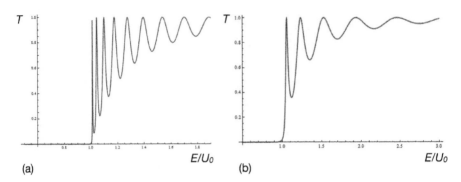

(a)   (b)

FIGURE 2.7   Plot of the transmission coefficient, T against (E/U_0) for a factor of (a) $\gamma = 30$ and (b) $\gamma = 13$.

In reality, quantum well structures can be fabricated using MBE grown semiconductor multilayers. A common example would be layers prepared using AlGaAs and GaAs. This is possible since the band gaps for the two materials are quite different; $\Delta_{GaAs} = 1.43$ eV and $\Delta_{AlGaAs} = 2$ eV, where we consider the alloy with equal quantities of Al and Ga. In Figure 2.8, we show a schematic illustration of the energy band diagram along the x direction along with the energy. Since the GaAs layer has a smaller bandgap, the conduction band forms a quantum (finite) well for electrons (e), where we illustrate three bound states. In a similar way, the valence band forms a quantum well for hole states (h). Since the confinement adds small energy to the well state energy with respect to that for a non-confined structure, the energy minimum for a transition from the valence to conduction band will correspond to $E_{min} \neq \Delta_{GaAs}$. This fact means that we can tune the energy transition just by controlling the thickness of the GaAs layer, t_{GaAs}. This has some very important implications for device functioning in quantum well structures. We will discuss this in more detail in the following chapter, when we consider the optical transitions in such devices.

2.5.2   *The Quantum Wire*

The quantum wire is a confined system, in which the electrons are permitted to move freely in only one direction. This situation has been illustrated above in the waveguide structure discussed in Section 2.3. It is a simple matter to extend the results for the quantum well for the quantum wire, where the energies of the bound states for an infinite well are given by:

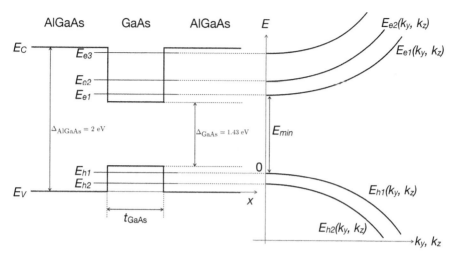

FIGURE 2.8 Schematic illustration of the energy band diagram for an Al-GaAs/GaAs quantum well structure. The energy diagram (right) shows the parabolic nature of the energy bands, marked as $E_{(e,h)n}(k_y, k_z)$. In fact, these could also be presented as the energy bands illustrated in Figure 2.4.

$$E_{n_x,n_y}(k_z) = \frac{\hbar^2 k_z^2}{2m^*} + \frac{\hbar^2 \pi^2}{8m^*}\left(\frac{n_x^2}{a^2} + \frac{n_y^2}{b^2}\right); \qquad n_{x,y} = 1,2,3,\dots \qquad (2.79)$$

where a and b denote the confinement lengths in the x an y directions, respectively, and n_x and n_y are the integers for the quantized states in these two directions, c.f. Eqs. (2.54) and (2.37). Equation (2.79) will be applied to the situation for the quantum point contact, which will be discussed in a later section in this chapter.

The wave functions of the various states will also display the same wave-like character along the wire, while having the node like structure in the transversal directions. In this case, the wave functions can be represented as a 2D membrane and can be expressed in the form:

$$\psi_{n_x,n_y}(x,y) = \frac{2}{\sqrt{ab}} \sin\left(\frac{n_x \pi x}{a}\right) \sin\left(\frac{n_y \pi y}{b}\right) \qquad (2.80)$$

Some examples are illustrated in Figure 2.9.

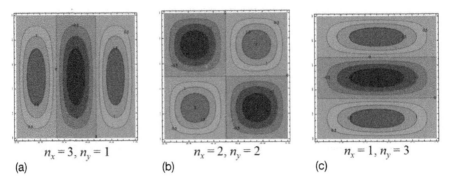

$$n_x = 3, n_y = 1 \qquad n_x = 2, n_y = 2 \qquad n_x = 1, n_y = 3$$

(a)   (b)   (c)

FIGURE 2.9   Illustration of the 2D wave function profiles for states with $n_x = 3, n_y = 1$, $n_x = 2, n_y = 2$, and $n_x = 1, n_y = 3$.

2.5.3   The Quantum Dot

The quantum dot refers to the case, where the electron/particle is confined in all three directions of space. Such structures can be formed lithographically or, more frequently, are simply nanoparticles typically of made of a semi-conducting material. We can extend the models we have discussed above for the quantum well and the quantum wire by considering the case of a box like structure. For the case of dimensions, (a,b,c), the energy equation for the infinite potential well is given by:

$$E_{n_x,n_y,n_z} = \frac{\hbar^2 \pi^2}{8m^*} \left(\frac{n_x^2}{a^2} + \frac{n_y^2}{b^2} + \frac{n_z^2}{c^2} \right); \qquad n_{x,y,z} = 1,2,3,\dots \qquad (2.81)$$

The corresponding wave functions take the form:

$$\psi_{n_x,n_y,n_z}(x,y,z) = \frac{2\sqrt{2}}{\sqrt{abc}} \sin\left(\frac{n_x \pi x}{a}\right) \sin\left(\frac{n_y \pi y}{b}\right) \sin\left(\frac{n_z \pi z}{c}\right) \qquad (2.82)$$

For the simple case of a cubic particle, the energy given by Eq. (2.81) becomes:

$$E_{n_x,n_y,n_z} = \frac{\hbar^2 \pi^2}{8m^* a^2} \left(n_x^2 + n_y^2 + n_z^2 \right); \qquad n_{x,y,z} = 1,2,3,\dots \qquad (2.83)$$

The quantum states are now fully discrete and expressed in terms of the three spatial quantum numbers n_x, n_y, n_z. The fourth quantum number remains the spin quantum number. In Figure 2.10, we illustrate the isosurface of the probability density function for the wave function, $\psi_{123}(x,y,z)$.

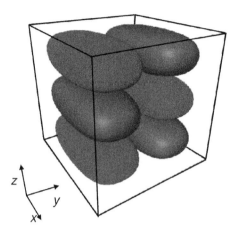

FIGURE 2.10   Illustration of the cubic quantum dot wave function profile for the state with $n_x = 1, n_y = 2, n_z = 3$.

For the case of spherical nanoparticles with radius R, the energy equation will be modified and is expressed as:

$$E_{nl} = \beta_{nl} \frac{\hbar^2}{8m^* R^2} \tag{2.84}$$

where β_{nl} are defined by Bessel functions. We will consider some further applications of the quantum dot when we deal with the optical properties of nanostructures in Chapter 3.

2.6   Density of States in Confined Systems

Dimensionality will clearly have an important effect on the allowed states in solid nanostructures. In this section, we will illustrate this using the so-called density of states function for different types of structure. It is instructive to consider a bulk (3D) structure, which we will assume to be a cube of side L. As we saw previously, the allowed wave vectors in say the x direction can be expressed as:

$$k_x = \frac{2n_x \pi}{L} \tag{2.85}$$

We can similarly express the wave vectors for the other directions as:

$$k_y = \frac{2n_y \pi}{L}; \quad \text{and} \quad k_z = \frac{2n_z \pi}{L} \tag{2.86}$$

c.f. Eqs. (2.31) and (2.32). The volume of k-space occupied per state will thus be:

$$V_k^{3D} = \frac{(2\pi)^3}{L^3} = \frac{(2\pi)^3}{V} \tag{2.87}$$

Each allowed state in reciprocal (k) space will therefore be evenly spaced, as illustrated in Figure 2.11.

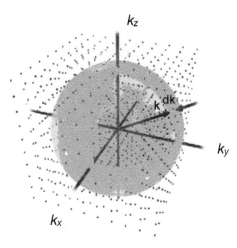

FIGURE 2.11 The allowed states in k-space. States are evenly spaced by $2\pi/L$, where L is the distance over which periodic boundary conditions apply to the free particles.

Assuming the states are closely spaced, we can evaluate the number of states in the thin shell between k and $k + dk$. The volume between these shells is given by:

$$V_k^{3D,\text{shell}}dk = 4\pi|k|^2dk \tag{2.88}$$

We can now express the number of states as twice the ratio of these volumes, which we can express as:

$$g(k)^{3D}dk = 2\frac{4\pi|k|^2dk}{(2\pi)^3/L^3} = L^3\frac{|k|^2dk}{\pi^2} \tag{2.89}$$

The extra factor two takes into account the spin of the electron. Using the fact that $E = \hbar^2k^2/2m^*$, we can express the density of states in terms of the energy E in the form:

$$g(E)^{3D}dE = \frac{1}{2\pi^2}\left(\frac{2m^*}{\hbar^2}\right)^{3/2}E^{1/2}dE \tag{2.90}$$

We can apply the same approach for a 2D system, where we have:

$$V_k^{2D} = \frac{(2\pi)^2}{L^2} \qquad (2.91)$$

and

$$V_k^{2D,\text{ring}} dk = 2\pi |k| dk \qquad (2.92)$$

and leads to the expression for the 2D density of states, see Figure 2.12:

$$g(k)^{2D} dk = 2\frac{2\pi |k| dk}{(2\pi)^2/L^2} = L^2 \frac{|k| dk}{\pi} \qquad (2.93)$$

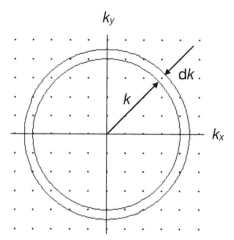

FIGURE 2.12 The allowed states in 2D k-space. States again are evenly spaced by $2\pi/L$, where L is the distance over which periodic boundary conditions apply to the free particles.

Making the change of variable from wave vector to energy as before, we obtain:

$$g(E)^{2D} dE = \frac{m^*}{\pi \hbar^2} dE \qquad (2.94)$$

Therefore, the 2D density of states does not depend explicitly on the energy. As the top of the energy gap is reached, there is a significant number of available states. Taking into account the other energy levels in the quantum well, the density of states takes on a staircase-like function given by:

$$g(E)^{2D} dE = \frac{m^*}{\pi \hbar^2} \sum_j \Theta(E - E_j) dE \qquad (2.95)$$

where $\Theta(E - E_j)$ is the so-called Heaviside function, which gives a step at the energy E_j.

In one dimension, two of the k-components are fixed, therefore the area of k-space becomes a length and the area of the annulus becomes a line. Now we find:

$$V_{\mathbf{k}}^{1D} = \frac{2\pi}{L} \tag{2.96}$$

and

$$V_{\mathbf{k}}^{1D,line} dk = 2dk \tag{2.97}$$

Therefore, the density of states per unit length in 1D can be expressed as:

$$g(\mathbf{k})^{1D} d\mathbf{k} = \frac{2L}{\pi} d\mathbf{k} \tag{2.98}$$

For one dimension, the density of states per unit volume at energy E is given by

$$g(E)^{1D} dE = \frac{1}{\pi} \left(\frac{2m^*}{\hbar^2} \right)^{1/2} \frac{dE}{E^{1/2}} \tag{2.99}$$

Using more than the first energy level, the density of states function becomes:

$$g(E)^{1D} dE = \frac{1}{\pi} \left(\frac{2m^*}{\hbar^2} \right)^{1/2} \sum_j \left[\frac{n_j \Theta(E - E_j)}{(E - E_j)^{1/2}} \right] dE \tag{2.100}$$

In a 0D structure, the values of k are quantized in all directions. All the available states exist only at discrete energies and can be represented by a delta function. In real quantum dots, however, the size distribution leads to a broadening of this line function. In Figure 2.13, we illustrate the form of the density of states for different cases.

2.7   Quantum Point Contacts

Quantum wires (QWR) are quasi-1D systems with a width of $W \simeq \lambda_F$. The wire is strictly 1D only, if the mode with the lowest energy is occupied. The QWR is designated as diffusive, if its length $L \gg l_e$, the elastic mean free path. In this case, the electrons suffer many collisions in their passage through the wire, see Figure 2.14(a). When the length of the wires is short enough such that $L \ll l_e$, then we say that the transport of electrons (or charge carriers) is in the *ballistic* regime and will generally not suffer any

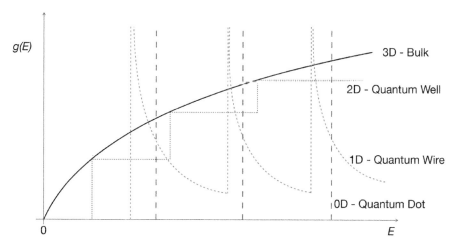

FIGURE 2.13 Density of states for bulk (3D black), quantum well (2D red), quantum wire (1D green) and quantum dot (0D blue).

scattering events in its passage through the device. This is illustrated in Figure 2.14(b). When the wire is very short, the QWR becomes a point contact or a quantum point contact (QPC), in which its width is comparable to its length; $W \simeq L \ll l_e$.

The quantum transport of interest can be evaluated by considering the current through the QPC, which we define as having a cross-section $ab =$

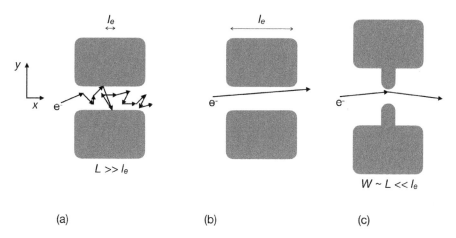

FIGURE 2.14 Transport regimes in quantum wires: (a) Diffusive regime; (b) Ballistic regime; (c) Quantum point contact.

A_{QPC}, where a and b are the lateral dimensions in the y and z directions. The current through the device $I = j_x/ab$ can be expressed in the form:

$$I = 2e \sum_m \int_{-\infty}^{\infty} \frac{dk_x}{2\pi} v_x(k_x) f_m(k_x) \qquad (2.101)$$

where $v_x = \hbar k_x/m^*$. The *filling factors*, $f_m(k_x)$, can be either open $(T = 1)$ or closed $(T = 0)$ for different current channels m. If a channel is closed, then all electrons or charge carriers passing through the cross-section from left are reflected from the barrier or device structure and will subsequently pass the same cross-section from the right. Therefore, for a closed channel, there will be the same amount of right- and left-going electrons, and the filling factors are the same for the two momentum directions; $f_m(k_x) = f_m(-k_x)$. Since their velocities are opposite, the contribution of closed channels to the net current will thus vanish.

For open channels, the filling factor for the two momentum directions is not the same. This leads us to the concept of reservoirs. Any nanostructure taking part in quantum transport is part of an electric circuit, i.e., connected to large macroscopic contact pads at a certain voltage (chemical potential, μ). These pads contain a large number of electrons at thermal equilibrium and are characterized by their respective filling factors. For our QPC, we can define two such reservoirs, the left reservoir (L) and the right reservoir (R). Electrons with $k_x > 0$ come from the left reservoir and will have filling factor: $f_L(E) \equiv f_F(E - \mu_L)$. For the electrons from the right, we have $k_x < 0$ and $f_R(E) \equiv f_F(E - \mu_R)$, see Figure 2.15.

Since the filling factors depend only on the energy, E, we can replace k_x in favor of E for each momentum direction. Since the velocity $v_x = \hbar^{-1}(\partial E/\partial k_x)$, we can write: $dE = \hbar v_x(k_x)dk_x$. We now substitute this into Eq. (2.101), and for open channels we have:

$$I = \frac{2e}{2\pi\hbar} \sum_{m;\text{open}} \int dE[f_L(E) - f_R(E)] \equiv \frac{2e}{2\pi\hbar} N_{\text{open}}(\mu_L - \mu_R) = G_Q N_{\text{open}} V$$

$$(2.102)$$

Here, the integration is made over the energy and yields a factor $(\mu_L - \mu_R)$. The factor 2 arises from the two spin states, so that for each m, there are in fact two channels. The simplest way to integrate is to consider vanishing temperature, where we have: $f_{L,R} = \Theta(\mu_{L,R} - E)$. The difference in the chemical potentials is simply related to the applied potential difference across the QPC: $V = (\mu_L - \mu_R)/e$. This voltage drives the current. Thermodynamic equilibrium occurs for $V = 0; I = 0$. The factor $(\mu_L - \mu_R)$ will be the same for

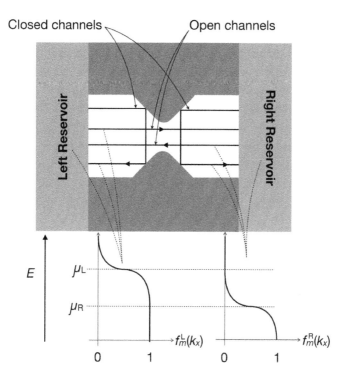

FIGURE 2.15 Conduction channels in a quantum point contact. Electrons are provided by the reservoirs to the left and right of the QPC.

all channels. Equation 2.102 thus illustrates that the current is proportional to the number of open channels, N_{open} and the applied voltage, V. The coefficient of proportionality, G_Q, is referred to as the quantum of conductance, and is defined as:

$$G_Q = \frac{2e^2}{2\pi\hbar} = \frac{2e^2}{h} \tag{2.103}$$

This has a numerical value of $G_Q = 7.7 \times 10^{-5}$ S, or a resistance equivalent of $R \sim 12.9$ kΩ. We note that the conductance of the systems can be expressed as I/V and appears to be quantized in units of G_Q. The quantum of conductance does not depend on the material parameters or properties, nor on the size or geometry of the nanostructure, it is in fact a fundamental constant. The current equation as expressed in Eq. (2.102) is known as the *Landauer* formula. We have derived it here for the case, when $T(E)$ is zero or one. We will revisit this shortly to discuss the general case. What is important to notice is that the variation of the current, as expressed in Eq. (2.102), will have a staircase-like appearance as the number of channels increases one at a time. This will be shown in an example later in this section.

The number of open channels, and consequently, the conductance are determined only by the narrowest portion of the "waveguide." Therefore, we can change the shape of the structure without changing its transport properties, provided that the narrowest cross-section remains unchanged. For a particular waveguide structure, there will be a finite number of channels at each energy and the spectrum will consist of discrete energy branches. In contrast, the number of transport channels approaching the QPC is infinite, with a continuous energy spectrum. Of all these channels, only a finite number can be transmitted through the constriction of the QPC.

The first QPC structures were prepared from a 2DEG fabricated from a quantum well structure of AlGaAs/GaAs. On top of the quantum well are two metallic electrodes which act as gates, see Figure 2.16. By applying an appropriate potential, U_g to the gate, it is possible to constrict the motion of the electrons in the 2DEG. This essentially consists of a depletion region resulting from the applied gate potential in the semiconducting layer. By adjusting the gate voltage, it is therefore possible to control the dimensions of the constricted region, d. Since the GaAs layer thickness is well-defined, it is possible to ensure that the depletion zone extends over the entire thickness of the GaAs (2DEG) region. The potential placed between the ends of the 2DEG ensures that the electrons move only in the x direction. Variation of the gate voltage will alter the extension of the depletion region (along the y direction) between the gates. In the constricted region, we can express the electron energy, where we consider the depletion region to be of infinite potential, in the following relation:

$$E_{k_x}^{QPC}(n_y, n_z) = \frac{\hbar^2 \pi^2}{2m^*} \left\{ \frac{k_x^2}{\pi^2} + \frac{n_y^2}{[L_y(x, U_g)]^2} + \frac{n_z^2}{L_z^2} \right\} \qquad (2.104)$$

where the minimum separation, $L_y^{min} \gg L_z$ in the 2DEG structure. The above equation shows that the electrons are free to move only in the x direction and are quantized in the y and z directions, with quantum numbers, n_y and n_z, respectively. Since the conductance along the x direction is governed by the constriction at its narrowest point, we can essentially ignore the x dependence of L_y. In Figure 2.17, we illustrate the cross-section of the QPC at its narrowest point. The distance between the gates is L_y^g, and the separation between

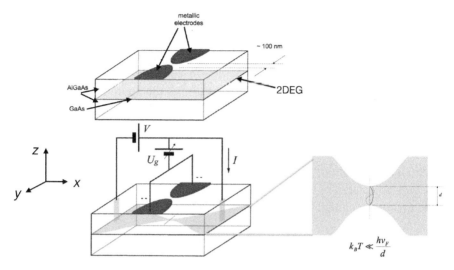

FIGURE 2.16 Quantum point contact fabricated from a 2DEG. An applied potential, V, is placed between the ends of the structure, which acts as the reservoirs, while a further potential, U_g, is placed on the gate structures to control the width, d, of the QPC.

the depletion zones, in which the electrons can pass, is $L_y^{\min}$ and is related to the gate potential through the expression:

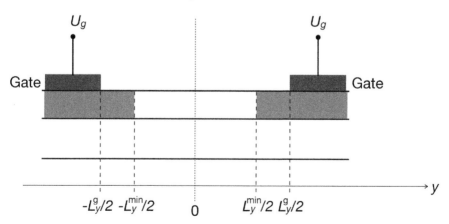

FIGURE 2.17 Quantum point contact fabricated from a 2DEG. An applied potential, V, is placed between the ends of the structure which acts as the reservoirs, while a further potential, U_g, is placed on the gate structures to control the width, d, of the QPC.

$$L_y^{\text{min}}(U_g) = L_y^g - 2L_y^{\text{depl}}(U_g) \tag{2.105}$$

where L_y^{depl} is the extension of the depletion zone below the gate in the y direction and depends on the gate voltage through the approximate relation:

$$L_y^{\text{depl}}(U_g) \simeq \sqrt{\frac{2\varepsilon_{\text{GaAs}}}{e} \frac{(U_{bi} - U_g + V)}{n_{\text{GaAs}}^{\text{2DEG}}}} \tag{2.106}$$

Here, $\varepsilon_{\text{GaAs}}$ is the dielectric constant of GaAs and $n_{\text{GaAs}}^{\text{2DEG}}$ represents the carrier density in the 2DEG of the GaAs layer. Once again, a staircase-like variation of the conductance will be expected for the QPC, as expressed by:

$$G(U_g) = G_Q N_{\text{open}}(U_g) \tag{2.107}$$

which arises from the appropriate Landauer formula with the change in the number of open transport channels, $N_{\text{open}}(U_g)$, which we can write as:

$$N_{\text{open}}(U_g) = \frac{k_F}{\pi} L_y^{\text{min}}(U_g) \tag{2.108}$$

where we assume that the thickness of the 2DEG is such that there is only one mode number (or channel) in the z direction; $N = (n_y, n_z) = (n_y, 1)$. If this is not the case, then further channels would be open and available for the transport of charge carriers. In Table 2.3, we give some of the modal numbers, patterns and number of available channels in the corresponding QPC.

Experimentally, the above observations have verified. In Figure 2.18(a), the schematic illustration of the experiment is illustrated, where an AFM has been adapted to measure the current flowing through a QPC device. Figure 2.18(b) illustrates the step-like conductance discussed above, while in Figure 2.18(c) we show the AFM imaging of the current density under varying values of the gate voltage (and consequently $L_y^{\text{min}}(U_g)$). The bright regions indicate the regions of electron flow. The cases for one, two and three channels are explicitly shown in images A, B, and C of Figure 2.18(c).

2.8   The Scattering Matrix and the Landauer Formula

In order to model the functioning of real nanostructures, we need to recognize that a number of imperfections can exist such as defects and fabrication errors. These will affect the quantum transport through the device. Given

TABLE 2.3   Mode Numbers and Mode Patterns for a Generalized QPC

n_y	n_z	$n_y^2 + n_z^2$	Mode Pattern	No. of Channels
1	1	2		1
1	2	5		
2	1			2
1	3	10		
3	1			3
2	2	8		4
1	4	17		
4	1			4
⋮	⋮	⋮	⋮	⋮

that such imperfections can influence the scattering processes of charge carriers passing through the system, we can describe the transport in terms of the transmission eigenvalues, which are derived from a characteristic scattering matrix.

As we mentioned in the previous section, a nanostructure which is under study will form part of a quantum transport device, which will be connected to electrical circuitry and to reservoirs of charge carriers, that we consider to be in thermal equilibrium at a fixed voltage. In this section, will consider only two connections/reservoirs, though a more general approach can consider any number of connections. Between the two reservoirs, we consider a scattering region, which is the nanostructure itself connected via ideal waveguides to the left (L) and right (R) reservoirs, as illustrated in Figure 2.19.

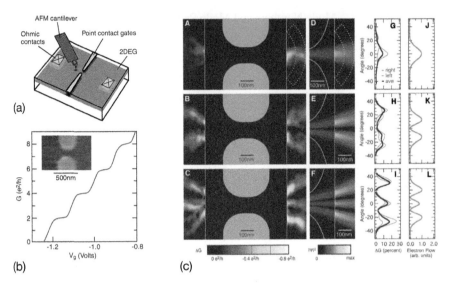

FIGURE 2.18 (a) Schematic diagram of the experimental setup. QPC conductance is measured as a function of the AFM tip position. (b) Point contact conductance G versus gate voltage V_g with no tip present at temperature $T = 1.7$ K. Plateaus at integer multiples of $2e/h$ are clearly seen. The inset shows a topographic image of the point contact device. (c) (A to C) Angular pattern of electron flow of individual modes of the QPC, comparing experiment with theory [(A) first mode, (B) second mode, (C) third mode (see text)]. (D to F) Calculated wave function $|\psi|^2$ for electrons passing from (D) the first mode, (E) the second mode, and (F) the third mode of the QPC (the areas in each simulation corresponding to areas not scanned in the experiment are dimmed). (G to I) Measured angular distribution of electron flow from (G) the first mode, (H) the second mode, and (I) the third mode. (J to L) Angular distribution of the wave function $|\psi|^2$ from (J) the first mode, (K) the second mode, and (L) the third mode. Reprinted by permission from Nature: M. A. Topinka, B. J. LeRoy, R. M. Westervelt, S. E. J. Shaw, R. Fleischmann, E. J. Heller, K. D. Maranowski, & A. C. Gossard, (2001). *Nature, 410*, 183, ©(2001).

The electronic wave functions can be fairly complicated in the scattering region, however, in the waveguides, they will always be of the form of plane waves. The left and right waveguides do not necessarily have the same axis or cross-section and we can define separate coordinates $(x_L < 0, y_L, z_L)$ and $(x_R > 0, y_R, z_R)$ for the left and right waveguides. Generally, a wave function at a fixed energy can be represented as a linear combination of plane waves, such that:

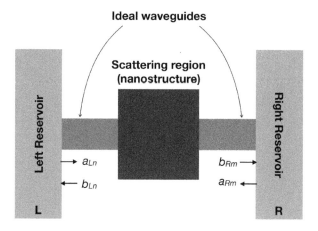

FIGURE 2.19 Idealized nanodevice for consisting of the scattering region, which represents the nanodevice itself and the external circuit made up of voltage source (reservoirs) and connectors (ideal waveguides).

$$\psi(x_L, y_L, z_L) = \sum_n \frac{1}{\sqrt{2\pi\hbar v_n}} \Phi_n(y_L, z_L) \left[a_{Ln} e^{ik_x^{(n)} x_L} + b_{Ln} e^{-ik_x^{(n)} x_L} \right] \quad (2.109)$$

and

$$\psi(x_R, y_R, z_R) = \sum_m \frac{1}{\sqrt{2\pi\hbar v_m}} \Phi_m(y_R, z_R) \left[a_{Rm} e^{-ik_x^{(m)} x_R} + b_{Rm} e^{ik_x^{(m)} x_R} \right] \quad (2.110)$$

In the above equations, we have labeled the transport channels in the left and right waveguides with the indices n and m. respectively. The corresponding transverse wave functions are $\Phi_n(y_L, z_L)$ and $\Phi_m(y_R, z_R)$, which have energies of the transverse motion given respectively by E_n and E_m. For a given transport channel, n or m, from the left or right, the energy will fix the value of the wave vector as $k_x^{(n,m)} = \sqrt{2m^*(E - E_{n,m})}/\hbar$. Since the transport must propagate as plane waves, $k_x^{(n,m)}$ must be real. Then, only a finite number of open channels, N_L to the left and N_R to the right, exist at a specific energy, E. The velocities of the charge carriers, $v_{n,m}$, are explicitly given in Eqs. (2.109) and (2.110) for each channel. This will ensure that the current density does not contain these factors and can be expressed in terms of the wave amplitudes from left and right; $a_{Ln}, b_{Ln}, a_{Rm}, b_{Rm}$. We note that the amplitudes a_{Ln} and a_{Rm} correspond to the waves emerging from the reservoirs, while b_{Ln} and b_{Rm} are the amplitudes of the transmitted/reflected waves from the scattering region, as illustrated in Figure 2.19. These coefficients are therefore not independent; the amplitude of the reflected wave from the scattering region will linearly depend on the amplitude of the incoming waves in all channels. We can therefore relate the a and b amplitudes as:

$$b_{\alpha,l} = \sum_{\beta=L,R} \sum_{l'} s_{\alpha,l} a_{\beta,l'}; \qquad \beta = L,R, \; l = n,m \qquad (2.111)$$

The coefficients of proportionality, $s_{\alpha,l}$, can be combined into a $(N_L + N_R) \times (N_L + N_R)$ scattering matrix, $\hat{s}$, which will have the following structure:

$$\hat{s} = \begin{pmatrix} \hat{s}_{LL} & \hat{s}_{LR} \\ \hat{s}_{RL} & \hat{s}_{RR} \end{pmatrix} \equiv \begin{pmatrix} \hat{r} & \hat{t}' \\ \hat{t} & \hat{r}' \end{pmatrix} \qquad (2.112)$$

The $N_L \times N_L$ reflection matrix, $\hat{r}$, describes the reflection of waves from the left. The element, $r_{nn'}$, corresponds to the amplitude of the process for electrons incident from the left via channel n' and reflected into channel n. This process has a probability represented by $|r_{nn'}|^2$. The $N_R \times N_R$ reflection matrix, $\hat{r}'$, describes the reflection of waves from the right. Similarly, we consider the transmission processes $\hat{t}$ and $\hat{t}'$, which are described by $N_R \times N_L$ and $N_L \times N_R$ matrices, respectively, for the transmission through the scattering region.

All scattering matrices must satisfy the unitary condition; $\hat{s}^{\dagger}\hat{s} = \hat{1}$. The diagonal element of $\hat{s}^{\dagger}\hat{s}$ is given by:

$$(\hat{s}^{\dagger}\hat{s})_{nn} = \sum_{n'} |r_{nn'}|^2 + \sum_m |t_{mn}|^2 = 1 \qquad (2.113)$$

This represents the total probability of an electron in channel n being either reflected or transmitted to any channel.

Returning to the consideration of the currents, we commence with the calculation of the current through a cross-section of the left waveguide; electrons have $k_x > 0$, originating from the left reservoir, with a filling factor $f_L(E)$. Electrons with $k_x < 0$ in a given channel n come from the scattering region, some of these electrons are the result of a reflection and will also have a filling factor $f_L(E)$. This fraction can be determined from the probability of being reflected to channel n from all possible starting channels n'; $R_n(E) = \sum_{n'} |r_{nn'}|^2$. Other electrons will be transmitted across the scattering region and will have a filling factor $f_R(E)$. The overall filling factor for electrons with $k_x < 0$ can therefore be represented by $R_n(E)f_L(E) + [1 - R_n(E)]f_R(E)$.

We now derive the current as:

$$I = 2e \sum_n \left\{ \int_0^\infty \frac{dk_x}{2\pi} v_x(k_x) f_L(E) + \int_{-\infty}^0 \frac{dk_x}{2\pi} v_x(k_x) \{ R_n(E) f_L(E) + [1 - R_n(E)] f_R(E) \} \right\}$$

$$= 2e \sum_n \int_0^\infty \frac{dk_x}{2\pi} v_x(k_x) [1 - R_n(E)] [f_L(E) - f_R(E)] \tag{2.114}$$

From the unitary relation, we have: $1 - R_n(E) = \sum_m |t_{mn}|^2 = (\hat{t}^\dagger \hat{t})_{nn}$. Changing the variable from k_x to E, we obtain:

$$I = \frac{2e}{2\pi\hbar} \int_0^\infty \mathrm{Tr}[\hat{t}^\dagger \hat{t}] [f_L(E) - f_R(E)] dE \tag{2.115}$$

where $\mathrm{Tr}[\hat{t}^\dagger \hat{t}] = \sum_n (\hat{t}^\dagger \hat{t})_{nn}$. Alternatively, the trace can be presented as a sum of eigenvalues, T_e, of the Hermitian matrix $\hat{t}^\dagger \hat{t}$. These are referred to as the transmission eigenvalues. As a consequence of the unitary relation, T_e are real numbers between zero and one, and will depend on the energy of the carriers. For low values of the applied voltage, in the linear regime, the transmission eigenvalues can be evaluated at the Fermi surface and the conductance is given by:

$$G = G_Q \sum_e T_e(\mu) \tag{2.116}$$

Calculations for the currents will be identical in both waveguides, since the current is conserved. Eq. (2.116) is referred to as the two-terminal Landauer formula.

To evaluate the transmission eigenvalues for a given nanodevice, we must solve the Schrödinger equation in the scattering region and match the asymptotes of the wave functions, $\psi(r_L)$ and $\psi(r_L)$, extracting the scattering matrix. This is a rather complex problem since a solution will depend on the specific details of the nanodevice under consideration.

In the simplest case, we can consider a scatterer that can only transmit via a single channel (for a given energy). This means that all other transmission channels are closed, with transmission eigenvalues of zero. In this case, we have a single transmission eigenvalue, T. This is the case for a potential barrier with a reflection coefficient of $R = 1 - T$. The scattering matrix will be a 2×2 matrix of the form:

$$\hat{s} = \begin{pmatrix} \sqrt{R} e^{i\Theta} & \sqrt{T} e^{i\eta} \\ \sqrt{T} e^{i\eta} & -\sqrt{R} e^{i(2\eta - \Theta)} \end{pmatrix} \tag{2.117}$$

The phases Θ and η do not manifest themselves in the transport in a single nanostructure of this type. However, these phases will be relevant, if we combine various nanostructures which produce quantum interference effects.

For the ideal systems, we have considered that the scattering does not mix between channels. For example, an electron in say channel n can either be reflected back and remain in the same channel or it can be transmitted through the scattering region into an identical channel on the other side. Therefore, the matrix of an ideal system is block diagonal, i.e., matrices $\hat{r}, \hat{r}', \hat{t}$ and $\hat{t}^\dagger \hat{t}$ are all diagonal. Thus, the transmission eigenvalues for these systems are just the transmission coefficients in these channels.

The operator formalism is often used when describing scattering problems. In this approach, the use of the creation and annihilation operators are of particular interest. An arbitrary wave function from the left waveguide can be represented as the sum of plane waves. Such plane waves do not, however, form a basis, since they only represent asymptotic expressions of wave functions, which have a complicated form in the scattering region and do not have to be orthogonal. It is convenient to use scattering states, which originate in the reservoirs as plane waves and are then partial transmitted through the scattering region and partially reflected. The scattering state arising in the left reservoir can be expressed as:

$$
\begin{aligned}
\psi_{Ln}(x_L, y_L, z_L) = {} & \frac{1}{\sqrt{2\pi\hbar v_n(E)}} \Phi_n(y_L, z_L) e^{ik_x^{(n)} x_L} \\
& + \sum_{n'} \frac{1}{\sqrt{2\pi\hbar v_{n'}(E)}} r_{nn'}(E) \Phi_{n'}(y_L, z_L) e^{-ik_x^{(n')} x_L}
\end{aligned}
\tag{2.118}
$$

in the left waveguide and:

$$
\psi_{Ln}(x_L, y_L, z_L) = \sum_m \frac{1}{\sqrt{2\pi\hbar v_m(E)}} t_{mn}(E) \Phi_m(y_R, z_R) e^{-ik_x^{(m)} x_R}
\tag{2.119}
$$

in the right hand waveguide. Analogous expressions can be established for the scattering states from the right reservoir for ψ_{Rm}.

For each of these states, we can introduce creation and annihilation operators; $\hat{a}_{Ln}^\dagger(E)$ and $\hat{a}_{Rm}^\dagger(E)$ are the creation operators responsible for "creating" electrons in scattering states with energy E, originating in the left reservoir in transport channel n and in the right reservoir in transport channel m, respectively. The conjugated operators $\hat{a}_{Ln}(E)$ and $\hat{a}_{Rm}(E)$ are responsible for the annihilation of electrons in the same states. Operators $\hat{a}^\dagger$ and $\hat{a}$ are sufficient for the quantum mechanical description of the system.

For convenience, we also introduce another set of operators; $\hat{b}^{\dagger}_{Ln\sigma}(E)$, which creates an electron with energy E and spin projection σ in the transport channel n of the left waveguide and propagating to the left. A corresponding creation operator can be written for the right hand waveguide for carriers moving to the right, $\hat{b}^{\dagger}_{Rm\sigma}(E)$. The annihilation operators will be of the form $\hat{b}_{Ln\sigma}(E)$ and $\hat{b}_{Rm\sigma}(E)$. As with the previous relations between the amplitudes, see Eq. (2.111), the $\hat{a}$ and $\hat{b}$ operators are linearly related via the scattering matrix:

$$\hat{b}_{\alpha l\sigma}(E) = \sum_{\beta=L,R}\sum_{l'} s_{\alpha l,\beta l'}(E)\hat{a}_{\beta l'\sigma}(E) \tag{2.120}$$

$$\hat{b}^{\dagger}_{\alpha l\sigma}(E) = \sum_{\beta=L,R}\sum_{l'} s_{\alpha l,\beta l'}(E)\hat{a}^{\dagger}_{\beta l'\sigma}(E) \tag{2.121}$$

Since electrons are fermions, the operators $\hat{a}$ obey anti-commutation relations:

$$\begin{aligned}
\hat{a}^{\dagger}_{\alpha l\sigma}(E)\hat{a}_{\beta l'\sigma'}(E') + \hat{a}_{\beta l'\sigma'}(E')\hat{a}^{\dagger}_{\alpha l\sigma}(E) &= \delta_{\alpha\beta}\delta_{ll'}\delta_{\sigma\sigma'}\delta(E-E'); \\
\hat{a}_{\alpha l\sigma}(E)\hat{a}_{\beta l'\sigma'}(E') + \hat{a}_{\beta l'\sigma'}(E')\hat{a}_{\alpha l\sigma}(E) &= 0; \\
\hat{a}^{\dagger}_{\alpha l\sigma}(E)\hat{a}^{\dagger}_{\beta l'\sigma'}(E') + \hat{a}^{\dagger}_{\beta l'\sigma'}(E')\hat{a}^{\dagger}_{\alpha l\sigma}(E) &= 0
\end{aligned} \tag{2.122}$$

Operators, which adhere to such relations, are said to form a basis. In the same way, in which operators describe left-moving electrons in the left waveguide, right moving electrons in the right waveguide also form a similar basis. We note that similar relations exist between the $\hat{b}$ and $\hat{b}^{\dagger}$ operators. This is not the case for mixed operations between $\hat{a}$ and $\hat{b}$ operators.

Considering the quantum mechanical averages using the creation and annihilation operators, and since right moving electrons in the left waveguide originate in the left waveguide, we can write:

$$\langle \hat{a}^{\dagger}_{\alpha l\sigma}(E)\hat{a}_{\beta l'\sigma'}(E')\rangle = \delta_{\alpha\beta}\delta_{ll'}\delta_{\sigma\sigma'}\delta(E-E')f_{\alpha}(E); \quad \alpha = L,R \tag{2.123}$$

The average value of two creation or two annihilation operators is always zero. We proceed by writing the field operators $\hat{\psi}_{\sigma}(\mathbf{r},t)$ and $\hat{\psi}^{\dagger}_{\sigma}(\mathbf{r},t)$, which annihilate and create the electron with a given spin projection, σ, at a specific position, $\mathbf{r}$ at a given time, t. For the left waveguide we have:

$$\hat{\psi}_{\sigma}(\mathbf{r},t) = \int dE\, e^{-iEt/\hbar}\sum_{n}\frac{\Phi_n(y_L,z_L)}{\sqrt{2\pi\hbar v_n(E)}}[\hat{a}_{Ln\sigma}e^{ik_x^{(n)}x_L} + \hat{b}_{Ln\sigma}e^{-ik_x^{(n)}x_L}] \tag{2.124}$$

$$\hat{\psi}_\sigma^\dagger(\mathbf{r},t) = \int dE e^{-iEt/\hbar} \sum_n \frac{\Phi_n^*(y_L,z_L)}{\sqrt{2\pi\hbar v_n(E)}} [\hat{a}_{Ln\sigma}^\dagger e^{ik_x^{(n)}x_L} + \hat{b}_{Ln\sigma}^\dagger e^{-ik_x^{(n)}x_L}] \quad (2.125)$$

From the form of the wave function, the quantum mechanical current density is obtained as:

$$\hat{I}(x_L,t) = \frac{e\hbar}{2im^*} \sum_\sigma \int dy_L dz_L \left[\hat{\psi}_\sigma^\dagger(\mathbf{r},t) \frac{\partial \hat{\psi}_\sigma(\mathbf{r},t)}{\partial x_L} - \frac{\partial \hat{\psi}_\sigma^\dagger(\mathbf{r},t)}{\partial x_L} \hat{\psi}_\sigma(\mathbf{r},t) \right]$$

$$(2.126)$$

To calculate the average current, we only require the time averaged current operator. To simplify the procedure, we imagine that all quantities are periodic in time, with a period $\mathcal{T} \to \infty$. The allowed values of the energy are then found from the condition that exponents of the form $e^{iEt/\hbar}$ are also periodic, hence $E = 2\pi q\hbar/\mathcal{T}$, where q is an integer. Consequently, we can replace $\int dE$ by $2\pi\hbar/\mathcal{T} \sum_n$, from which obtain $\langle e^{i(E-E')t/\hbar} \rangle_t = \delta_{qq'}$, where $\langle...\rangle_t$ denotes the time average. This means that in Eq. (2.126), both field operators must be evaluated at the same energy. From this, we obtain:

$$\langle \hat{I} \rangle_t = \frac{G_Q}{e} \left(\frac{2\pi\hbar}{\mathcal{T}} \right)^2 \sum_{n\sigma} \sum_E [\hat{a}_{Ln\sigma}^\dagger(E)\hat{a}_{Ln\sigma}(E) - \hat{b}_{Ln\sigma}^\dagger(E)\hat{b}_{Ln\sigma}(E)] \quad (2.127)$$

This equation is interpreted as follows: the current in the left waveguide is the number of electrons moving to the right (represented by $\hat{a}^\dagger\hat{a}$) minus the number of electrons moving to the left (represented by $\hat{b}^\dagger\hat{b}$), summed over all channels and all energies. Using expressions (2.120) and (2.121), we can write:

$$\langle \hat{I} \rangle_t = \frac{G_Q}{e} \left(\frac{2\pi\hbar}{\mathcal{T}} \right)^2 \sum_{n\sigma} \sum_{\alpha\beta,ll'} \sum_E \hat{a}_{\alpha l\sigma}^\dagger(E)\hat{a}_{\alpha l\sigma}(E)[\delta_{\alpha L}\delta_{\beta L}\delta_{nl}\delta_{nl'} - s_{\alpha l,Ln}^*(E)s_{\beta l',Ln}(E)]$$

$$(2.128)$$

The final step is to perform the quantum mechanical average to obtain the average current. This may appear odd, since the average of the product of two operators at the same energy will be infinity. However, for a system with discrete energies, we can replace the delta function with the Kronecker delta: $\delta(E-E') \to (\mathcal{T}/2\pi\hbar)\delta_{qq'}$. This cancels one of the factors of $\mathcal{T}$. We can now take the limit $\mathcal{T} \to \infty$, and from the discrete sum, we come back to the integral over energy. Taking into account that the average procedure yields $\alpha = \beta$, $l = l'$ and using the unitary condition, we obtain the Landauer formula.

The scattering matrix approach is well-known in many branches of physics and can be applied to more complex situations. Indeed, nanodevices can be connected to more than two contacts such as gates, etc., which can be used to control the transport. Such multi-terminal devices can also be treated with the scattering matrix. Such a treatment is more complex than the two-terminal nanodevice we have discussed above, however, the physical principles and mathematical formalism will be the same. Indeed, the Landauer formula is also valid for any number of terminals. Such a situation was considered by Büttiker. In this approach, the Landauer formula is extended to sum over all contacts. Using the expression for the two-terminal Landauer formula of the form: $I = (2e/2\pi\hbar)\bar{T}[\mu_1 - \mu_2]$, c.f. Eq. (2.102), summing over all terminal pairs, which are indexed as p and q, we can write:

$$I_p = \frac{2e}{2\pi\hbar}\sum_q[\bar{T}_{q\leftarrow p}\mu_p - \bar{T}_{p\leftarrow q}\mu_q] \tag{2.129}$$

Here, the arrows indicate the direction of current flow, but are generally dropped. We note that $\bar{T}_{p\leftarrow q} = M_{p\leftarrow q}T_{p\leftarrow q}$ is the transmission function, where $T_{p\leftarrow q}$ is the transmission probability from contact p to contact q, and $M_{p\leftarrow q}$ is the number of transverse modes between them. Substituting $V = \mu/e$, we can write:

$$I_p = \sum_q[G_{qp}V_p - G_{pq}V_q] \tag{2.130}$$

where $G_{pq} = (2e/2\pi\hbar)\bar{T}_{q\leftarrow p}$. Using the sum rule, which ensures that the current is zero, when the potentials are equal, and is expressed as: $\sum_q G_{qp} = \sum_q G_{pq}$, we can write:

$$I_p = \sum_q G_{pq}[V_p - V_q] \tag{2.131}$$

Such multi-terminal devices are very important, since they allow the determination of the currents in a range of devices, from diodes to transistors and multiple coupled quantum dot structures. We will discuss some of these devices in the following sections.

2.9 Resonant Tunneling

The resonant tunneling diode is a specific type of nanodevice with particular transport properties. In the following, we will consider a device based on a double barrier resonant tunneling diode or RTD. This structure is created using a double barrier heterostructure based on the AlGaAs/GaAs quantum well system and is illustrated schematically in Figure 2.20(a). The outer layers of the structure are heavily n-doped, such that the Fermi level sits within the conduction band and effectively act as contacts. The thickness of the barriers, formed of the AlGaAs layers, and the central quantum well region (GaAs) can be varied. In this latter, the discrete quantized levels exist and act as quasi-bound states, the formation of which has been discussed above in Section 2.5.1.

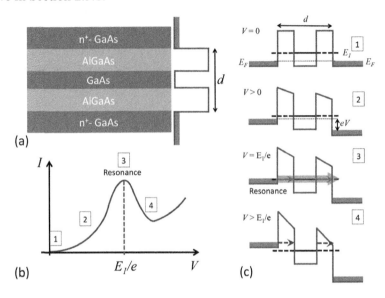

FIGURE 2.20   Resonant tunnel diode (RTD). (a) Basic design using semicon-ductors GaAs and AlGaAs. (b) Current–voltage characteristics and (c) Band dia-grams (conduction band) for different applied voltages, with points indicated (1–4) in (b). As the applied bias increases, the bands of the conduction band shift and the Fermi-level of the left contact approaches the level of the quasi-bound state, E_1. This is accompanied by a gradual increase in the current through the device. Once the Fermi-level reaches the same level as E_1, the transmission probability will reach a maximum. This is the point of resonance. Once the bottom of the conduction band of the contact reaches the level of the quasi-bound state, with further increase of V, the transmission probability will rapidly decrease as no free states are level with E_1.

The resonant tunneling diode can be considered to be a system with two contacts with 3D states, while the QW has 3D electron states. These three subsystems are weakly coupled via the quantum tunneling between them. In Figure 2.20(c), we demonstrate the sequence of events, when we increase the potential, V, between the two contact (n^+-GaAs) layers. For $V = 0$, the sys-tem is in equilibrium and the QW is shown as having a single quasi-bound state, indicated as E_1. This is the lowest of the 2D subbands as in-plane mo-tion is possible within the QW in the transverse directions. We note that the number of quasi-bound states and their energy separation will depend on the details of the heterostructure; well depth and thickness, etc. The diode parameters are chosen such that in the non-biased state, the quasi-bound level, E_1, lies above the Fermi energy, E_F, of the contact layers. Once a small potential is applied to the contacts, the well states are shifted. For electrons with arbitrary energy, a certain tunneling probability through the barrier will exist, though it will be small, since the tunneling from one contact to another is very unlikely due to the broadness of the effective barrier width, which in this case corresponds to the region of width d, see Figure 2.20(a). The structure is indeed designed to prevent thermal transfer. The only situation, which is favorable for the transmission of electrons through the structure, is when the quasi-bound state lies below the Fermi-level and above the bottom of the conduction band, E_C, of the contacts. Here, the electrons from the emitter, or left contact, whose kinetic energy of in-plane motion, $E_\perp = \hbar^2 k_z^2 / 2m^*$, coincides with E_1 are transmitted through the structure with a finite probability. This is the resonant-tunneling process, as illustrated in Figure 2.20(b) at point number 3. There is a small window of enhanced electron transfer, which corresponds to the situation, where electrons in the emitter are level with the energy E_1. This corresponds to the resonant state. Further increase of the applied bias will shift the bands further up on the left and the tunneling probability with fall off rapidly. This region of the $I - V$ characteristic corresponds to a region of negative differential resistance and is reminiscent of the action of the Esaki (or tunnel) diode, though with an important difference. In this case, the tunneling occurs via a well-defined quasi-bound state in a quantum well.

Recalling that the $I - V$ characteristics are related to the resistance, R, according to the relation: $I = V/R$, if a conductor has a non-linear $I - V$ characteristic, we can introduce the so-called differential resistance, R_d:

$$R_d = \left(\frac{dI}{dV} \right)^{-1} \tag{2.132}$$

Negative differential resistance is used to denote the situation where $R_d < 0$, which corresponds to the unusual effect of a decrease in current with increasing applied voltage. In RTDs, where there are further quasi-bound levels, further increase of the bias voltage will allow these states to become resonant, with the corresponding peaks in the current.

The resonant transmission of electrons through a double barrier diode, as we have seen will depend on the relative bias applied to the device. In general, we can describe the transmission process via a consideration of the single electron wave functions through the entire quantum structure. In both emitter (cathode) and collector (anode), we can define these wave functions as a plane wave. For the emitter, the incident plane wave can be expressed as:

$$\psi_{em}(\mathbf{r}, z) = A_{em} e^{i(\mathbf{k}\cdot\mathbf{r} + k_z^{em} z)} \tag{2.133}$$

The outgoing wave that has passed through the QW structure can be expressed in similar fashion as:

$$\psi_{col}(\mathbf{r}, z) = B_{col} e^{i(\mathbf{k}\cdot\mathbf{r} + k_z^{col} z)} \tag{2.134}$$

where $\mathbf{k}$ and $\mathbf{r}$ are 2D in-plane wave vectors and z is perpendicular to the layers, in the direction of transmission. k_z^{em} and k_z^{col} are the z-components of the wave vector in the emitter and collector, respectively. In the ideal case, where $\mathbf{k}$ is conserved, we can write:

$$E_{\perp}^{em} - E_{\perp}^{col} = \frac{\hbar^2 (k_z^{em})^2}{2m^*} - \frac{\hbar^2 (k_z^{col})^2}{2m^*} = eV \tag{2.135}$$

The coefficients A_{em} and B_{col} can be found from the solution of the Schrödinger equation with the potential corresponding to the double barrier structure under the voltage bias and using the appropriate boundary conditions. The ratio of these coefficients will define the transmission coefficient:

$$T(E_{\perp}) = \frac{|B_{col}|^2}{|A_{em}|^2} \tag{2.136}$$

As discussed above, the resonant tunneling effect will occur as a sharp peak in the tunneling probability, $T(E_{\perp})$, in the vicinity of the resonance energies, defined by the quasi-bound levels in the QW, E_n. For a structure with symmetric barriers, we can approximate these peaks with the relation:

$$T(E_{\perp}) = \frac{1}{1 + [(E_{\perp} - E_n)/\Delta E]^2} \tag{2.137}$$

where $2\Delta E$ is the full width half-maximum (FWHM) of the transmission peak. This relation assumes that the tunneling probability is unity when $E_{\perp} = E_n$, as illustrated in Figure 2.21. We also introduce the probability of tunneling of the electrons out of the well per second, Γ, which is related to the linewidth as:

$$\Gamma = \frac{\Delta E}{\hbar} \tag{2.138}$$

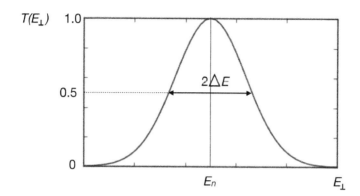

FIGURE 2.21 Idealized resonance peak illustrating the full width half-maximum.

The lifetime of electrons on the quasi-bound level on the quantum well, between the barriers is then given as:

$$\tau = \frac{1}{2\Gamma} \qquad (2.139)$$

The tunneling probability per unit time can be expressed as the product of the attempt rate, $v_z/2L_w$, and the probability of tunneling through a single barrier with one attempt, T:

$$\Gamma = \frac{T v_z}{2L_w} \qquad (2.140)$$

Here, L_w denotes the width of the QW, and v_z is the transverse velocity of the electron in the QW, which is estimated as: $v_z = \sqrt{2E_n/m^*}$. For a quantum well of width 10 nm and barrier width 4 nm with a barrier height of 0.3 eV and $E_1 = 50$ meV, we can estimate $\Gamma = 10^{11}$ s^{-1}, which gives an electron lifetime on the QW as $\tau = 5$ ps.

For the case of an asymmetric barrier structure, it is necessary to calculate the transmission coefficients separately for the left (L) and right (R) barriers. This gives a total transmission coefficient, which can be approximated as:

$$T(E_\perp) = \frac{4T_L T_R}{(T_L + T_R)^2} \frac{1}{1 + [(E_\perp - E_n)/\Delta E]^2} \qquad (2.141)$$

where $\Delta E = \hbar\Gamma$ and $\Gamma = (T_L + T_R)v_z/2L_w$. The transmission even at resonance, $E_\perp = E_n$, will be less than unity.

In order to calculate the current of tunneling electrons, we need to account for all the electrons, which tunnel from the emitter to the collector. We assume a Fermi distribution for the electrons entering the double barrier. If we apply a bias V, the difference between the Fermi energies of emitter, E_F^{em}, and collector, E_F^{col}, will simply be: $E_F^{em} - E_F^{col} = eV$. To obtain the Landauer formula for the RTD, we need to obtain the equation for the net current in the device; $I = I^{em} - I^{col}$. These currents can be expressed in terms of the electron concentrations for the two contacts; $n^{em}(E_\perp)$ and $n^{col}(E_\perp)$:

$$I^{em,col} = \frac{e}{2\pi\hbar} \int dE_\perp T(E_\perp) n^{em,col}(E_\perp) \qquad (2.142)$$

The integration runs over energies $E_\perp$ above the bottom of the conduction band for emitter and collector. For low temperatures, the concentration of electrons takes the form:

$$n^{em}(E_\perp) = \frac{m^*}{\pi\hbar}(E_F - E_\perp) \qquad (2.143)$$

The transverse electron energy is limited by $E_C^{em} < E_\perp < E_F^{em}$. The current is then evaluated as:

$$I = \frac{em^*}{2\pi^2\hbar^3} \int_{E_F^{em}-eV}^{E_F^{em}} dE_\perp (E_F^{em} - E_\perp) T(E_\perp); \qquad eV < E_F^{em} \qquad (2.144)$$

This takes into account the fact that, since all states in the collector with energy $E_\perp < E_F^{em} - eV$ are occupied, tunneling is possible only for those electrons with energies satisfying the following conditions: $E_F^{em} > E_\perp > E_F^{em} - eV$. For $eV > E_F^{em}$, all emitter electrons can tunnel through the barriers and we obtain:

$$I = \frac{em^*}{2\pi^2\hbar^3} \int_{E_C^{em}}^{E_F^{em}} dE_\perp (E_F^{em} - E_\perp) T(E_\perp); \qquad eV > E_F^{em} \qquad (2.145)$$

It has been assumed that the Lorentzian linewidths of $T(E_\perp)$ are narrow with respect to E_F^{em}. The integration over $E_\perp$ yields the following current:

$$I = \frac{em^* v_z}{2\pi^2\hbar^3 L_w}[E_F^{em} - E_n(V)]\frac{T_L T_R}{(T_L + T_R)} \qquad (2.146)$$

Eq. 2.146 represents the tunnel current through the bound state with an energy $E_n(V)$, which depends on the applied bias. The peak value of the current will occur at the resonance energy, with:

$$I = \frac{em^* v_z}{2\pi^2\hbar^3 L_w} E_F \frac{T_L T_R}{(T_L + T_R)} \qquad (2.147)$$

When the bound state energy E_n falls below the bottom of the emitter conduction band, E_C^{em}, the current decreases rapidly and drops to a value determined by the off-resonance tunneling. This value can be estimated from the off-resonance transmission coefficients as:

$$I = \frac{em^*}{2\pi^2\hbar^3}E_F^2 T_L T_R \qquad (2.148)$$

Since for the transmission coefficients the inequality $T_L T_R \ll 1$ is valid, the resonant current will be much greater than that in the off-resonance condition.

2.10   Coulomb Blockade and Single Electron Transfer

The transport of charge carriers in macroscopic systems is generally described without the inclusion of interparticle interactions, despite the fact that Coulomb repulsion will always be a factor. This is because, in a large majority of cases, these effects are negligible and are conveniently ignored. This simplifies the treatment of transport phenomena, which is generally discussed in terms of the scattering of charge carriers in their transit through the medium. However, when we discuss mesoscopic and nanoscopic systems, interactions between the charge carriers become significant and charge effects must be taken into account because of the spatial confinement inherent in such cases. Such effects are referred to as the *Coulomb blockade* regime, where oscillations of the conductance are observed and charge transport in a device can occur via the movement of a single electron. The physics of such phenomena is based on the charge quantization discussed in previous sections for the quantum dot, which is isolated from its surroundings via tunnel junctions. Often, such a structure is referred to as an island.

The accumulation of charges on isolated structures gives rise to charging energy, denoted as E_c, and this is associated with charge blocking effects. We can describe a nanostructure in terms of an equivalent capacitance in a circuit, which means that the island is capacitively coupled to the circuit and charge transfer occurs via weak tunneling coefficients, whereby a single electron will pass from the lead (reservoir) to the quantum dot and between the quantum dot and the leads. It is common to distinguish the nature of the island, being either metallic or semiconducting, since this will influence the number of electrons stored on the island and the extent, to which quantum confinement affects the allowed electron states. In semiconductors, spatial confinement effects can be very significant and lead to the concept of the *artificial atom*, whose stationary electronic states can be manipulated by specific nanostructuring in the three spatial dimensions. This can be clearly seen, for example, with reference to Eq. 2.81, which explicitly considers the dimensions of the nanostructure in the three spatial directions for a quantum dot. Metallic systems have much higher concentrations of electrons, with a mean free path of the order of a few nm. As such, they are more bulk-like, though tunneling processes are important in both types of structures to couple to the external electric circuit.

We will consider an island, which is isolated in space and has a specific charge, that can be either positive or negative. In either case, this must be an integer value of the elementary electronic charge, i.e., $Q_i = N_i e$, where N_i is the number of excess or deficient electrons on the island. The transfer of an electron from (or onto) one of the reservoirs onto (or from) an almost isolated island will cause a rearrangement of the charge and an alteration in the electrostatic potential on the island. For sufficiently small quantum dots, the change in potential may be greater than the available thermal energy, $k_B T$, particularly at low temperatures. The shift of electrostatic potential can be particularly important for a single electron and may result in a gap in the energy spectrum at the Fermi energy. This results in a Coulomb blockade and will inhibit any further tunneling of electrons until the additional energy is overcome by an applied bias voltage. From electrostatics, we can represent the electrostatic potential energy due to a capacitance C in the form:

$$E_c = \frac{1}{2}QV = \frac{Q^2}{2C} \qquad (2.149)$$

where Q is the total charge stored on the capacitor. Expressing this energy in terms of the discrete charge of a number of electrons, we have:

$$E_c = \frac{N_i^2 e^2}{2C} = E_c N_i^2 \qquad (2.150)$$

where $E_c = e^2/2C$ is the excess energy due to one electron. For a spherical island of radius r, the capacitance is given by: $C = 4\pi\varepsilon\varepsilon_0 r$, thus the potential required to charge the island can be expressed as:

$$V = \frac{Q}{C} = \frac{e}{8\pi\varepsilon\varepsilon_0 r} \qquad (2.151)$$

If we take $\varepsilon = 1$, the voltage for an island of radius 10 nm will be about 72 mV. As the island increases, in size, this value will decrease and Coulomb blockade effects would only be observable at low temperatures. We note that at room temperature, the thermal energy is around 26 meV. The transfer of a single electron to the island will block the additional transfer of electrons due to their mutual Coulomb repulsion, at least until a suitable potential is applied via an external bias. In island structures of this type, charging effects lead to an increase of the low-temperature resistance, which has an activated conductance, as is the case for semiconductors, of the form: $\sigma \sim e^{-E_c/K_B T}$. This similarity with semiconductors has given rise to the term *Coulomb gap* in activated tunneling transport.

One of the simplest devices that can be considered is a single insulated quantum dot separated from two metallic contacts. There are a number of ways, in which such a set-up can be studied. One example would be using an STM tip as one of the leads, which is connected to an external voltage source and then to the metallic substrate. Deposited on the substrate are an insulating oxide layer and a quantum dot. The QD acts as the island and the oxide layer and the gap between the QD and the STM tip form two tunnel junctions (TJ). This is illustrated in Figure 2.22(a). In Figure 2.23, we show the experimental and theoretical $I - V$ characteristics for an In quantum dot coupled to an external circuit using the STM system to make the contacts. For low applied voltages, we note that the current remains zero. This region, for both positive and negative applied voltages, corresponds to the Coulomb blockade regime. As the potential increases, the energy shift is sufficient to overcome the barrier or Coulomb gap, and a current can pass, via tunneling, from the emitter, to the QD, and to the collector. The direction being governed by the polarity of the bias voltage. The staircase nature of the $I - V$ characteristics are typical. A small asymmetry is evident in the experimental data and arises due to uncontrolled background charging effects.

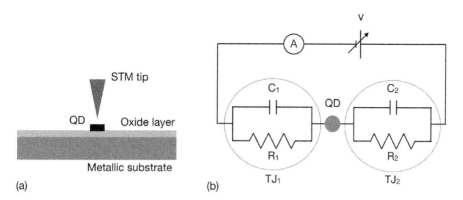

FIGURE 2.22 (a) Schematic illustration of a quantum dot weakly coupled to an external circuit via two tunnel junctions (TJ) using the STM as a measurement system. (b) Equivalent circuit of the set-up in (a) showing an external voltage source. Each tunnel junction is characterized by a capacitance, C, and a resistance, R.

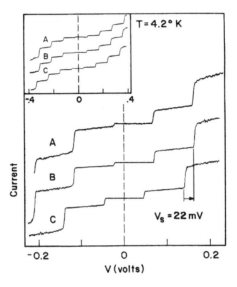

FIGURE 2.23 Experimental data and theoretical fit for an In droplet with Al substrate, which is oxidized to form one of the tunnel junctions. The other TJ is formed using an Al STM tip. Curve A is an experimental $I-V$ characteristic from the In droplet in a sample with an average droplet size of 30 nm. The peak-peak current is 1.8 nA. Curve B is a theoretical fit to the data for $C_1 = 3.5 \times 10^{-19}$ F, $C_2 = 1.8 \times 10^{-18}$ F, $R_1 = 7.2$ MΩ, and $R_2 = 4.4$ GΩ. The obvious asymmetric features in curve A require a voltage shift $V_s = 22$ mV. Curve C, calculated for $V_s = 0$, shows the (seldom observed) symmetric case. A small quadratic term was added to the computed tunneling rate for each junction. Inset: A wider voltage scan for this same structure; again, the topmost curve is experimental data. Reprinted figure with permission from R. Wilkins, E. Ben-Jacob & R. C. Jaklevic, (1989). *Phys. Rev. Lett.*, *63*, 801, Copyright (1989) by the American Physical Society.

The equivalent circuit of this experimental set-up is shown in Figure 2.22(b), where each tunnel junction is depicted as a capacitor and a resistance in parallel. The capacitor acts as the insulating gap between lead and QD, while the resistance represents a tunnel probability. For this circuit, we can write the capacitor charges as: $Q_1 = C_1V_1$ and $Q_2 = C_2V_2$. The net charge on the island will be the difference between Q_1 and Q_2. In the absence of tunneling, the difference in charge would be zero and the island will have neutral charge. tunneling allows an integer number of excess electrons to accumulate on the island, such that $Q = Q_1 - Q_2 = -ne$, where $n = n_1 - n_2$ is the net number of excess electrons on the island. This can be positive or negative, convention is such that an increase in either n_1 or n_2 corresponds to increasing either junction charge Q_1 or Q_2, respectively. The sum of the junction potentials is simply the applied voltage, V_a, where we can now write:

$$V_1 = \frac{1}{C_{eq}}(C_2 V_a + ne)$$ (2.152)

$$V_2 = \frac{1}{C_{eq}}(C_1 V_a - ne)$$ (2.153)

with $C_{eq} = C_1 + C_2$ being the equivalent island capacitance. The electrostatic energy stored by the capacitors being:

$$E_S = \frac{Q_1^2}{2C_1} + \frac{Q_2^2}{2C_2} = \frac{(C_1 C_2 V_a^2 + Q^2)}{2C_{eq}}$$ (2.154)

The corresponding work done by the voltage source in the transfer of an electron via tunneling can be expressed by:

$$W_S = \int dt V_a I(t) = V_a \Delta Q$$ (2.155)

ΔQ being the charge transferred due to tunneling and charge accumulation due to the change of electrostatic potential on the island. We note that the change in the charge on the island due to an electron tunneling through barrier (TJ$_2$), such that $n_2' = n_2 + 1$, $Q' = Q + e$ and $n' = n + 1$, causes a change in potential across junction 1 of $V_1' = V_1 + e/C_{eq}$, and as a result, a polarization charge flows from the voltage source, with $\Delta Q = -eC_1/C_{eq}$, to compensate. The work done through the two junctions will therefore be:

$$W_S(n_1) = -n_1 e V_a \frac{C_2}{C_{eq}}$$ (2.156)

and

$$W_S(n_2) = -n_2 e V_a \frac{C_1}{C_{eq}}$$ (2.157)

The total energy of the circuit can thus be written as:

$$E(n_1, n_2) = E_S + W_S = \frac{(C_1 C_2 V_a^2 + Q^2)}{2C_{eq}} + \frac{e V_a}{C_{eq}}(C_1 n_2 + C_2 n_1)$$ (2.158)

The energy change of the system due to a particle tunneling event (in the second junction) can now be expressed as:

$$\Delta E_2^{\pm} = E(n_1, n_2) - E(n_1, n_2 \pm 1) = \frac{Q^2}{2C_{eq}} - \frac{(Q \pm e)^2}{2C_{eq}} \mp \frac{e V_a C_1}{C_{eq}}$$

$$= \frac{e}{C_{eq}} \left[-\frac{e}{2} \pm (ne - V_a C_1) \right] \tag{2.159}$$

The corresponding energy change for a tunnel event in junction 1, will be

$$\Delta E_1^{\pm} = E(n_1, n_2) - E(n_1 \pm 1, n_2) = \frac{e}{C_{eq}} \left[-\frac{e}{2} \mp (ne + V_a C_1) \right] \tag{2.160}$$

For the case, where the island was initially neutral, with $n = 0$, we can write:

$$\Delta E_{1,2}^{\pm} = -\frac{e^2}{2C_{eq}} \mp \frac{eV_a}{C_{eq}} C_{2,1} > 0 \tag{2.161}$$

For all possible transitions to and from the island, the leading term for the Coulomb energy of the island causes ΔE to be negative until V_a exceeds the threshold, which depends on the lesser of the two capacities. In the case of $C_1 = C_2 = C$, this condition reduces to:

$$|V_a| > \frac{e}{C_{eq}} \tag{2.162}$$

tunneling will occur above this threshold and hence the initial flat region of the $I - V$ curve for low applied voltages. This region of Coulomb blockade is a direct result of the additional Coulomb energy $e^2/2C_{eq}$, which must be expended by an electron in order to be able to tunnel to or from the island. For large area junctions, C_{eq} is large and consequently the Coulomb regime will not be observed.

The equilibrium band diagram for the double tunnel-junction (lead1 - island–lead2) is illustrated in Figure 2.24 (left), for the symmetric case; $C_1 = C_2 = C$. The Coulomb gap of width e^2/C_{eq} opens at the Fermi energy of metal, such that no states are available for electrons to tunnel into the island from either electrode. Electrons on the island have no available states to tunnel until the blockade regime is overcome by the appropriately applied potential, as shown in Figure 2.24 (right). For the situation illustrated, with $C_1 = C_2 = C$, we have $V_a > e/2C$ and one electron can tunnel onto the island. Once this occurs, the Fermi energy of the quantum dot is raised by a quantity $e^2/2C$ and a gap again appears, prohibiting further tunneling on to the island from the right electrode. This will only be overcome when the applied potential reaches $V_a > 3e/2C$, and so on. It is now clear that the step-like $I - V$ behavior is a result of the periodic condition imposed on the tunneling effect.

The existence well-defined structure in the transport properties multi-junction devices, due to Coulomb charging effects, depends on the magnitude

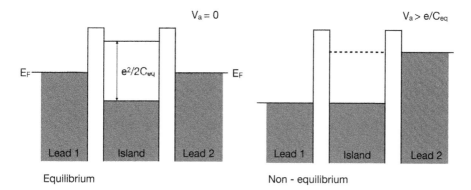

FIGURE 2.24 Band diagram for a double barrier for the case of a symmetrical structure, with $C_1 = C_2 = C$. The Coulomb blockade regime is illustrated in the equilibrium state. tunneling can only occur for the condition, $V_a > e/C_{eq}$, and is illustrated on the right.

of the Coulomb gap, e^2/C_{eq}, compared to the thermal energy. So to observe a well-defined Coulomb blockade staircase structure, for a given temperature, the following condition must be satisfied:

$$\frac{e^2}{C_{eq}} > k_B T \qquad (2.163)$$

Furthermore, we must consider the quantum fluctuations which can occur as a result of the Heisenberg uncertainty principle; $\Delta E \Delta t \geq \hbar$, where we equate ΔE with E_c, and the time required for charge transfer will depend on the characteristic time of the circuit; $\Delta t \simeq R_t C_{eq}$, with R_T being the smaller of the two tunnel resistances. This allows us to establish the condition:

$$\frac{e^2}{C_{eq}} R_t C_{eq} > h \qquad (2.164)$$

Our requirement for Coulomb charging effects now can be expressed in terms of the tunnel resistance, which needs to be sufficiently large (to be considered a tunnel junction), such that: $R_t > h/e^2 = 25.8 \text{ k}\Omega$.

2.11 Single-Electron Transistor

The formation of a single-electron transistor (SET) can be envisaged as the addition of further contact to the island structure we considered in the previous section. This contact must be isolated from the power supply, from which electrons are supplied for single-electron tunneling to and from the island. This additional contact forms a gate, with a "gate" potential, V_g, and an associated capacitance, C_g, allowing the control and tuning of the tunneling process between the island and the electrodes. A schematic illustration of the structure and equivalent circuit are illustrated in Figure 2.25. As shown, the gate voltage is controlled separately from that applied across the island, being coupled to the island via an ideal capacitor, C_g. This additional potential will modify the charge balance on the island, and we now write:

$$Q_g = C_c(V_g - V_2) \tag{2.165}$$

and the island charge becomes:

$$Q = Q_2 - Q_1 - Q_g = -ne + Q_p \tag{2.166}$$

where Q_p accounts for unintentional background potential charges, which typically exist in real experiments due to the work function differences and random charges trapped near junctions. These background potentials are manifested as asymmetries in the $I - V$ characteristics, as shown in Figure 2.23.

The potential differences across the two tunnel junctions will be modified by the gate voltage, and now we must express these potentials as:

$$V_1 = \frac{1}{C_{eq}}[(C_g + C_2)V_a - C_gV_g + ne - Q_p] \tag{2.167}$$

$$V_2 = \frac{1}{C_{eq}}[C_1V_a + C_gV_g - ne + Q_p] \tag{2.168}$$

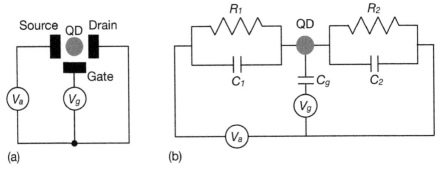

FIGURE 2.25 (a) Schematic illustration of the basic structure of the single electron transistor. (b) The equivalent circuit for the SET.

Here, $C_{eq} = C_1 + C_2 + C_g$. The modified electrostatic energy will also now include the energy of the gate capacitor, $e^2/2C_g$, such that:

$$E_S = \frac{1}{2C_{eq}}[C_1 C_g (V_a - V_g)^2 + C_1 C_2 V_a^2 + C_g C_2 V_g^2 + Q^2] \tag{2.169}$$

The work done in the electron tunneling through the two junctions will also be modified by the gate voltage, and are expressed by the relations:

$$W_S(n_1) = -n_1 \left[\frac{C_2}{C_{eq}} eV_a + \frac{C_g}{C_{eq}} e(V_a - V_g) \right] \tag{2.170}$$

and

$$W_S(n_2) = -n_2 \left[\frac{C_1}{C_{eq}} eV_a + \frac{C_g}{C_{eq}} eV_g \right] \tag{2.171}$$

The total energy of the charged state (n_1, n_2) will take a modified form and we express the energy change across the two junctions as follows: for junction 1 we have:

$$\Delta E_1^{\pm} = E(n_1, n_2) - E(n_1 \pm 1, n_2) = \frac{Q^2}{2C_{eq}} - \frac{(Q \pm e)^2}{2C_{eq}} \mp \frac{e}{C_{eq}}[(C_2 + C_g)V_a - C_g V_g]$$

$$= \frac{e}{C_{eq}} \left\{ -\frac{e}{2} \mp [ne - Q_p + (C_2 + C_g)V_a - C_g V_g] \right\} \tag{2.172}$$

The corresponding energy change for a tunnel event in junction 2, will be:

$$\Delta E_2^{\pm} = E(n_1, n_2) - E(n_1, n_2 \pm 1) = \frac{Q^2}{2C_{eq}} - \frac{(Q \pm e)^2}{2C_{eq}} \mp \frac{e}{C_{eq}}(C_1 V_a + C_g V_g)$$

$$= \frac{e}{C_{eq}} \left[-\frac{e}{2} \pm (ne - Q_p - C_1 V_a - C_g V_g) \right] \tag{2.173}$$

From these relationships, we can see that the gate voltage allows a change in the effective charge on the island and therefore a shift of the Coulomb blockade region with V_g. The stable region of Coulomb blockade can therefore be obtained for $n \neq 0$. As previously, the tunneling condition at low temperature will be $\Delta E_{1,2} > 0$, such that the total energy of the system will be reduced by tunneling. The effect of the random polarization will offset the gate voltage so that we can write a new gate voltage, $V_g' = V_g + Q_p/C_g$. The condition for forward and reverse tunneling can now be expressed as:

$$-\frac{e}{2} \mp [ne + (C_2 + C_g)V_a - C_g V_g'] > 0 \tag{2.174}$$

and

$$-\frac{e}{2} \pm (ne - C_1 V_a - C_g V'_g) > 0 \qquad (2.175)$$

These equations describe the four possible tunneling events that can occur on the SET, as illustrated in Figure 2.26, for transfer between leads (L) and island (I), in either direction: IL_1, IL_2, LI_1, and LI_2.

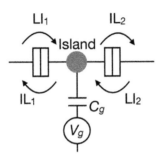

FIGURE 2.26   Illustration of the four possible tunneling events on a SET.

The state of Coulomb blockade is that for which the above conditions are NOT satisfied, and therefore all tunneling channels are blocked and no electron transfer can take place. This situation is illustrated in Figure 2.27. Here, the dashed lines indicate that the four tunneling pathways are blocked in the equilibrium state. A shift in either the bias between source and drain or on the gate can modify this situation and open one or more of the tunneling channels. In the case, where the applied potential, V_a, is modified, this will alter the electrochemical potentials on the leads, giving a relative difference in the potential and with the gate potential. A change in the applied gate potential will alter the island states with respect to the electrodes.

In Figure 2.28(a), we show the process whereby the channel LI_1, between lead 1 and the island is open due to the application of a potential V_a, between the leads 1 and 2. In Figure 2.28(b), the channel IL_2, between the island and lead 2, is open. The transfer of an electron from lead 1 to the island, $n \rightarrow n+1$, changes the island potential such that the tunneling channel IL_2 opens and an electron can transfer from the island to lead 2. All other channels remain closed. This illustrates a sequential transfer of a single electron from lead 1 to lead 2.

If we only alter the gate voltage V_g, we can open channels for the transfer of electrons only onto the island. This is illustrated in Figure 2.29. Here, we see that for a voltage $V_g < e/C_{eq}$, two tunneling channels, LI_1 and LI_2, are open. These channels only allow the passage of electrons onto the island,

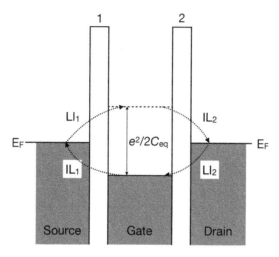

FIGURE 2.27 Illustration of the band diagram showing the four possible tunneling events on a SET, as shown in Figure 2.26.

which would cause it to charge, but no current will flow between the leads. However, there will be conductance between the island and the gate. This can lead to conductance oscillations as a function of the gate voltage. If now we apply a small bias voltage, however, we can expressly control the current flow between the two leads. The bias voltage can be quite small, and smaller than e/C_{eq}, as shown in Figure 2.30.

We can construct a stability diagram, which shows the limits of tunneling for the four channels and can be depicted in the V_a - V_g plane. This shows the stable regions for each value of n and the limits, where a tunneling process can occur $n \rightarrow n \pm 1$. In Figure 2.31, we demonstrate the simple case for which $C_g = C_2 = C$ and $C_1 = 2C$. The diagonal lines represent the boundaries for the onset of tunneling. In fact, there are four lines, which are defined by setting the inequalities indicated in Eqs. 2.174) and (2.175) to equalities, which form the boundaries for tunneling. This set of four equations will then form the four channels for tunneling. In Figure 2.31, the blue diamonds indicate regions, where there are no solutions to the tunneling equations and the islands have a stable charge, being fixed in the Coulomb blockade regime. Each region corresponds to a different integer number of electrons on the island. The charge stability will also depend on thermal fluctuations and may require low temperatures. The gate voltage then allows the tuning between the stable regions, where an electron is added or removed, one at a time, from the island. This is referred to as a single electron transfer. By adjusting the

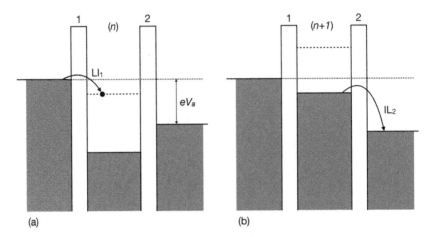

FIGURE 2.28 (a) The application of a potential $V_a < e/C_{eq}$ will shift the relative positions of the energy bands, opening the tunneling channel LI_1. (b) The transfer of an electron from lead 1 onto the island changes its potential, opening channel LI_2, and allows the electron to transfer onto lead 2. This illustrates the process by which a single electron can be transferred from lead 1 to lead 2, via the intermediary of the island.

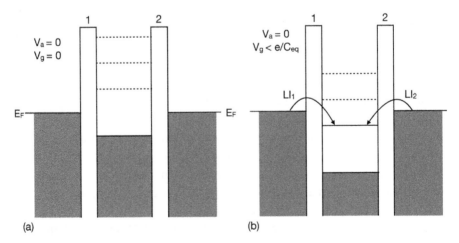

FIGURE 2.29 Adjusting the gate voltage alone will not lead to current flow between the leads, but charging of the island.

applied and/or gate voltage, it is possible to attain the necessary conditions for further charge transfer of two more electrons.

One of the consequences of the Coulomb diamond formation is the capacity for the construction of the single-electron transistor (SET), which exhibits

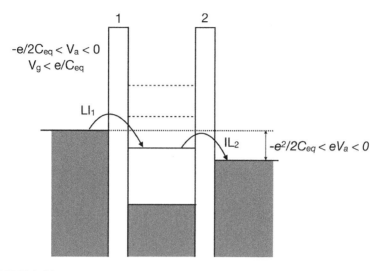

FIGURE 2.30   Adjusting the gate voltage and the bias voltage can lead to current flow between the leads.

Coulomb oscillations in the conductance, when the gate voltage is varied for a small bias in the source–drain voltage. For a given S–D bias, V_a, over which Coulomb blockade occurs, an increase of the gate voltage will allow the passage of electrons through successive transfer and blockade regions. As can be seen from the stability diagram, the maximum blockade occurs for $C_g V_g = me$, where $m = 0, \pm 1, \pm 2, \ldots$ As $C_g V_g$ approaches half-integer values of the electronic charge, the width of the Coulomb blockade region vanishes and tunneling can occur. Therefore, for small values of V_a, current peaks will occur for a narrow range of V_g around these half-integer values. The separation of peaks will be limited by thermal broadening and smear out at temperatures greater than the energy width of the Coulomb blockade regime. The conductance linewidth can be evaluated with regard to thermal fluctuations. The variation of conductance between source and drain as a function of the gate voltage is illustrated in Figure 2.32, along with the energy band diagram for the double junction SET device. It should be noted that the peaks in the conductance curve correspond with the steps in the staircase structure of the $I - V$ characteristics of the confined quantum dot systems. This can be seen simply from the relation between the current and the voltage, which gives the conductance as the derivative of the current with respect to the voltage; $G = dI/dV$.

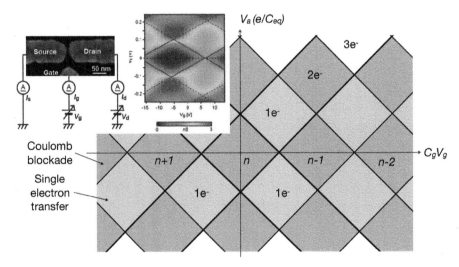

FIGURE 2.31   Stability diagram, showing the Coulomb blockade diamonds. The current in the blue diamond regions is blocked. Within these areas, the number of electrons *n* remains constant. Conductances peaks/oscillations occur along the C_gV_g axis, at points where the neighboring Coulomb diamonds touch. Applying a bias voltage will lead to electron transfer from the leads to the island. As long as there is a small bias, a shift in gate voltage can lead to a single electron transfer. The inset shows a real SET device along with the $V_a - V_g$ characteristics. Indeed, we see a very good agreement with the expected behavior for the device at a low temperature. Reprinted with permission from Y. Azuma et al., (2010). *Jpn. J. Appl. Phys., 49,* 090206, ©2010 The Japan Society of Applied Physics.

2.12   Coupled Quantum Dots

The use of coupled quantum dots or islands can give rise to even more complex and interesting transport behavior due to Coulomb blockade effects. In this case, the coupling occurs electrostatically between the electrons on different islands; the excess discrete charge on the island *i* changes the energy of the charge state on the island *j*. Charging effects in many-island systems can be analyzed by using the equivalent capacitive circuits, as discussed in the case for the single-electron transistor. The state of any Coulomb blockade system is characterized by the number of excess charges on the island. The electrostatic energy of a system of *N* islands can be written as:

$$E^{(N)} = \sum_{i,j \neq i}^{N} E_{ij}^{(C)} \left(N_i - \frac{q_i}{e} \right) \left(N_j - \frac{q_j}{e} \right) \tag{2.176}$$

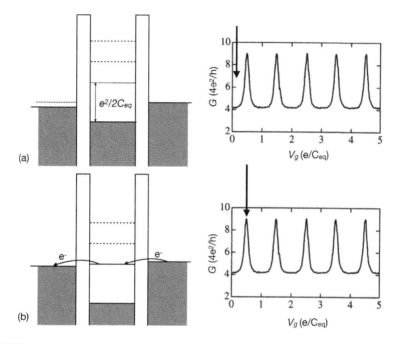

(a)

(b)

FIGURE 2.32 Energy band diagram for the double junction SET, illustrating the occurrence of conductance peaks as a result of Coulomb blockade. (a) System in the off-resonance state. (b) Resonance condition occurs when the excited state due to Coulomb charging lies between the chemical potentials of the root and left electrodes. Arrows indicate the position of the conductance for each case.

where i, j are the island labels for two coupled islands, N_i is the discrete charge on island i and q_i is the charge induced in the same island by any electrodes. The term $E_{ij}^{(C)}$ is the Coulomb energy of the form $e^2/2C_{ij}$. It will be noted that the charging energy is a matrix proportional to the inverse capacitance matrix of the islands. The diagonal elements give the energy expenditure due to the adding of excess electrons to a given island. Off-diagonal elements denote the repulsive interactions between the quantized charges on different islands.

We will now consider the general case of a capacitance circuit with gate electrodes, labeled as k, for the islands and biases given by $V_k^{(g)}$. We also take into account the source electrode among the gates. There is a capacitive coupling between each island i and the gate electrode k, which has a capacitance of $C_{ik}^{(g)}$. Additionally, the islands are interconnected by capacitors, $C_{ik}^{(c)}$. The island biases are designated as V_i. The full energy includes the capacitor energies and the work done by the voltage source, such that we can write:

$$E = \frac{1}{2}\sum_{i>j}C_{ij}^{(c)}(V_i - V_j)^2 + \frac{1}{2}\sum_{i,k}C_{ik}^{(g)}(V_i - V_k^{(g)})^2 - \sum_{i,k}q_{ik}V_k^{(g)} \qquad (2.177)$$

where q_{ik} is the charge in capacitance $C_{ik}^{(c)}$, so that:

$$q_{ik} = C_{ik}^{(g)}(V_k^{(g)} - V_i) \qquad (2.178)$$

We now write:

$$\begin{aligned}
E &= \frac{1}{2}\sum_{i>j}C_{ij}^{(c)}(V_i - V_j)^2 + \frac{1}{2}\sum_{i,k}C_{ik}^{(g)}V_i^2 - \frac{1}{2}\sum_{i,k}C_{ik}^{(g)}(V_k^{(g)})^2 \\
&= \frac{1}{2}\sum_{i>j}C_{ij}^{(c)}(V_i - V_j)^2 + \frac{1}{2}\sum_{i,k}C_{ik}^{(g)}[V_i^2 - (V_k^{(g)})^2]
\end{aligned} \qquad (2.179)$$

For a single electron box, we neglect the final group of terms since they do not depend on the charge state of the system. To proceed, we note that the full charge on island i is the sum over charges accumulated in all capacitors connected to the island, as expressed by:

$$Q_i = |e|N_i = \sum_j q_{ij} + \sum_k q_{ik} = \sum_j C_{ij}^{(c)}(V_i - V_j) - \sum_k C_{ik}^{(g)}(V_i - V_k^{(g)}) \quad (2.180)$$

This can be expressed in terms of the capacitance matrix in the form:

$$eN_i - q_i = \sum_j C_{ij}V_j \qquad (2.181)$$

where $q_i = -\sum_k C_{ik}^{(g)}V_k^{(g)}$ is the offset charge on island i due to all gate electrodes. Diagonal elements of the capacitance matrix are the sums of all the capacitances connected to the island, while the off-diagonal elements are cross-capacitances having a negative sign. We can express this as:

$$C_{ij} = \begin{cases} \sum_j C_{ij}^{(c)} + \sum_k C_{ik}^{(g)}; & i = j \\ -C_{ij}^{(c)}; & i \neq j \end{cases}$$

Inverting this matrix, we can express the voltage in terms of the island charge as:

$$V_i = \sum_j (C_{ij})^{-1}(eN_j - q_j) \qquad (2.182)$$

We can now express the charging energy in the form:

$$E_{ij}^{(c)} = \frac{e^2}{2}(C_{ij})^{-1} \qquad (2.183)$$

We will now illustrate the example of a double quantum dot system, which is shown in Figure 2.33. Each island is controlled by its own gate, so charges induced at each island can be expressed as: $q_{1,2} = C_{1,2}^{(g)} V_{1,2}^{(g)}$. There is a capacitance, C_{12}, between the two quantum dots. This capacitance will determine the coupling or interaction between the quantum dots and is what makes the system different from two independent quantum dots.

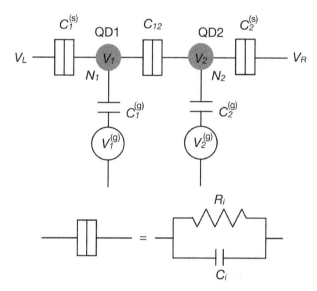

FIGURE 2.33 Double quantum dot system coupled in series. The figure illustrates the equivalent circuit for the double island structure.

Considering the charges on the two islands, we can establish the following expressions:

$$Q_1 = C_1^{(s)}(V_1 - V_L) + C_1^{(g)}(V_1 - V_1^{(g)}) + C_{12}(V_1 - V_2)$$
$$Q_2 = C_2^{(s)}(V_2 - V_R) + C_2^{(g)}(V_2 - V_2^{(g)}) + C_{12}(V_2 - V_1) \tag{2.184}$$

where, $V_{1,2}$ is the electrostatic potential on island (1, 2) and $V_{L,R}$ refers to the applied potential to the left and right of the 2 QD assembly. These two equations can be written in the form:

$$\begin{pmatrix} Q_1 + C_1^{(s)} V_L + C_1^{(g)} V_1^{(g)} \\ Q_2 + C_2^{(s)} V_R + C_2^{(g)} V_2^{(g)} \end{pmatrix} = \begin{pmatrix} C_1 & -C_{12} \\ -C_{12} & C_2 \end{pmatrix} \begin{pmatrix} V_1 \\ V_2 \end{pmatrix} \tag{2.185}$$

where $C_{1,2} = C_{12} + C_{1,2}^{(g)} + C_{1,2}^{(s)}$. This can be re-arranged to obtain:

$$\begin{pmatrix} V_1 \\ V_2 \end{pmatrix} = \frac{1}{(C_1 C_2 - C_{12}^2)} \begin{pmatrix} C_2 & C_{12} \\ C_{12} & C_1 \end{pmatrix} \begin{pmatrix} Q_1 + C_1^{(s)} V_L + C_1^{(g)} V_1^{(g)} \\ Q_2 + C_2^{(s)} V_R + C_2^{(g)} V_2^{(g)} \end{pmatrix} \tag{2.186}$$

To establish the electrostatic energy, we can use the following relation employing the capacitance matrix:

$$E = \frac{1}{2} \vec{V} (C_{ij}) \vec{V} = \frac{1}{2} \vec{Q} (C_{ij})^{-1} \vec{Q} \tag{2.187}$$

To simplify the analysis, we will consider the case where there is no applied potential, $V_L = V_R = 0$. Using $Q_{1,2} = |e| N_{1,2}$, we can now express the double quantum dot electrostatic energy as:

$$E(N_1, N_2) = \frac{1}{2} N_1^2 E_{C1} + \frac{1}{2} N_2^2 E_{C2} + N_1 N_2 E_{C12} + E(V_1^{(g)}, V_2^{(g)}) \tag{2.188}$$

where we have expressed the energy in terms of the charging energies of the individual quantum dots; $E_{C1} = (e^2 C_2)/(C_1 C_2 - C_{12}^2)$ and $E_{C2} = (e^2 C_1)/(C_1 C_2 - C_{12}^2)$, the electrostatic coupling energy; $E_{C12} = (e^2 C_{12})/(C_1 C_2 - C_{12}^2)$ and a term which accounts for the coupling to the gate potentials:

$$E(V_1^{(g)}, V_2^{(g)}) = \frac{1}{-|e|} \left[(N_1 E_{C1} + N_2 E_{C12}) C_1^{(g)} V_1^{(g)} + (N_2 E_{C2} + N_1 E_{C12}) C_2^{(g)} V_2^{(g)} \right]$$
$$+ \frac{1}{e^2} \left[\frac{1}{2} E_{C1} (C_1^{(g)} V_1^{(g)})^2 + \frac{1}{2} E_{C2} (C_2^{(g)} V_2^{(g)})^2 + C_1^{(g)} V_1^{(g)} C_2^{(g)} V_2^{(g)} E_{C12} \right] \tag{2.189}$$

For the case, where the coupling vanishes, $C_{12} = 0; E_{C12} = 0$, Eq. 2.188, reduces to:

$$E(N_1, N_2) = E(N_1) + E(N_2) = \frac{[-N_1 e + C_1^{(g)} V_1^{(g)}]^2}{2C_1} + \frac{[-N_2 e + C_2^{(g)} V_2^{(g)}]^2}{2C_2} \tag{2.190}$$

i.e., the energy of two independent quantum dots. When the coupling is very strong and C_{12} becomes the dominant capacitance, the electrostatic energy becomes:

$$E(N_1, N_2) = E(N_1 + N_2) = \frac{[-(N_1 + N_2)e + C_1^{(g)} V_1^{(g)} + C_2^{(g)} V_2^{(g)}]^2}{2(\tilde{C}_1 + \tilde{C}_2)} \tag{2.191}$$

where $\tilde{C}_{1,2} = C_{1,2} - C_{12}$. Eq. (2.191) has the form of the electrostatic energy of a single dot with a charge $(N_1 + N_2)e$ and a capacitance $\tilde{C}_1 + \tilde{C}_2$. The strong coupling between the two dots effectively makes the ensemble act as a large single dot.

The electrochemical potential $\mu_{1,2}(N_1, N_2)$ of dot (1, 2) is defined as the energy required to add the $N_{1,2}^{\text{th}}$ electron to dot (1, 2), while having $N_{2,1}$ electrons on dot (2, 1). From the expression (2.188), the electrochemical potential for the two dots are given by:

$$\mu_1(N_1, N_2) \equiv E(N_1, N_2) - E(N_1 - 1, N_2)$$

which gives:

$$\mu_1(N_1, N_2) = \left(N_1 - \frac{1}{2}\right) E_{C1} + N_2 E_{C12} - \frac{1}{|e|}(C_1^{(g)} V_1^{(g)} E_{C1} + C_2^{(g)} V_2^{(g)} E_{C12})$$
(2.192)

and

$$\mu_2(N_1, N_2) \equiv E(N_1, N_2) - E(N_1, N_2 - 1)$$

yielding:

$$\mu_2(N_1, N_2) = \left(N_2 - \frac{1}{2}\right) E_{C2} + N_1 E_{C12} - \frac{1}{|e|}(C_1^{(g)} V_1^{(g)} E_{C12} + C_2^{(g)} V_2^{(g)} E_{C2})$$
(2.193)

It will be noted from these expressions, that if we add an electron to one of the dots, at fixed gate voltages, the change in electrochemical potential will be: $\Delta\mu_1 = \mu_1(N_1 + 1, N_2) - \mu_1(N_1, N_2) = E_{C1}$ for dot 1, and $\Delta\mu_2 = \mu_2(N_1, N_2 + 1) - \mu_2(N_1, N_2) = E_{C2}$ for dot 2. Furthermore, $\mu_1(N_1, N_2 + 1) - \mu_1(N_1, N_2) = \mu_2(N_1 + 1, N_2) - \mu_2(N_1, N_2) = E_{C12}$.

We will now construct the stability diagrams for the double quantum dot system based on the electrochemical potentials expressed in Eqs. 2.192) and (2.193). These illustrate the number of electrons (N_1, N_2) on the islands as a function of the gate voltages $V_1^{(g)}$ and $V_2^{(g)}$. In zero bias, we set the electrochemical potentials of the leads to zero, $\mu_L = \mu_R = 0$. Hence at equilibrium, the charges on the dots are the largest values of N_1 and N_2, for which both $\mu_1(N_1, N_2)$ and $\mu_2(N_1, N_2)$ are less than zero. If either is less than zero, electrons will escape to the leads. The various constraints and conditions give rise to a specific shape of the stability diagram in the $(V_1^{(g)}, V_2^{(g)})$ domain. Inside each cell, the charge configuration is stable. In the following, we will discuss the various coupling regimes.

In the simplest case, we can discuss the situation, where there is no coupling between the two quantum dots; $C_{12} = 0; E_{C12} = 0$. The stability diagram is illustrated in Figure 2.34(a). Here, the stability diagram shows that increasing either of the gate voltages, with other remaining constant, will increase the number of charges, one at a time, on the corresponding island. For very strong coupling, at the other extreme, the boundaries between tunneling events go from vertical/horizontal to diagonal. The boundaries between the dots are now indicated as dashed lines and effectively, the number of total electrons on the two islands remains stable between the tunnel boundaries (solid lines), see Figure 2.34(b). This means that the electrons can begin in either one of the two islands. Again, a gradual increase of either gate voltage will increase, in a discrete manner, the number of electrons on the islands.

For intermediate coupling, the stability diagram has a hexagonal appearance, where we can consider that, from the decoupled state, as the coupling increases, the boundaries become inclined, until they align in the strong coupling regime. This is illustrated in Figure 2.34(c). Here, the vertices of the square stability domains split into two distinct vertices or triple points. In order to obtain a measurable current, the tunnel barrier should be sufficiently transparent, however, not too much, so that a stable number of electrons can be defined on each island. For the configuration of quantum dots, as illustrated in Figure 2.33, the condition for a conduction resonance will be encountered, when electrons can tunnel through both quantum dots. Such a condition is met, whenever three charge states become degenerate, this occurs at the vertices. A close-up of this situation is shown in Figure 2.34(d), where we now distinguish between the two types of vertices. At the points marked (•), the islands cycle thought the sequence: $(N_1, N_2) \rightarrow (N_1 + 1, N_2) \rightarrow (N_1, N_2 + 1) \rightarrow (N_1, N_2)$. This shuttles an electron through the system, as illustrated by the anti-clockwise path of the electron (e). tunneling, in this case, occurs from left to right. For the vertices marked (○), the sequence corresponds to: $(N_1 + 1, N_2 + 1) \rightarrow (N_1 + 1, N_2) \rightarrow (N_1, N_2 + 1) \rightarrow (N_1 + 1, N_2 + 1)$. This clockwise path corresponds to a hole (h) moving in the opposite direction to the electron motion.

In Figure 2.35, we show a scanning electron micrograph (SEM) image of a double quantum well structure, details are explained in the figure caption. Each dot has roughly a physical dimension of 120 nm by 80 nm, the connecting ribbon structure is of 100 nm length and 35 nm width. Specifically, the gates LP (RP) and LB (RB) are designed to control the electrochemical potential of the left dot (right dot) and left barrier (right barrier), respectively.

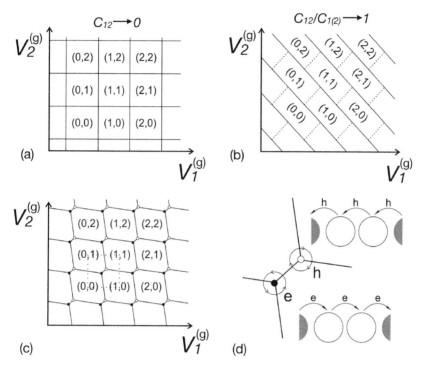

FIGURE 2.34 Schematic stability diagram of the double-dot system for (a) small, (b) large, and (c) intermediate inter-dot coupling. The equilibrium charge on each dot in each domain is denoted by (N_1, N_2). The two kinds of triple-points corresponding with the electron transfer process ($\bullet$) and the whole transfer process ($\circ$) are illustrated in (d). The region in the dotted square in (b) is depicted in more detail in (d). Reprinted figure with permission from B. J. van Wees, H. van Houten, C. W. J. Beenakker, J. G. Williamson, J. P. Kouwenhoven, D. van der Marel, & C. T. Foxon, *Phys. Rev. Lett., 60*, 848 (1988). Copyright (1988) by the American Physical Society.

The middle gate (marked as M) is primarily used to tune the capacitive and the tunneling coupling of the two adjacent dots and gate Q is for the adjustment of the QPC's working point for optimum sensitivity.

2.12.1 The Electron Turnstile

We now consider the case of an array of quantum dot islands in a chain, each coupled capacitively on either side with the end islands being connected to the left and right electrodes. We can manipulate the conductances of the various junctions by switching them on and off, using a gating voltage. We

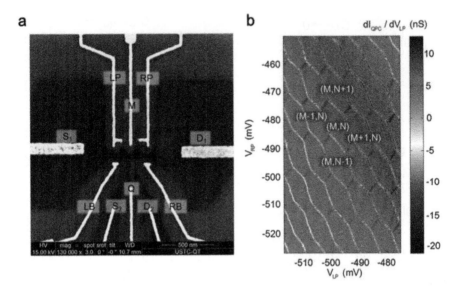

FIGURE 2.35 (a) Scanning electron micrograph in the false-color of the device. Dark regions are the graphene base structure consisting of a double quantum dot (100 nm × 100 nm for each dot, 35 nm × 100 nm for each ribbon) and an integrated QPC channel serving as in-situ charge detector. Gray area is the etched-away region that shows the SiO_2 substrate surface. The yellow regions are the metal gates used for control and as source/drain electrodes. (b) A typical QPC charge sensor stability diagram of the double quantum dot: differential charge sensor current dI_{QPC}/dV_{LP} as a function of two gate voltages V_{LP} and V_{RP}. The structure shows a well-defined double quantum dot with characteristic honeycomb patterns. Charge state is defined by (M, N), where M and N are the electron number in the left and right dot, respectively. Reprinted from D. Wei, H.-O. Li, G. Cao, G. Luo, Z.-X. Zheng, T. Tu, M. Xiao, G.-C. Guo, H.-W. Jiang, & G.-P. Guoi, (2013). *Scientific Reports, 3*, 3175, under the Creative Commons Attribution 4.0 International License.

can envisage a cycle, in which we start with a particular cell in the state "0." Turning on the left double junction and maintaining the right double in the off-state, we can take an electron from the left electrode. This process corresponds to writing a state "1" on our cell. Now, we switch the right junction on and set the left junction to off. We can now remove the electron from the cell to the right electrode. In doing so, we have written the "0" state on the cell and we return to our starting point. Such a cycle describes a controlled transfer of a single electron from the left electrode to the right electrode.

Instead of just controlling the conductances, as we considered previously, we can exploit the Coulomb blockade to do this for us. This device is referred

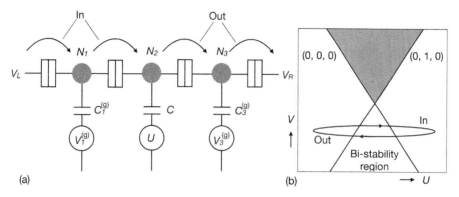

FIGURE 2.36 (a) Schematic illustration of a three-quantum dot single-electron turnstile. The cycling operation is controlled using an applied potential to the central island, U and the device is biased using a voltage, V. (b) The cycle crosses a region where both (0, 0, 0) and (0, 1, 0) are stable. The tunneling "in" and "out" occurs in two separate steps after crossing in the boundary to the bi-stability region. The gray area at the top indicates the region where uncontrolled tunneling would occur.

to as an electron turnstile. In its simplest form, we can consider a three island array, as illustrated in Figure 2.36. The turnstile is operated by changing the charge induced on the central island. The region of operation is where the Coulomb diamond shards of stability for the state (0, 1, 0) and (0, 0, 0) overlap. Region (0, 1, 0) is bounded by alone that corresponds to (0, 0, 1) through junction 3. The system does not remain long in this (0, 0, 1) state as it is not stable in the operational region. The electron tunnels to the right, bringing the system to the stable (0, 0, 0). Similarly, the (0, 0, 0) shards bound by the line of transition to (1, 0, 0). This state is not stable, switching to (0, 1, 0). The working cycle starts deep within the (0, 0, 0) region. When a boundary is crossed, an electron is transferred in two jumps, from the left electrode to the central island. In crossing to the (0, 1, 0) region, the electron, the electron can then be released to the right electrode. In the region of operation, it is not possible to reverse the direction of electron motion by changing the cycle. In fact, the direction of motion is set by the bias applied to the turnstile. This acts as a protection to the turnstile operation.

To demonstrate the operation, a time-dependent signal is applied to the central gate with a frequency, $f = \omega/2\pi$. If the turnstile works in an ideal manner, the current will be precisely: $I = ef = e\omega/2\pi$, depending on neither the modulation amplitude nor on the bias voltage. Indeed, the current is controlled by the frequency applied to the potential, clearly being the number of electrons per unit time. The accuracy of more complicated turnstiles allows their use in metrology as an accurate current and capacitance standard. In Figure 2.37, we again illustrate the quantized nature of the electron turnstile for a fixed frequency with variable gate settings. In this particular example, the turnstile structure is fabricated using a tunnel device of a superconductor (S) and a normal metal (N) which sandwich the insulating (I) layer. In fact, two such junctions are used in the SINIS device.

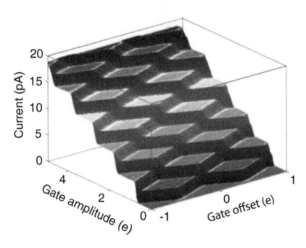

FIGURE 2.37   Current plateaus at $I = nef$ of a hybrid SINIS single-electron turnstile measured at f = 20 MHz. Reprinted by permission from Nature: J. P. Pekola, J. J. Vartiainen, M. Möttönen, O.-P. Saira, M. Meschke, & D. V. Averin, (2008). *Nature Physics, 4*, 120–124, ©(2008).

2.12.2   *The Single-Electron Pump*

An alternative single charge transfer device is the so-called *single-electron pump*, in which the charge transfer is quasi-adiabatic. In the simplest case, we can consider the double quantum dot structure illustrated above, see Figure 2.33, with a stability diagram as shown in Figure 2.34(c). For small bias voltages, $V \simeq 0$, stable configurations exist for the excess electron numbers on each island, N_1 and N_2, for various combinations of the gate voltages, $V_1^{(g)}$ and $V_2^{(g)}$. We can set the bias voltage to sit near one of the electron triple points, Figure 2.34(d), and then it is possible to traverse the region around the triple point in either a clockwise or counter-clockwise direction using phase-shifted ac biases on $V_1^{(g)}$ and $V_2^{(g)}$, applied on top of the dc biases already imposed to move close to the triple point, such that; $\tilde{V}_i^{(g)} = V_i^{(g)} + \tilde{v}_i^{(g)}$, with $i = 1, 2$. If, for example, $\tilde{v}_2^{(g)}$ lags $\tilde{v}_1^{(g)}$ by a factor of $\pi/2$, the system moves from state (0, 0) to state (1, 0) as $\tilde{v}_1^{(g)}$ increases, as it crosses the domain boundary. In this passage, an electron moves from the left electrode onto island 1, during the first positive part of the ac cycle. Then, as $\tilde{v}_2^{(g)}$ increases, a second domain will be crossed, thus taking the system from the state (1, 0) to state (0, 1), i.e., the electron shifts from island 1 to island 2. Finally, during the negative-going part of the $\tilde{v}_1^{(g)}$ cycle, the system crosses back down into the original domain for the state (0, 0), which corresponds to the electron tunneling out onto electrode 2. The net effect is a negative current of magnitude ef. If the phase of the two ac gate voltages is reversed, such that $\tilde{v}_1^{(g)}$ lags $\tilde{v}_2^{(g)}$ by $\pi/2$, the path around the triple point will be reversed. This will then give rise to a positive current, with an electron passing from right to left through the double QD structure. If, however, we bias the system close to the h triple point, a counter clock-wise motion results in a current, which is opposite to that produced by an electron triple point.

Despite the fact that semiconductor quantum dots contain a much smaller number of electrons than metallic islands, tunnel barriers can be easily formed by electrostatic potentials and controlled externally via gating. The height of the barrier is also easily controlled and devices can be tailor-made to good precision. As an example, a double barrier single-electron pump system is schematically illustrated in Figure 2.38 and is based on a Si MOSFET, as conceptualized by Ono and Takahashi (2003). In panel (b), the energy diagram is shown for the case where an electron is shuttled across the device in the cycle described above. In the state I, the system resides in the Coulomb blockade regime, containing N electrons. Closing the left channel with a negative bias on gate 1, brings the system to state II. Maintaining the island potential nearly constant by applying a positive bias to gate 2, the electron is then ejected to the right channel and the island potential is raised and ends up in a new Coulomb blockade state with $N - 1$ electrons, in position III. The next state (IV) is reached by again opening the left channel and simultaneously closing the right, which keeps the island in an almost constant state. The cycle is closed in state V by lowering the island potential so that an electron can enter from the left. The process is illustrated on the stability diagram in panel (c) of Figure 2.38.

In Figure 2.39 we show a single-electron pumping device (a) along with the conduction energy-band profile for a pumping cycle (b). Once again, we observe the current as being set by the frequency of the applied signals, $I = ef$, depending on whether the phase shift between the gate ac

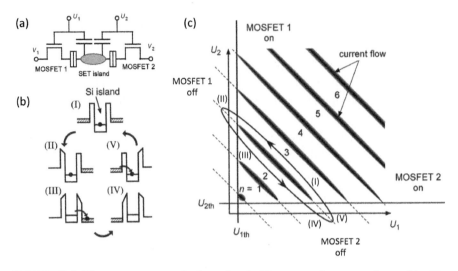

FIGURE 2.38   (a) Device equivalent circuit. The gate voltages are denoted by U_1 and U_2. (b) Potential-profile diagrams of transfer steps. (c) Stability diagram in the gate-voltage plane. The vertical and horizontal lines indicate the threshold voltages of MOSFET-1 and -2, respectively. Three bold lines indicate open states with non-zero conductivity, and attached dotted lines indicate hidden open states with vanishing conductivity due to the closed MOSFET(s). The regions between these lines are Coulomb blockade states. The oval shows the trajectory of the two gate voltages. The states marked by Roman numerals correspond to those in (b). Reprinted by permission from Y. Takahashi, Y. Ono, A. Fujiwara, K. Nishiguchi, & H. Inokawa, (2009). *Device Applications of Silicon Nanocrystals and Nanostructures*, N. Koshida (Ed.), by Springer Nature: Springer, ©(2009).

biases is positive or negative and as independent of the static bias voltage applied across the junctions. The quantization of the current, and its polarity dependence on the phase difference of the two ac sources, illustrates nicely the pump principle (Zhao et al., 2017). The operation of the single-electron pump is a single-electron analog of charge-coupled devices, which are extensively used in memory and image processing applications. The capacitive coupling strength of the quantum dot to each gate is obtained from the period of the corresponding Coulomb blockade oscillations. Then, the two gate voltages that have the strongest capacitive coupling to the dot potential, namely, V_{B1} and V_{PL}, are selected to be the main sweep parameters. Sinusoidal excitation is set up with a relatively low frequency, starting from 500 MHz, is applied to B1. The rf-drive power P_{B1} is gradually increased until a plateau structure appears in the $V_{B1} - V_{PL}$ plane, see Figure 2.39 (c). The ef plateaus are well pronounced, shown in Figure 2.39 (d), up to 2 GHz, with no change in the gate voltages or in the rf power. Note the linear dependence in the pump current, I_P.

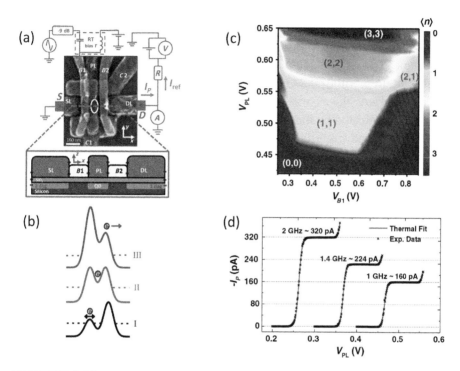

FIGURE 2.39 (a) False-color SEM image of an electron pump similar to the one used. The yellow circle highlights the approximate region where a quantum dot is formed. A schematic of the measurement setup as well as an illustrative cross-sectional view of the metal-oxide-semiconductor structure is also shown. The drain contact is connected to the reference current source used in the high-accuracy measurements. It consists of a temperature-controlled 1-GΩ-thick film resistor and a voltage source monitored by a high-accuracy voltmeter. (b) Sketch of the conduction-band energy profile (solid lines) and Fermi level (dashed lines) during a pumping cycle. (c) Coarsely tuned plateaus for $f = 500$ MHz, measured using the normal-accuracy setup. In the notation (m, n), m (n) represents the ideal number of captured (ejected) electrons. Here, $V_{SL} = V_{B2} = 1.5$V, $V_{DL} = 1.75$V, $V_{C1} = 1$V, $V_{C2} = 0.2$V, and $P_{B1} = 2$ dBm. (b) Normal-accuracy measurements (black crosses) of pumped current as a function of the plunger gate voltage for different pumping frequencies and fits the thermal model (red solid lines). Data have been horizontally shifted for clarity. Parameter settings: $V_{SL} = V_{B2} = 1.5$ V, $V_{B1} = 0.45$ V, $V_{DL} = 1.9$V, $V_{C1} = ?1.04$V, $V_{C2} = 0.187$V, and $P_{B1} = 3$dBm. Reprinted figure with permission from R. Zhao et al., (2017). *Phys. Rev. Appl.*, 8, 044021, (2017), ©(2017) by the American Physical Society.

Improved precision in the pumping operation can be found by increasing the number of quantum dots. A single-electron pump consisting of a series array of five ultra-small metallic tunnel junctions is illustrated in Figure 2.40(a). On the right side, the chain of tunnel junctions is connected to a metallic electrode, the "island," which has a small total capacitance of about $C_\Sigma = 20$ fF. With a fast train of voltage pulses on the gate electrodes of the pump (V1-4, bottom figure), within 0.25 s an electron is pumped through the chain onto the island. Here, the excess electron causes a change in potential of about 8 V, which is detected by the single-electron transistor being capacitively coupled to the island (right-hand side). After a wait time t_w, the excess electron is removed from the island by a pulse sequence in the opposite direction. In the "shuttle pumping mode" demonstrated, the successive charging and discharging of the island are repeated periodically. In the "hold mode," the gate voltages on the single-electron pump are not modulated. Then, the charge state of the island remains constant, since tunnel processes through the junction chain are suppressed by the Coulomb blockade effect.

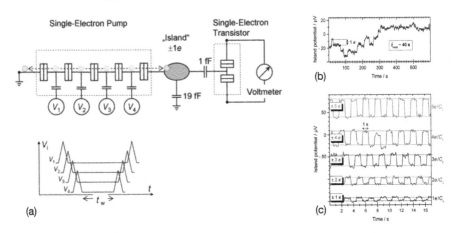

FIGURE 2.40 (a) The single-electron pump based on a series of tunnel junctions and controlled using four gating voltages. (b) The single-electron pump in the "hold mode." (c) Signal time series in the "shuttle pumping mode"; charge packets of one up to five electrons are shuttled across the device [H. Scherer, https://www.ptb.de/cms/en/gateways/the-economy/news/single-news.html (accessed on 30 March 2020)].

Measurements of the output signal of the single-electron transistor, connected to the electrical potential of the island, can be made in a controlled manner. Figure 2.40(b) shows the time series of the signal when the single-electron pump is kept in the "hold mode": Apart from noise intrinsic to the single-electron detector, the signal is constant for several seconds. Unwanted tunnel processes of single electrons through the chain of junctions cause step-like quantized fluctuations of the island potential. Changes of the island charge by one electron corresponds to a potential change of 8 V. The mean time interval between these unwanted "error" events on the island was 40 s. Figure 2.40(c) shows the signal time series in the "shuttle pumping mode," when charge packets of one up to five electrons where shuttled between the island and the opposite pump side with a clock time of $t_w = 1$ s. Obviously, the detected potential changes follow the second beat of island charging and discharging, and the signal amplitude is proportional to the number of excess electrons on the island.

2.13   The Electronic Properties of Graphene and Carbon Nanotubes

Graphene is a pure carbon structure—a single layer of graphite—whose atoms are arranged in a 2D hexagonal array, and is the basis of many different allotropes of carbon. Stacking graphene sheets will form the graphite structure, while rolling a graphene sheet into a cylinder will produce a carbon nanotube (CNT). Fullerenes can be formed by introducing pentagons, which cause a curvature of the sheet into a spherical like surface. These are illustrated in Figure 2.41. Graphene was known to have exotic linear dispersion, bearing a close resemblance to that of a massless relativistic particle. However, this was not thought to have much relevance to real systems until 2004, when the group at the University of Manchester, led by Andre Geim and Konstantin Novoselov, discovered a novel method for obtaining fairly large areas of graphene layers from graphitic samples. The principal findings of the Manchester group are related to the observation of a subtle optical effect of a single graphene layer and led to the identification of the dispersion relation for graphene, which explains the experimental studies of the quantum Hall effect and electron transport in high magnetic fields. Indeed, the earlier findings of Wallace (1947), who studied the band theory for graphite, illustrated the correct treatment for the case of a single layer of graphite. The work of the Manchester group led to the awarding of the 2010 Nobel prize in physics to Geim and Novoselov.

In terms of bonding, the underlying structure of graphene arises from the hybridization of the s and p_x, p_y atomic states into sp^2 molecular orbitals in

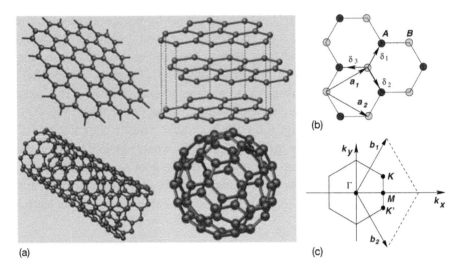

FIGURE 2.41 (a) Graphene (top left) is a honeycomb lattice of carbon atoms. Graphite (top right) can be viewed as a stack of graphene layers. Carbon nanotubes are rolled-up cylinders of graphene (bottom left). Fullerenes C_{60} are molecules consisting of wrapped graphene by the introduction of pentagons on the hexagonal lattice (bottom right). (b) Lattice structure of graphene, made out of two interpenetrating triangular lattices (a_1 and a_2 are the lattice unit vectors, and δ_i, $i = 1, 2, 3$ are the nearest-neighbor vectors). (c) Brillouin zone for graphene. The Dirac cones are located at the K and K' points. Reprinted figure with permission from A. H. Castro Neto, F. Guinea, N. M. R. Peres, K. S. Novoselov, & A. K. Geim, *Rev. Mod. Phys.*, *81*, 109–162, (2009), ©(2009) by the American Physical Society.

the plane of graphene, framing directed σ-bonds between nearest neighbors. There are three nearest neighbors in the plane, giving rise to the hexagonal lattice structure, as illustrated in Figure 2.41(b). The remaining p_z orbitals are relatively weakly bound. They contribute one electron per atom and form the π bands in the graphene lattice, which is half-filled. These π bands are primarily responsible for the higher-lying conduction and valence bands associated with transport and the position of the Fermi energy.

The primitive vectors of the graphene hexagonal structure are associated with the two atom basis of carbon atoms at each point. That is, we consider two sub lattices (A and B). Any equivalent point within the graphene lattice is obtained by a translation in terms of integer multiples of these vectors, as given by:

$$\mathbf{a}_1 = \frac{a_0}{2}(3\hat{x} + \sqrt{3}\hat{y}) \qquad (2.194)$$

and

$$\mathbf{a}_2 = \frac{a_0}{2}(3\hat{x} - \sqrt{3}\hat{y}) \tag{2.195}$$

where $a_0 = \sqrt{3}a_{cc}$, and a_{cc} is the nearest neighbor bonding distance, of magnitude 1.42 Å. The corresponding reciprocal lattice can be generated from the reciprocal lattice vectors obtained from the condition:

$$\mathbf{a}_i \cdot \mathbf{b}_j = 2\pi\delta_{ij} \tag{2.196}$$

from which we obtain:

$$\mathbf{b}_1 = \frac{b_0}{2}(\hat{x} + \sqrt{3}\hat{y}) \tag{2.197}$$

$$\mathbf{b}_2 = \frac{b_0}{2}(\hat{x} - \sqrt{3}\hat{y}) \tag{2.198}$$

where $b_0 = 4\pi/3a_0$. The reciprocal lattice is shown in Figure 2.41(c). High symmetry points correspond to the positions indicated as Γ, M and K, the latter of which are situated at the vertices of the hexagonal first Brillouin zone in k-space. The points of high symmetry, located at K and K', are of particular interest, at the corners of the Brillouin zone. These are known as the Dirac points, as will be discussed later, and are positioned in reciprocal space at:

$$\mathbf{K} = \frac{2\pi}{3a_0}\left(\hat{x} + \frac{\hat{y}}{\sqrt{3}}\right) \tag{2.199}$$

and

$$\mathbf{K}' = \frac{2\pi}{3a_0}\left(\hat{x} - \frac{\hat{y}}{\sqrt{3}}\right) \tag{2.200}$$

The nearest neighbor positions are expressed by the vectors δ_i, $i = 1, 2, 3$, illustrated in Figure 2.41(b), and given as:

$$\delta_1 = \frac{a_0}{2}\left(\hat{x} + \frac{\hat{y}}{\sqrt{3}}\right), \ \delta_2 = \frac{a_0}{2}\left(\hat{x} - \frac{\hat{y}}{\sqrt{3}}\right), \ \delta_3 = -a_0\hat{x} \tag{2.201}$$

Within the tight-binding model, the electronic states of the π-bonds are independent of those of the σ bonds and the latter being tightly bound are negligible to the electronic properties. The atom A (B) is uniquely defined by one orbital per atom site and since p_z orbitals are responsible for the contributing electrons, $\Phi_{p_z}(\mathbf{r} - \mathbf{r}_A)$ ($\Phi_{p_z}(\mathbf{r} - \mathbf{r}_B)$), we can use these to derive the electronic spectrum from the Schrödinger equation by applying the Bloch theorem. As such, we can express the wave function in the form:

$$\Psi_{\mathbf{k}}(\mathbf{r}) = u_A(\mathbf{k})\tilde{\Phi}^A_{p_z}(\mathbf{k},\mathbf{r}) + u_B(\mathbf{k})\tilde{\Phi}^B_{p_z}(\mathbf{k},\mathbf{r}) \qquad (2.202)$$

where

$$\tilde{\Phi}^A_{p_z}(\mathbf{k},\mathbf{r}) = \frac{1}{\sqrt{N_{\text{cell}}}} \sum_{\mathbf{l}} e^{i\mathbf{k}\cdot\mathbf{l}} \Phi_{p_z}(\mathbf{r} - \mathbf{r}_A - \mathbf{l})$$

$$\tilde{\Phi}^B_{p_z}(\mathbf{k},\mathbf{r}) = \frac{1}{\sqrt{N_{\text{cell}}}} \sum_{\mathbf{l}} e^{i\mathbf{k}\cdot\mathbf{l}} \Phi_{p_z}(\mathbf{r} - \mathbf{r}_B - \mathbf{l})$$

N_{cell} being the number of unit cells in the graphene sheet and $\mathbf{l}$ is the position vector. The energy spectrum is solved using the Schrödinger equation, which we can reduce to matrix form as:

$$\begin{vmatrix} \mathcal{H}_{AA} - E & \mathcal{H}_{AB} \\ \mathcal{H}_{BA} & \mathcal{H}_{BB} - E \end{vmatrix} = 0 \qquad (2.203)$$

The matrix elements being defined by:

$$\mathcal{H}_{AA}(\mathbf{k}) = \frac{1}{N_{\text{cell}}} \sum_{\mathbf{l},\mathbf{l}'} e^{i\mathbf{k}\cdot(\mathbf{l}-\mathbf{l}')} \langle \Phi^{A,\mathbf{l}}_{p_z} | \mathcal{H} | \Phi^{A,\mathbf{l}'}_{p_z} \rangle \qquad (2.204)$$

$$\mathcal{H}_{AB}(\mathbf{k}) = \frac{1}{N_{\text{cell}}} \sum_{\mathbf{l},\mathbf{l}'} e^{i\mathbf{k}\cdot(\mathbf{l}-\mathbf{l}')} \langle \Phi^{A,\mathbf{l}}_{p_z} | \mathcal{H} | \Phi^{B,\mathbf{l}'}_{p_z} \rangle \qquad (2.205)$$

with $\Phi^{A(B),\tau}_{p_z} = \Phi_{p_z}(\mathbf{r} - \mathbf{r}_{A(B)} - \tau)$. Restricting the interactions to just the nearest neighbors allows us to simplify this to:

$$\mathcal{H}_{AB}(\mathbf{k}) = \langle \Phi^{A,0}_{p_z} | \mathcal{H} | \Phi^{B,0}_{p_z} \rangle + e^{-i\mathbf{k}\cdot\mathbf{a}_1} \langle \Phi^{A,0}_{p_z} | \mathcal{H} | \Phi^{B,-\mathbf{a}_1}_{p_z} \rangle + e^{-i\mathbf{k}\cdot\mathbf{a}_2} \langle \Phi^{A,0}_{p_z} | \mathcal{H} | \Phi^{B,-\mathbf{a}_2}_{p_z} \rangle$$

$$= -\hat{t}\alpha(\mathbf{k}) \qquad (2.206)$$

Here, $\hat{t}$ represents the transfer integral between nearest neighbor π - orbitals, $\alpha(\mathbf{k}) = 1 + e^{-i\mathbf{k}\cdot\mathbf{a}_1} + e^{-i\mathbf{k}\cdot\mathbf{a}_2}$, and we have $\langle \Phi^{A,0}_{p_z} | \mathcal{H} | \Phi^{A,0}_{p_z} \rangle = \langle \Phi^{B,0}_{p_z} | \mathcal{H} | \Phi^{B,0}_{p_z} \rangle = 0$, which sets the reference energy for the system. This allows the dispersion relation to be evaluated as:

$$E^{\pm}(\mathbf{k}) = \pm t \{3 + 2\cos(\mathbf{k}\cdot\mathbf{a}_1) + 2\cos(\mathbf{k}\cdot\mathbf{a}_2) + 2\cos[\mathbf{k}\cdot(\mathbf{a}_1 - \mathbf{a}_2)]\}^{1/2}$$

$$(2.207)$$

This can be rewritten as:

$$E^{\pm}(\mathbf{k}) = \pm t \left[1 + 4\cos\left(\frac{3k_x a_0}{2}\right)\cos\left(\frac{\sqrt{3}k_y a_0}{2}\right) + 4\cos^2\left(\frac{\sqrt{3}k_y a_0}{2}\right) \right]^{1/2}$$

(2.208)

The $\mathbf{k} = (k_x, k_y)$ vectors belong to the first hexagonal Brillouin zone (BZ) and constitute an ensemble of available electronic momenta. These dispersion relations can be represented along with the directions of high symmetry in the BZ; points Γ, M and K, to give the band structure. These are illustrated as 2D and 3D plots in Figure 2.42.

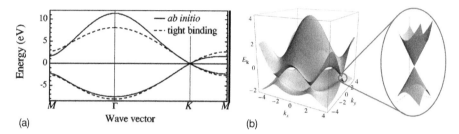

(a)   (b)

FIGURE 2.42   (a) Comparison of the tight-binding energy dispersion curve with an ab-initio calculation. (b) Electronic dispersion in the honeycomb lattice. Left: energy spectrum in units of t for finite values of t, with $t = 2.7$ eV. Right: zoom-in of the energy bands close to one of the Dirac points. Reprinted figure with permission from A. H. Castro Neto, F. Guinea, N. M. R. Peres, K. S. Novoselov, & A. K. Geim, (2009). *Rev. Mod. Phys., 81*, 109–162, ©(2009) by the American Physical Society.

In the 3D plot, we observe that the high-symmetry K-points occur at k-space positions:

$$\mathbf{K} = \left(n \pm \frac{1}{3}\right)\mathbf{b}_1 + \left(m \pm \frac{1}{3}\right)\mathbf{b}_2$$

(2.209)

where m, n are integers. Substituting these vectors into the expression for the energy, we find that $E^{\pm}(\mathbf{k})$ goes to zero. These are the Dirac points, where the two solutions become degenerate, and hence there is no energy gap, with the Fermi energy lying at the point of intersection between the upper and lower branches or bands. We note that the upper branch forms the conduction band, while the lower forms the valence band. Expanding the energy around one of the equivalent Dirac points we can write:

$$E^{\pm}(\mathbf{k}) \simeq \pm \frac{3}{2}a_0 t |\mathbf{k} - \mathbf{K}| + O[(k)^2] = \pm v_F |\mathbf{q}| + O[(q/K)^2]$$

(2.210)

where $\mathbf{q} = \mathbf{k} - \mathbf{K}$, is the momentum measured relative to the Dirac point and v_F is the Fermi velocity. Using the definition for the group velocity: $v_g = \nabla_{\mathbf{k}}[E^{\pm}(\mathbf{k})]/\hbar$, we can estimate the Fermi velocity as $v_F \simeq 3a_o t/2\hbar \simeq 1 \times 10^6$ m s^{-1}, where we have used $t = 2.8$ eV.

One further important point regards the effective mass of the electrons in graphene. From most textbooks on solid-state physics, we can find that the relation between the effective mass and the energy dispersion can be expressed as:

$$m^* = \hbar^2\{\nabla_{\mathbf{k}}^2[E^{\pm}(\mathbf{k})]\}^{-1} \qquad (2.211)$$

In bulk solids, we typically find a parabolic relation for the energy bands, in which case we find a finite value for the effective mass. At Brillouin zone boundaries, the effective mass diverges strongly, where we meet the diffraction condition associated with the Bragg condition and the forbidden energies of the electron, where the bandgap originates. In graphene, the dispersion relation at the K points is linear with the wave vector and therefore, it also appears that the effective mass also diverges. This is a confusing result at first sight, since the electrons in graphene are said to be "massless," which appears in strong contradiction to the above. The problem actually arises from the definition of the effective mass, as expressed in Eq. (2.211). This definition assumes a parabolic energy dispersion and is invalid for the case of graphene.

We can address this problem by considering the crystal momentum, as expressed by $p = \hbar k = m^* v_g$. This allows us to establish an effective mass in a different form from Eq. (2.211). We can express this as:

$$m^* = m^*(E,k) = \frac{p}{v_g} = \hbar^2 k\{\nabla_{\mathbf{k}}[E^{\pm}(\mathbf{k})]\}^{-1} \qquad (2.212)$$

This is often referred to as the optical effective mass. This can alternatively be expressed as: $m^* = \hbar^2 k(\partial E/\partial k)^{-1}$. The electrons in the energy bands near the K-points resemble those for relativistic particles. The energy, in this case, should be expressed as:

$$E^2(k) = (pc)^2 + (m_0 c^2)^2 = (\hbar k c)^2 + (m_0 c^2)^2 \qquad (2.213)$$

where m_0 represents the rest mass. For a massless particle like the photon, we have $E(k) = \pm\hbar kc$, c being the speed of light. This dispersion relation is linear in k, which is also the case for electrons in graphene near the K-points. It is in this sense that the electrons (and holes) in graphene near the

Fermi energy are considered massless. The excitation of an electron from the valence to the conduction band will create an electron-hole pair as it does in a semiconductor. However, in this case, all the energy of the excitation will be in the form of kinetic energy, since there is no rest mass to overcome, which is reflected in the fact that the bands meet at $k = 0$. In the case of the solid, where there is a finite rest mass at $k = 0$, the energy between the bands ($+$ and $-$ energies at this momentum) will be $2m_0c^2$, which represents the band gap from the relativistic point of view. Interestingly, using the definition of the effective mass in Eq. (2.212) along with the relativistic energy, Eq. (2.213), returns an expression:

$$m^* = \frac{\sqrt{(\hbar kc)^2 + (m_0c^2)^2}}{c^2} \tag{2.214}$$

In this case, we obtain an effective mass of m_0 for $k = 0$.

We previously noted that the carbon nanotube (CNT) can be obtained by rolling up a sheet of graphene. The manner, in which the sheet is rolled up, can have important consequences to the form of the CNT (this is discussed in more detail in Chapter 9 of Volume 1). We can define a lattice vector for a carbon atom on the graphene sheet as:

$$\mathbf{c} = n\mathbf{a}_1 + m\mathbf{a}_2 \tag{2.215}$$

or simply refer to it as (n, m), where n and m are integers. We can use such a vector to define the rolled sheet of graphene, which must join the two edges of the sheet at the appropriate crystallographic equivalent sites. The length of the vector $\mathbf{c}$ corresponds to the circumference of the CNT, which will have a diameter:

$$d = \frac{|\mathbf{c}|}{\pi} = \frac{a_0}{\pi}(n^2 + mn + m^2)^{1/2} \tag{2.216}$$

There are two special cases that we can consider; (i) the zig-zag CNT, where $m = 0$, and (ii) $m = n$ which is referred to as armchair CNT, see Fig 2.43(a).

The closure of the graphene sheet into a cylindrical CNT imposes periodic boundary conditions on the wave functions of the electron. This means that the projection of the wave vector along the chiral vector $\mathbf{c}$ will be quantized according to the relation:

$$\mathbf{k} \cdot \mathbf{c} = 2\pi w \tag{2.217}$$

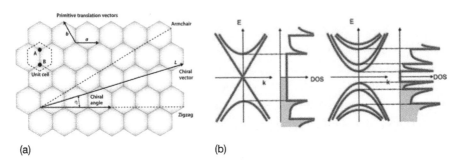

(a)   (b)

FIGURE 2.43   (a) The honeycomb lattice of graphene. The hexagonal unit cell contains two carbon atoms (A and B). The chiral vector determining the structure of a carbon nanotube is given by **L**, and its length gives the circumference. The chiral angle is denoted by η, with $\eta = 0$ corresponding to zig-zag nanotubes and $\eta = \pi/6$ to armchair nanotubes. (b) Schematic of the band structure and density of states of a metallic nanotube (left) and semiconducting carbon nanotube. (a) Reprinted by permission from Nature: T. Ando, (2009). *NPG Asia Materials, 1*, 17, ©(2009); (b) Reprinted from S. Nanot, E. H. Hroz, J.-H. Kim, R.H. Hauge, & J. Kono, (2012). *Adv. Mater., 24*, 4977, under the Creative Commons Attribution 4.0 International License (Wiley Open Access Articles).

where w must also be an integer quantum number. Electron motion in a direction perpendicular to the chiral vector will be unconstrained, leading to a quasi-1D system, while the transverse modes are defined by the above condition, Eq. (2.217).

To evaluate the effect of this on the $E(\mathbf{k})$ dispersion relation, we can consider one of the Dirac points around K:

$$E^{\pm}(\mathbf{k}) \simeq \pm \frac{3}{2} a_0 t |\mathbf{k} - \mathbf{K}| = \pm \frac{3}{2} a_0 t \sqrt{k_{c,w}^2 + k_t^2} \qquad (2.218)$$

where we have written $k_{c,w}$ as the chiral wave vector parallel to **c** and k_t as the transversal wave vector. The form can be expressed as:

$$k_{c,w} = \frac{(\mathbf{k} - \mathbf{K}) \cdot \mathbf{c}}{|\mathbf{c}|} = \frac{2\pi w - 2\pi(n-m)/3}{d\pi} = \frac{3w - (n-m)}{3d} \qquad (2.219)$$

The minimum energy in $E(\mathbf{k})$ defines the bandage for the nanotube. If $(n - m)$ is a multiple of 3, the minimum energy will be zero and the dispersion relation corresponds to a metallic nanotube with a linear 1D dispersion around K, equivalent to that for graphene. This will occur for the case of an armchair CNT, since $n = m$. A schematic illustration of the band diagram for the metallic and semiconducting examples are illustrated in Figure 2.43(b). A more detailed illustration of the band structure for armchair and zig-zag

CNT is shown in Figure 2.44. In the case of the armchair CNT, the band

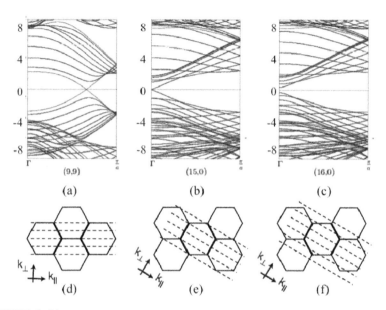

FIGURE 2.44 (a) The band structure of a (9,9) armchair nanotube. One of the quantization lines bands crossing the Fermi energy stems from the quantization line in (d), which crosses the K-points and makes this SWCNT metallic. (b) The band structure of a (15,0) zig-zag nanotube. It has a doubly degenerate band crossing the Fermi level in (e). When the curvature effects are taken into account, the K-points move away from the corners of the hexagon, which makes metallic zig-zag nanotubes to be small band gap semiconductors. (c) and (f) Some tubes are semi-conducting with a bigger band gap as the (16,0) zig-zag tube. Reprinted from B. Xu, J.Yin and Z. Liu, in *Physical and Chemical Properties of Carbon Nanotubes*, S. Suzuki (Ed.), Intech Open (2013), under the Creative Commons Attribution 4.0 International License.

diagram shows no gap, as predicted for the lowest subband, which will have metallic like behavior with linear dispersion, Figure 2.44(a). For the zig-zag CNT, where n is a multiple of 3, the CNT also has a degenerate band with no gap, Figure 2.44(b). However, when n is not a multiple of 3, a non-zero band gap results and the CNT will be semiconducting, as illustrated in Figure 2.44(c). The minimum value of $k_{c,w}$ is $k_{c,w} = 2/3d$. From this, the band gap minimum at K can be estimated as:

$$\Delta_g = \frac{2\sqrt{3}a_0t}{d} \simeq \frac{0.8(eV)}{d} \qquad (2.220)$$

where d is given in nm. The semiconducting band gap in a zig-zag CNT is inversely proportional to the diameter of the nanotube.

2.14   Summary

The electronic properties of nanomaterials are of particular importance. They form some of the principal applications for metallic and semiconducting materials and are the basis for a large volume of the applications in nanotechnologies. We can probably say without fear of contradiction that it is the most developed area of nanotechnologies and has been the driving force for research and development from the very beginning of this subject. Indeed, much experimental and theoretical research in materials, as well as the testing of many concepts in solid-state physics and quantum mechanics, can be traced back to the principles which have given birth to the nanotechnology revolution. This is intimately tied up with the developments in the microelectronics and semiconductor industries.

The length scales of importance in the consideration of the electrical properties of nanomaterials and low dimensional systems can be expressed in terms of the Fermi wavelength and other physical quantities related to the Fermi energy of metals. Another important and guiding principle in the length scale is related to the scattering of charge carriers. Once the dimensions of a physical object reach these orders of length, then we can expect to observe quantization effects for the former, and ballistic transport in the latter.

Quantum confinement is a well-known phenomenon and can lead to a number of modifications of the transport behavior of electrons and charge carriers and also has important implications for the optical properties of low dimensional structures. The effects of the spatial confinement of electrons have been known since the early days of the development of quantum mechanics. The fabrication of reduced dimensionality in solid-state physics allows the direct testing of these basic quantum principles. Much research was performed in semiconductors to determine the physical properties related to quantum wells, wires, and dots. These have subsequently formed the basis of the development of many devices, most notably the solid-state semiconducting laser.

Reduced physical dimensions produce an energy quantization of allowed electron states. This affects the density of states in the systems and can be evaluated in one, two and three dimensions, where confinement of charge carriers occurs to limit their motion in a plane (well), a line (wire) or in all directions (dot). These electronic systems are said to be 2D, 1D, and 0D, respectively. This spatial confinement, which is considered in terms of the confining potentials, can be used to evaluate the allowed energies of the charge carriers and in the case of infinite potential barriers, analytical formulae can be readily derived. For finite barriers, the calculations are more complex, but the calculations can be made computationally.

There are a large number of potential devices that can be made from low-dimensional systems. One of the simplest is the quantum point contact or QPC. This relies on making at least one of the physical dimensions continuous to allow conduction in one direction, as in a quantum wire. Once the length of this wire is reduced sufficiently it is possible to establish ballistic transport. The lateral dimensions of the device will determine the number of conducting channels that can conduct. The system is considered as having reservoirs of charge carriers at either end of the QPC. Adjusting the lateral dimensions of the QPC results in discrete jumps in the device conductivity. These jumps are related to the number of conducting channels and have a height, which is proportional to the quantum of conductance, defined as $G_Q = 2e^2/h$, and is a fundamental constant for quantum transport.

A more generalized approach to the consideration of the transmission of charge carriers across a device can be constructed using the scattering matrix formalism. This can be used for any type of quantum transport device with any number of contacts or reservoirs. The quantum current can then be evaluated in terms of the transmission probabilities between the different reservoirs and the applied potentials. This leads to the well-known Landauer formula for the conductance and the Landauer–Büttiker relation for the current of the device.

Resonant tunneling is a specific condition in which the current across a quantum well is enhanced due to the matching of the applied potential with the quantum levels on the well structure. The current–voltage characteristics of such a resonant tunneling diode shows current maxima at specific potentials corresponding to the bound states of the QW. There are also regions of negative resistance. The current can be evaluated in terms of the transmission coefficients from the reservoirs at either side of the device and conform to the Landauer–Büttiker formalism.

Isolated island structures are of special interest since they can be charge by single electron transfer processes via appropriately applied potentials. This charging can effectively block further charge transfer to the island and lead to a phenomenon known as Coulomb blockade. These blockade regimes lead to a step-like current–voltage characteristic. A consideration of the coupling capacitances to the island via the contacts allows the calculation of the different energy states of the system, depending on the number of electrons that are transferred to it. Gating the island with a third electrode allows the device to be controlled in what is called a single-electron transistor. A stability diagram allows the device action to be understood in terms of the transfer of discrete quantities of electronic charge (electrons).

The capacitive coupling of further islands in a chain can lead to further operations and devices. Stability diagrams help to understand their operation and are based on the considerations of the energy and capacitances between the island and the electrodes. Applications of these systems have to lead to the single-electron turnstile and the single-electron pump.

As we have seen in previous chapters, carbon nanotubes and graphene have attracted enormous attention in recent decades due to their unique mechanical and electronic properties. These latter are related to the electronic energy bands which are derived from the hexagonal periodic honeycomb structure of the graphene sheet. These systems display cone structures at the Dirac point in their energy band structure and give rise to effective massless fermions at the Fermi energy. They exhibit very large electric mobilities, making them ideal for electronic devices and applications. Since CNTs are also based on the same hexagonal periodic arrays of carbon atoms, their electron properties are also of interest, and they also display metallic and semiconducting transport behavior depending on the orientation of the CNT. Different behavior being noted for armchair and zig-zag structures.

2.15 Problems

(1) (a) How does the frequency of an electronic device scale with its size?
   (b) How large would a device need to be for it to function at microwave frequencies at room temperature? Consider only the thermal velocity of the electron.
(2) Consider a sample of crystalline GaAs with an electronic effective mass of $m^* = 0.067\,m_0$ and a mobility of $m = 10^5$ cm^2 V^{-1} s^{-1} at liquid nitrogen temperatures. Calculate the following parameters:

   (a) de Broglie wavelength λ
   (b) Scattering time τ_e
   (c) Thermal electron velocity v_T

(d) Mean free path λ_e

(e) Diffusion coefficient D $(\alpha = 3)$.

Determine the transport regime for devices with feature sizes of $L_x = 0.05, 0.5$ and 5 mm.

(3) Show that the low temperature Fermi wave vector for a two dimensional electron gas (2DEG) can be expressed as:

$$k_F = \sqrt{2\pi n_s}$$

where n_s is the equilibrium electron density.

(4) A particle of mass m^* is confined to move in a quantum well with infinite barriers and a width of d. Show that the energy separation of the first two levels is equal to $k_B T/2$, when d is given by:

$$d = \sqrt{\frac{3h^2}{4m^* k_B T}}$$

Calculate d for electrons of effective mass $m^* = m_0$ and $m^* = 0.1\, m_0$ at 300 K. Hence, show that the value of d is smaller than Δx (given by the Heisenberg uncertainty principle) by a factor of $\sqrt{3\pi^2}$.

(5) Estimate the emission wavelength of a 15 nm GaAs quantum well laser at room temperature. Use the following data for GaAs: $m_{el}^* = 0.067\, m_e$, $m_{hh1}^* = 0.5\, m_e$, $D_{GaAs}^{300\,K} = 1.424$ eV.

(6) Show that the density of states for an electron with:

$$E = \frac{h^2 k^2}{2m_e}$$

is given by:

$$g(E) = \frac{1}{2\pi^2} \left(\frac{2m^*}{h^2} \right)^{\frac{3}{2}} E^{\frac{1}{2}}$$

Determine the corresponding result for a quantum well structure.

(7) (a) Calculate the excitonic Rydberg and Bohr radius for GaAs, for which we have: $\varepsilon_r = 12.8$; $m_e^* = 0.067\, m_e$ and $m_h^* = 0.2\, m_e$.

(b) GaAs has a lattice constant of around 0.56 nm. Estimate the number of unit cells contained in the $n = 1$ orbit.

(c) Estimate the highest temperature at which it is possible to observe stable excitons in GaAs.

(8) How can quantum dot be defined? What characteristic wavelength plays the main role in its definition? Please explain why. Compare proper-

ties of quantum dots with properties of atoms. What holds electrons together in the quantum dot? How can the number of electrons in quantum dot be changed? What are the main applications of quantum dots?

(9) Explain what is meant by *Coulomb blockade* and explain the origin of the formation of the *Coulomb Diamond* structure experimentally observed in *single electron transfer* devices. Further show that a potential of e/C is required to transfer a single electron across a tunnel junction.

(10) Determine the energy of the Coulomb barrier for islands of 2, 5 and 10 nm in diameter. Assume that the capacitance of the island contact junctions to be 1 fF.

(11) Consider a semiconductor of width W and length L connected to two regions (contacts) of 2DEGs of the same material with an effective mass of $0.07\,m_e$. In the narrow region of the device, it is necessary to take account of the discreteness of transversal modes. Plot the electron density as a function of the Fermi energy for $W = 100$ nm assuming a hard wall potential.

(12) What is the shape of conductance oscillations in QD? Does it resemble Coulomb blockage oscillations? Could this effect be described by the diamond stability diagram for the Coulomb-blockade single-electron transistor?

(13) Can the double-island system be used as the current standard? What is its current accuracy? How accurate is the number of electrons pumped? What is the co-tunneling and how can it be prevented?

(14) How does the stability diagram of two-island, two-gate single-electron transistor change in the presence of nonzero bias voltage? What is the principle of electron pump operation? What is the link between the current on the plateau of the diagram and the frequency of the AC signal added to the gate voltage?

References and Further Reading

Ando, T., (2009). *NPG Asia Materials, 1,* 17.

Azuma, Y., Yasutake, Y., Kono, K., Kanehara, M., Teranishi, T., & Majima, Y., (2010). *Jpn. J. Appl. Phys., 49,* 090206.

Bandyopadhyay, S., (2012). *Physics of Nanostructured Solid State Devices,* Springer, New York.

Beenakker, C. W. J., & van Houten, H. (1991). *Semiconductor heterostructures and nanostructures,* Ehrenreich, H., & D. Turnbull (Eds.), *Solid State Physics,* volume *44,* 1, Academic, New York.

Blömers, C., Schäpers, Th., Richter, T., Calarco, R., Lüth, H., & Marso, M., (2008). *App. Phys. Lett., 92*, 132101.

Büttiker, M. (1986). *Phys. Rev. Lett., 57*, 1761.

Castro Neto, A. H., Guinea, F., Peres, N. M. R., Novoselov, K. S., & Geim, A. K., (2009). *Rev. Mod. Phys., 81*, 109–162.

Chiu, K.-L., & Xu, Y., (2016). arXiv.org–cond-mat–arXiv:1601.00986v2.

Datta, S., (1995). *Electronic Transport in Mesoscopic Systems*, Cambridge University Press, Cambridge.

Dresselhaus, M. S., & Dresselhaus, G., (2002). *Adv. Phys., 51*, 1.

Ferry, D. K., Goodnick, S. M., & Bird, J., (2009). *Transport in Nanostructures*, Second edition, Cambridge University Press, Cambridge.

Fischetti, M. V., & Vandenberghe, W. G., (2016). *Advanced Physics of Electron Transport in Semiconductors and Nanostructures*, Springer, Berlin.

Fujisawa, T., Oosterkamp, T. H., van der Wiel, W. G., Broer, B. W., Aguado, R., Tarucha, S., & L.Kouwenhoven, P., (1998). *Science, 282*, 932.

Geim, A. K., & Novoselov, K. S., (2007). *Nature Mater., 6*, 183.

Geim, A. K., & Grigorieva, I. V., (2013). *Nature, 499*, 419.

Guttinger, J., Stampfer, C., Libisch, F., Frey, T., Burgdorfer, J., Ihn, T., & Ensslin, K., (2009). *Phys. Rev. Lett., 103*, 046810.

Heikkilä, T. T. (2013). *The Physics of Nanoelectronics*, Oxford University Press, Oxford.

Heinzel, T., (2006). *Mesoscopic Electronics in Solid State Nanostructures*, Wiley-VCH.

Ihn, T., (2009). *Semiconductor Nanostructures*, Oxford University Press, Oxford.

Kaplan, S. B., & Hartstein, A., (1986). *Phys. Rev. Lett., 56*, 2403.

Kastner, M. A., (1993). *Physics Today, 46*, 24.

Koch, H., & Lubbig, H., (1992). *Single-Electron Tunneling and Mesoscopic Devices*, Springer-Verlag, Berlin.

Kouwenhoven, L. P., Oosterkamp, T. H., Danoesastro, M. W. S., Eto, M., Austing, D. G., Honda, T., & Tarucha, S., (1997). *Science, 278*, 1788.

Landauer, R., (1957). *IBM J. Res. Dev., 1*, 223.

Landauer, R., (1988). *IBM J. Res. Dev., 32*, 306.

Landauer, R., (1989). *Phys. Today, 42*, 119.

Mello, P. A., & Kumar, N., (2004). *Quantum Transport in Mesoscopic Systems: Complexity and Statistical Fluctuations*, Oxford University Press, Oxford.

Mitin, V. V., Kochelap, V. A., & Stroscio, M. A., (2008). *Introduction to Nanoelectronics: Science, Nanotechnology, Engineering and Applications*, Cambridge University Press, Cambridge.

Morozov, S. V., Novoselov, K. S., & Geim, A. K., (2008). *Phys.-Usp., 51*, 744.

Nanot, S., H‡roz, E. H., Kim, J.-H., Hauge, R. H., & Kono, J., (2012). *Adv. Mater., 24*, 4977.

Ono, Y., & Takahashi, Y., (2003). *App. Phys. Lett., 82*, 1221.

Ouisse, T., (2008). *Electron Transport in Nanostructures and Mesoscopic Devices: An Introduction*, ISTE J. Wiley and Sons, London and New York.

Pekola, J. P., Vartiainen, J. J., M. Möttönen, O.-Saira, P., Meschke, M., & Averin, D. V., (2008). *Nature Physics, 4*, 120–124.

Pekola, J. P., Saira, O.-P., Maisi, V. F., Kemppinen, A., Möttönen, M., Pashkin, Y. A., & Averin, D. V., (2013). *Rev. Mod. Phys., 85*, 1421–1469.

Reed, M., & Kirk, W. P. (Eds.), (1989). *Nanostructure Physics and Fabrication*, Academic, New York.

Reed, M., (1993). *Scientific American, 268*, 118.

Scherer, H., Lotkhov, S. V., G.-Willenberg, D., & Camarota, B., (2009). *IEEE Trans. Instr. Meas., 58*, 997.

Schmidt, T., Haug, R. J., Klitzing, K. V. Fšrster, A., & LŸth, H., (1997). *Phys. Rev. Lett., 78*, 1544.

Skocpol, W. J., Mankiewich, P. M., Howard, R. E., Jackel, L. D., Tennant, D. M., & Stone, A. D., (1986). *Phys. Rev. Lett., 56*, 2865.

Takahashi, Y., Ono, Y., Fujiwara, A., Nishiguchi, K., & Inokawa, H., (2009).*Device Applications of Silicon Nanocrystals and Nanostructures*, N. Koshida (Ed.), Springer Science + Business Media, New York.

Topinka, M. A., LeRoy, B. J., Shaw, S. E. J., Heller, E. J., Westervelt, R. M., Maranowski, K. D., & Gossard, A. C., (2000). *Science, 289*, 2323.

Topinka, M. A., LeRoy, B. J., Westervelt, R. M., Shaw, S. E. J., Fleischmann, R., Heller, E. J., Maranowski, K. D., & Gossard, A. C., (2001). *Nature, 410*, 183.

van Wees, B. J., van Houten, H., Beenakker, C. W. J., Williamson, J. G., Kouwenhoven, J. P., van der Marel, D., & Foxon, C. T., (1988). *Phys. Rev. Lett., 60*, 848.

van der Wiel, W. G., De Franceschi, S., Elzerman, J. M., Fujisawa, T., Tarucha, S., & Kouwenhoven, L. P., (2003). *Rev. Mod. Phys. 75*, 1–22.

Wallace, P. R., (1947). *Phys. Rev., 71*, 622.

Waugh, F. R., Berry, M. J., Mar, D. J., Westervelt, R. M., Campman, K. L., & Gossard, A. C., (1995). *Phys. Rev. Lett., 75*, 705.

Wei, D., Li, H.-O., Cao, G., Luo, G., Zheng, Z.-X., Tu, T., Xiao, M., G.-Guo, C., H.-Jiang, W., & G.-Guoi, P., (2013). *Scientific Reports, 3*, 3175.

Wilkins, R., Ben-Jacob, E., & Jaklevic, R. C., (1989). *Phys. Rev. Lett., 63*, 801.

Xu, B., Yin, J., & Liu, Z., (2013) *Physical and Chemical Properties of Carbon Nanotubes*, S. Suzuki (Ed.), Intech Open, DOI: 10.5772/51451.

Zhao, R., Rossi, A., Giblin, S. P., Fletcher, J. D., Hudson, F. E., Möttönen, M., Kataoka, M., & Dzurak, A. S., (2017). *Phys. Rev. Appl., 8*, 044021.

Chapter 3

Optical Properties of Nanostructures

3.1 Overview of Optical Properties of Solids

The subject of light–matter interaction is a vast area of physical science, covering a broad range of topics, from classical and quantum electrodynamics to the study of black holes and neutron stars. For our study, we will limit the discussion to that which concerns the optical properties of nanomaterials and confined structures. In particular, we will consider how the nature of the interactions between light and matter is governed by the physical dimensions and nature of the material in question. There are a number of physical phenomena that we shall consider, such as the plasmonics and photonics, which have emerged in recent decades as topics of growing importance. It is also the objective of this chapter to address the optical properties of semiconductor materials, which, as we saw in the previous chapter, hold a special interest due to the enormous range of applications of this class of material. We will see the strong link between the optical and electronic properties of solids and build on some of the topics that were addressed in the previous chapter.

Take a look around you and you will notice, that light interacts with different materials in apparently different ways. Metals have a shiny appearance, water and glass are transparent, gemstones and stained glass having specific colors, which are transmitted through their bulk, some liquids are cloudy, because incoming light is scattered in all directions by particles, while other objects transmit no light and have a particular color due to the absorption of certain wavelengths of light and the reflection of others. Indeed, there are a number of processes, which cause all objects to have a particular appearance. We can classify this in terms of a number of physical phenomena. The simplest way to do this is via the phenomenology of *reflection, propagation, and transmission*. We schematically illustrate this in Figure 3.1, where we show a light beam incident upon a medium. A proportion of the light is shown to be reflected from the front surface of the object, while the rest is transmitted across the interface into the medium itself. During the propagation of light in the medium, some of the light may be absorbed by the atoms of the medium, some may be scattered and some may reach the back surface and again suffer a reflection. Emerging from the material will generally be a smaller proportion of the light that enters it in the form of the transmitted light. Even this simple classification illustrates a number of processes that can occur, when light enters into contact with an object. Clearly, the conservation of energy will require that the total number of photons is conserved. For example, at the interface between media, the intensity of light must follow the general rule, $I = R + T$, where I, R and T are the intensities of the incident, reflected and transmitted beams. This is indeed the case for all wave phenomena, as for example the case of a beam of electrons incident on a potential barrier.

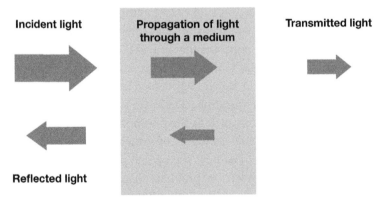

FIGURE 3.1   Illustration of the reflection, propagation, and transmission of light from an incident beam on an optical medium.

Between the front and rear surfaces of the medium, the propagation of light can undergo various physical processes, such as: (i) *refraction*, which causes the light to propagate at a lower velocity than in free space. This typically leads to the deviation of the propagation direction of the light within the medium with respect to that upon its arrival at the surface. This is described by Snell's law; $n_1 \sin \theta_1 = n_2 \sin \theta_2$, where the refractive index $n_i = c/v_i$ is the ratio of the velocity of light in free space with that in medium i. The angle of incidence/propagation with respect to the surface normal is expressed by $\theta_{1,2}$. This will not affect the light intensity as it propagates. (ii) *Absorption* can occur during the propagation of light through the medium, if the energy (frequency, wavelength) of the light is resonant with the component atoms of the medium. This process will reduce the light intensity as a function of its depth within the medium. In certain compounds, a selective absorption of specific wavelengths of light will result in the coloration of the medium, as with gemstones such as ruby and emerald. The absorption of light is typically quantified by the absorption coefficient, α, and can be expressed as the attenuation of light intensity using the Beer–Lambert law:

$$I = I_0 e^{-\alpha x} \tag{3.1}$$

where I_0 is the intensity of light at $x = 0$, and x is the direction of propagation. (iii) The phenomenon of *luminescence* concerns the spontaneous emission of light by excited atoms within the medium. It is common that this excitation arises via the promotion of electrons from the ground state to an excited state in the absorption process. Luminescence will accompany the propagation of light within the medium. The de-excitation, therefore, is the process, by which the energy of the excited state is reduced (in taking the atom back to a lower and typically its ground state) and converted into the emission of a photon. The frequency of the emitted light is different from that of the incoming radiation and is non-directional, i.e., emitted in all directions. It should be noted, that not all absorbed radiation is re-emitted as light. The time between absorption and spontaneous emission is finite. This means, that some of the energy absorbed can be converted into other processes, such as heating the solid via the generation of phonons. (iv) Another process, which commonly accompanies the propagation of light, is *scattering*. This process will change the direction of propagation and frequently also its wavelength. The simple fact, that the light changes direction, will mean an effective attenuation of the intensity of the propagating light. As we have just noted, scattering can change the energy/wavelength/frequency of the light. In doing so, energy is partially converted to or from some other form and the scattering is said to be *inelastic*. If the energy decreases, then energy must be transferred to the medium, and the process is referred to as *Stokes scattering*. Typically, this type of process corresponds to the emission of phonons. In the opposite process, where the frequency of radiation increases, there must already be phonons in the medium, which can be converted into light, thus lowering the temperature of the medium. This process is referred to as *Anti-Stokes scattering*. Such effects are typically studied using Brillouin light scattering and Raman spectroscopy techniques. If the frequency of the scattered light is unchanged, then no energy conversion takes place and the scattering is referred to as *elastic*. (v) More complex optical effects can occur for higher incident intensities, and typically can be observed when using higher power lasers.

Such effects include second harmonic generation (SHG), where the incident frequency is doubled, and other nonlinear effects.

The phenomena of refraction and absorption can be jointly described using the *complex refractive index*. This quantity is defined as follows:

$$\tilde{n} = n + i\kappa \tag{3.2}$$

The real part of this quantity is the refractive index we described above using the ratio of velocities in free space and the medium in question, while the imaginary part, κ, is referred to as the extinction coefficient. In fact, this is directly related to α, the absorption coefficient. Electromagnetic theory and in particular the Maxwell equations allows us to establish the wave equation for electromagnetic radiation, such as light. From this, we can determine the group velocity of electromagnetic waves as:

$$v = \frac{1}{\sqrt{\mu_0 \mu_r \varepsilon_0 \varepsilon_r}} \tag{3.3}$$

In free space, the relative permittivity, ε_r, and permeability, μ_r, are unity, giving the velocity of light in free space as: $c = 1/\sqrt{\mu_0 \varepsilon_0}$. This yields the expression for the velocity of light in a medium as $v = c/\sqrt{\mu_r \varepsilon_r}$. Therefore, from the definition of the refractive index, $n = c/v$, and the fact that at optical frequencies we can write $\mu_r = 1$, we have

$$n = \sqrt{\varepsilon_r} \tag{3.4}$$

which shows that the dielectric constant of a material is directly related to its refractive index.

Let us consider the propagation of the electric field of the electromagnetic wave, which we express as:

$$\mathbf{E} = \mathbf{E}_0 e^{i(kx - \omega t)} \tag{3.5}$$

where k is the wave vector and ω denotes the angular frequency. In the case of a non-absorbing medium, we can show that, since the frequency remains unchanged, the light will change wavelength to accompany the change in velocity in the medium, i.e., $\lambda' = \lambda/n$. In this case, we can express the wave vector as:

$$k = \frac{2\pi n}{\lambda} = \frac{n\omega}{c} \tag{3.6}$$

Generalizing this result for the complex refractive index, for an absorbing medium, we have:

$$k = \frac{\omega}{c}\tilde{n} = \frac{\omega}{c}(n + i\kappa) \qquad (3.7)$$

Substituting k into the exponent for the propagating electric field, Eq. (3.5), we find:

$$\mathbf{E} = \mathbf{E}_0 e^{-\kappa\omega x/c} e^{i\omega(nx/c-t)} \qquad (3.8)$$

This clearly shows, that a non-zero extinction will lead to attenuation of the propagating wave as it passes through the medium of a form given by the Beer–Lambert law, Eq. (3.1), with an attenuation coefficient of: $\alpha = 2\omega\kappa/c = 4\pi\kappa/\lambda$. Here, λ refers to the free space wavelength and not the wavelength of the light propagating in the medium. The real part of the refractive index determines the phase of the propagating wave.

We demonstrated that the refractive index can be expressed in terms of the dielectric constant, Eq. (3.4). Therefore, we can also represent the dielectric constant as a complex quantity:

$$\tilde{\varepsilon}_r = \varepsilon_1 + i\varepsilon_2 \qquad (3.9)$$

From the relation between the complex refractive index and the complex relative dielectric constant, $\tilde{n}^2 = \tilde{\varepsilon}_r$, we can relate the real and imaginary parts of these quantities as:

$$\varepsilon_1 = n^2 - \kappa^2 \qquad (3.10)$$

$$\varepsilon_2 = 2n\kappa \qquad (3.11)$$

The above parameters are also useful in determining the reflectivity of a material. The Fresnel equation, which relates the reflectivity with the real and imaginary parts of the refractive index, can be easily derived from the Maxwell equations. For normal incidence, this is expressed as:

$$R = \left|\frac{\tilde{n}-1}{\tilde{n}+1}\right|^2 = \frac{(n-1)^2 + \kappa^2}{(n+1)^2 + \kappa^2} \qquad (3.12)$$

For transparent materials, i.e., where the absorption is weak such as in glasses, the refractive index can be seen to essentially be real.

In the introductory paragraph to this chapter, we mentioned the various optical properties of certain types of material. We will provide a very brief summary of some of the more common optical properties of solids as it will allow us to introduce some of the principal concepts that will serve us as we go through some of the specialized topics in this chapter. These will mainly concern metals, semiconductors, insulators, and glasses. As this list might suggest, and as we mentioned in Chapter 2, there is a strong connection between the electrical and optical properties of solids.

On the highly conductive side of the electronic materials, metals are considered to be systems, in which the free electrons can be treated as a gas that is referred to as a plasma. We noted, that metals have a shiny appearance. Indeed, metals such as silver and aluminum are commonly used for making mirrors, and hence have high reflectivity in the visible range of the electromagnetic spectrum, see Figure 3.2. The high reflectivity in metals is due to the interaction of light with the free electrons in the metal.

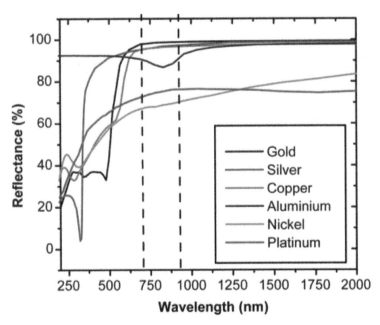

FIGURE 3.2 Reflectance of commonly used metals as a function of wavelength from 200 to 2000 nm. Reprinted from B. Wang & L. Gallais, (2013). *Opt. Exp., 21,* 14698, under the Creative Commons Attribution 4.0 International License.

To understand the underlying physics, we usually apply the simplified model due to Drude and Lorentz. In this model, we consider the oscillations of the free electrons as induced by the oscillating component of the electric field of the incident electromagnetic radiation, $E(t)$. The equation of motion for the electron is represented as a function of its displacement x and can be expressed in the form:

$$m_0 \frac{d^2x}{dt^2} + m_0 \gamma \frac{dx}{dt} = -eE(t) = -eE_0 e^{-i\omega t} \tag{3.13}$$

Here, ω refers to the frequency of the light and E_0 represents the amplitude of the electric field of the electromagnetic wave. In Eq. (3.13), the first term concerns the acceleration of the electron, while the second term is the frictional damping force as characterized by γ. The right hand side of the equation represents the driving force for motion due to the electric field. The solution to the equation of motion can be obtained using the general term for the oscillatory motion of the electron as given by: $x = x_0 e^{-i\omega t}$. Substituting into Eq. (3.13) yields the expression of the time-dependent displacement, given by:

$$x(t) = \frac{eE(t)}{m_0(\omega^2 + i\gamma\omega)} \tag{3.14}$$

Since the electric displacement can be expressed as:

$$D = \varepsilon_r \varepsilon_0 E = \varepsilon_0 E + P \tag{3.15}$$

we can use the fact that $P = -Nex$ to derive the relative dielectric function as:

$$\varepsilon_r(\omega) = 1 - \frac{Ne^2}{\varepsilon_0 m_0(\omega^2 + i\gamma\omega)} = 1 - \frac{\omega_p^2}{(\omega^2 + i\gamma\omega)} \tag{3.16}$$

where the plasma frequency is given by:

$$\omega_p = \sqrt{\frac{Ne^2}{\varepsilon_0 m_0}} \tag{3.17}$$

For lightly damped systems, we can take $\gamma \simeq 0$, which yields:

$$\varepsilon_r(\omega) = 1 - \frac{\omega_p^2}{\omega^2} \tag{3.18}$$

The complex refractive index that we introduced earlier can be expressed in terms of the dielectric function as: $\tilde{n} = \sqrt{\varepsilon_r(\omega)}$. Therefore, for low frequencies, $\omega < \omega_p$, $\tilde{n}$ will be imaginary, while for high frequencies, $\omega > \omega_p$, $\tilde{n}$ will be positive. At the frequency $\omega = \omega_p$, $\tilde{n}$ will be precisely zero. The reflectivity of the metal will be expressed by Eq. (3.12), which means that for $\omega \leq \omega_p$, the reflectivity will be unity, while for $\omega > \omega_p$, R will rapidly decrease and tends to zero with increasing frequency. This indicates that we expect the reflectivity of the metal to be almost perfect for low frequencies up to the plasma frequency. Indeed, the experiment supports the principal conclusion, as illustrated in Figure 3.2, where the data is plotted as a function of the wavelength, which corresponds to high reflectivity at high wavelengths. The differences between the metals will arise from the electron density N. Added to this, we should also note that certain metals have characteristic colors. For example, copper has a pinkish hue, while gold appears to be yellowish. These deviations from the normal silvery appearance are due to inter-band electronic transitions, which absorb a certain portion of the visible spectrum, hence not all wavelengths are totally reflected, hence producing a slight coloration.

The optical properties of semiconductors are a rather different story. In Figure 3.3, we show the absorption spectra for a number of common semiconducting bulk materials. In the energy range illustrated, we observe that there is a sharp cut-off in the absorption at a specific energy, which corresponds to the absorption edge. Below this energy, the semiconductor is effectively transparent. The cut off energy occurs at the energy of the bandgap, where the photon energy is sufficient to excite an electron from the valence to the conduction band. It is worth stating in simple terms what this exactly means. When we say that the electron passes from the valence to the conduction band, it means that an atomic electron is delocalized from the atomic site. In physical terms, the bandgap corresponds to the binding energy. In this case, the binding energy will be for the most weakly bound electrons localized on the atomic site. The cut-off energies or absorption edges, shown in Figure 3.3, can now be understood as the point at which energy is absorbed from the incident radiation in the liberation of an electron from its localized state on an atom. Once the electron is removed from its bound state, it can move within the periodic potential of the crystal lattice and hence contribute to electrical conduction and is said to be in the conduction band. In the case, where a conduction electron is captured by an atom, which is missing an electron, then energy will be liberated from the excess energy, either as a photon or some other non-radiative transition. It should now be clear that this process corresponds to the transition of the electron from the conduction band to the valence band. We understand then, that the electron-free state of the atom corresponds to a hole state.

The binding energy of the electron is temperature dependent, as can be seen from the shift in the absorption edge, shown for Ge at room temperature and 77 K. Clearly, for low temperatures the binding energy is stronger, which translates into an increase in the bandgap at low temperatures. So while band

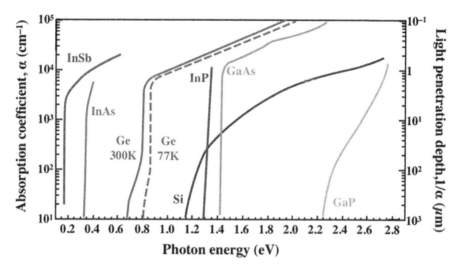

FIGURE 3.3 Absorption coefficient and penetration depth values as a function of photon energy for different semiconductors. Reprinted from L. Ferre Llin and D. J. Paul, *Thermoelectrics, Photovoltaics and Thermal Photovoltaics for Powering ICT Devices and Systems, ICT–Energy Concepts for Energy Efficiency and Sustainability*, G. Fagas (Ed.), InTech, DOI: 10.5772/65983 (2017), under the Creative Commons Attribution 4.0 International License.

theory is not entirely indispensable for the understanding of phenomena such as radiative transitions, it is a useful conceptual tool, which makes the treatment of the electronic and optical properties of semiconductors less complex.

The optical properties of insulating materials can be understood in much the same way, with similar physical mechanisms. Obviously, the difference being the size of the energy gap. Since the energy requirements in insulating materials are greater than in semiconductors, this translates as an increase of the forbidden energy gap and hence a correspondingly higher binding energy of the electron states in the valence band.

Glasses form an important class of optical materials and have been known for many centuries, where their use in windows, glassware and optical instruments is well-known. More recent applications include optical fibers. What distinguishes glasses as a material is their microstructure. While most of the materials we have discussed are usually considered from the point of view of their crystalline structures, glasses are essentially amorphous and have no long-range order of their constituent atoms. This has the important effect of making their properties isotropic. Most materials with crystalline structure show various forms of anisotropy, which could be with respect to the magnetic, electronic or optical properties. Clearly, glasses are well-known for their transparent optical properties in the visible region of the electromagnetic spectrum, and with most glasses being derivatives of silica SiO_2, their transparency ranges from around 200 nm in the UV to well into the infrared region at around 2000 nm.

3.2   Optical Transitions in Semiconductor Heterostructures

The optical absorption coefficient, α, is determined by the quantum mechanical transition rate, $W_{i \to f}$, for the excitation of an electron from an initial quantum state, ψ_i, to a final quantum state, ψ_f, via the absorption of a photon with angular frequency ω. This is usually expressed using the transition rate given by *Fermi's golden rule*, which we can write in the form:

$$W_{i \to f} = \frac{2\pi}{\hbar} |M|^2 g(\hbar\omega) \tag{3.19}$$

The transition rate will thus depend on; (i) the matrix element, M, and (ii) the density of states, $g(\hbar\omega)$. The matrix elements describe the effect of an external perturbation caused by photons, of energy $\hbar\omega$, on the electrons and can be expressed by the relation:

$$M = \langle f | \mathcal{H}' | i \rangle = \int \psi_f^*(\mathbf{r}) \mathcal{H}' \psi_i(\mathbf{r}) d^3\mathbf{r} \tag{3.20}$$

where $\mathcal{H}'$ is the perturbation associated with the light wave. Using a semiclassical approach (electrons from quantum theory and photons from classical electromagnetic theory), we can write the perturbation Hamiltonian as: $\mathcal{H}' = -\mathbf{p} \cdot \mathbf{E}$, where $\mathbf{p}$ is the dipole moment of the particle and $\mathbf{E}$ is the electric field of the perturbation (light). These can be expressed as:

$$\mathbf{p} = -e\mathbf{r} \tag{3.21}$$

and

$$\mathbf{E}_{ph}(\mathbf{r}) = \mathbf{E}_0 e^{\pm i\mathbf{q} \cdot \mathbf{r}} \tag{3.22}$$

from which we obtain:

$$\mathcal{H}' = e\mathbf{E}_0 \cdot \mathbf{r} e^{\pm i\mathbf{q} \cdot \mathbf{r}} \tag{3.23}$$

The electron states in a crystalline solid can be represented by Bloch functions, which allows us to express the initial and final wave functions respectively as:

$$\psi_i(\mathbf{r}) = \frac{1}{\sqrt{V}} u_i(\mathbf{r}) e^{i\mathbf{k}_i \cdot \mathbf{r}} \tag{3.24}$$

$$\psi_f(\mathbf{r}) = \frac{1}{\sqrt{V}} u_f(\mathbf{r}) e^{i\mathbf{k}_f \cdot \mathbf{r}} \tag{3.25}$$

where $u_{i,f}$ are the appropriate envelope functions for the initial and final states and V is the normalization volume. $\mathbf{k}_{i,f}$ express the initial and final state wave vectors. We can now substitute the wave functions and the Hamiltonian into the expression for the matrix elements in the form:

$$M = \frac{e}{V} \int u_f^*(\mathbf{r}) e^{-i\mathbf{k}_f \cdot \mathbf{r}} (\mathbf{E}_0 e^{\pm i\mathbf{q}\cdot\mathbf{r}}) u_i(\mathbf{r}) e^{i\mathbf{k}_i \cdot \mathbf{r}} d^3\mathbf{r} \tag{3.26}$$

where the integration is over the volume of the crystal. Conservation of momentum demands that the change in crystal momentum for the electron must be equal to that of the photon, which we can express as:

$$\hbar\mathbf{k}_f - \hbar\mathbf{k}_i = \pm\hbar\mathbf{q} \tag{3.27}$$

We therefore note that, from Eq. (3.26), the phase factor must be zero. If this were not the case, then there would be a phase difference for the different unit cells of the crystal and the integral would sum to zero. The Bloch theorem requires that $u_i(\mathbf{r})$ and $u_f(\mathbf{r})$ are periodic functions with the same periodicity as the crystal lattice. As such, the integral over the crystal can be represented for one unit cell (of volume Ω_{uc}), since they are equivalent and in phase. We now write:

$$M \propto \int_{\Omega_{uc}} u_f^*(\mathbf{r}) x u_i(\mathbf{r}) d^3\mathbf{r} \tag{3.28}$$

We have defined the light as being polarized along the x-axis.

The envelope functions can be derived from the atomic orbitals of the constituent elements of the crystal. We can further simplify by considering the wave vectors of the electrons and photons. For the photons, we write: $k = 2\pi/\lambda$, where λ is the wavelength of the light. Thus, for optical frequencies, we have $k \simeq 10^7$ m^{-1}. For electrons, the wave vector will be much larger since $k \simeq \pi/a$, (which is related to the size of the Brillouin zone) where a is the lattice parameter. Since $a \simeq 10^{-10}$ m, we find $k \simeq 10^{10}$ m^{-1}. Therefore, we can neglect the photon momentum, i.e., $\mathbf{k}_f = \mathbf{k}_i$. We can thus note that a direct optical transition leads to a negligible change of the electron wave vector. In the energy diagram, we can therefore express such a transition as a vertical arrow to represent the absorption process in a direct optical transition, as represented in the $E - k$ diagram, see Figure 3.4.

The factor $g(\hbar\omega)$ in the expression for the transition rate, Eq. (3.19), represents the joint density of states, evaluated at the photon energy. This accounts for the fact that both the initial and final electron states lie within the continuous energy bands. For electrons in a band, the density of states can be expressed by:

$$g(E)dE = 2g(k)dk \qquad (3.29)$$

The factor 2 arises from the two electron spin states for each value of k. Therefore; we may write:

$$g(E) = \frac{2g(k)}{dE/dk} \qquad (3.30)$$

Clearly, dE/dk represents the gradient of the dispersion curve $(E - k)$. The momentum density of states, $g(k)$, is calculated from the number of states (k) in the incremental volume between shells in k-space, of radii k and $k + dk$. Thus, we can write:

$$g(k)dk = \frac{4\pi k^2 dk}{(2\pi)^3} \qquad (3.31)$$

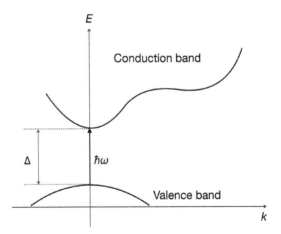

FIGURE 3.4   Direct optical transition due to the absorption of a photon, of energy $\hbar\omega$, resulting in the excitation of an electron from the valence band to the conduction band.

or

$$g(k) = \frac{k^2}{2\pi^2} \tag{3.32}$$

For an electron in a parabolic band with effective mass m^*, we obtain:

$$g(E) = \frac{1}{2\pi^2}\left(\frac{2m^*}{\hbar^2}\right)^{3/2} E^{1/2} \tag{3.33}$$

We recognize this as the standard expression for the density of states.

The joint density of states is finally obtained by evaluating $g(E)$ at E_i and E_f, where they are related to $\hbar\omega$ through the details of the band structure. This can be done as follows:

We consider the bands as determined from the effective masses of the charge carriers; $m_e^*, m_{hh}^*, m_{lh}^*$. The energy dispersion relations $E(k)$ for the parabolic bands are expressed as:

$$E_C(k) = \Delta + \frac{\hbar^2 k^2}{2m_e^*}; \quad \text{for electrons in the conduction band} \tag{3.34}$$

$$E_{hh}(k) = -\frac{\hbar^2 k^2}{2m_{hh}^*}; \quad \text{for holes in the (heavy hole) valence band} \tag{3.35}$$

$$E_{lh}(k) = -\frac{\hbar^2 k^2}{2m_{lh}^*}; \quad \text{for holes in the (light hole) valence band} \tag{3.36}$$

Transitions for light (lh) and heavy holes (hh) are illustrated in Figure 3.5.

From the conservation of energy during a light hole or heavy hole transition, we have:

$$\hbar\omega = \Delta + \frac{\hbar^2 k^2}{2m_e^*} + \frac{\hbar^2 k^2}{2m_h^*} \tag{3.37}$$

where $m_h^* = m_{hh}^*$ or m_{lh}^*. We can define the reduced electron–hole mass as:

$$\frac{1}{\mu} = \frac{1}{m_e^*} + \frac{1}{m_h^*} \tag{3.38}$$

which means, we can express Eq. (3.37) in the simplified form:

$$\hbar\omega = \Delta + \frac{\hbar^2 k^2}{2\mu} \tag{3.39}$$

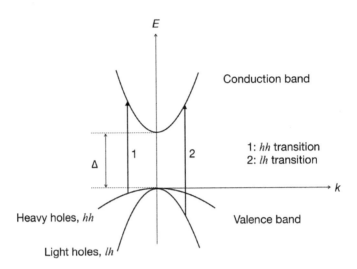

FIGURE 3.5   Optical transitions from light hole (lh) and heavy hole (hh) valence band states to the conduction band.

As we stated earlier, we are interested in evaluating $g(E)$ at the energy $E = \hbar\omega$. Thus, we can write: For $\hbar\omega < \Delta$, we have:

$$g(\hbar\omega) = 0 \qquad (3.40)$$

while for $\hbar\omega \geq \Delta$, we have:

$$g(\hbar\omega) = \frac{1}{2\pi^2}\left(\frac{2\mu}{\hbar^2}\right)^{3/2}(\hbar\omega - \Delta)^{1/2} \qquad (3.41)$$

Therefore, we see that the joint density of states factor increases as $(\hbar\omega - \Delta)^{1/2}$ for photon energies greater than the band gap, Δ. We thus expect, that the absorption coefficient will have the following properties: For $\hbar\omega < \Delta$:

$$\alpha(\hbar\omega) = 0 \qquad (3.42)$$

while for $\hbar\omega \geq \Delta$:

$$\alpha(\hbar\omega) \propto (\hbar\omega - \Delta)^{1/2} \qquad (3.43)$$

i.e., no absorption for $\hbar\omega < \Delta$ and the absorption increases as $(\hbar\omega - \Delta)^{1/2}$ for photon energies greater than the band gap of the semiconductor. In addition to this, transitions with reduced mass μ will give rise to stronger absorption due to the $\mu^{3/2}$ factor.

Let us now consider the case of a quantum well structure with a well thickness d in the z-direction, see Figure 3.6(a). We consider the case, where

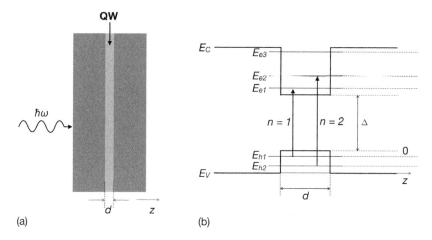

FIGURE 3.6 (a) Schematic illustration of a quantum well structure with an incident photon. (b) Band diagram for the quantum well, of thickness d, showing the absorption transitions for $n = 1$ and $n = 2$.

the polarization vector is in the $x - y$ plane. In this case, the matrix element takes the form: $M = \langle f|x|i \rangle$. This differs from the previous case, since for the quantum well we have: $\langle f|x|i \rangle = \langle f|y|i \rangle \neq \langle f|z|i \rangle$.

For transitions from the valence band to the conduction band in the quantum well, as shown in Figure 3.6(b), we illustrate the specific transitions for $n = 1$ and $n = 2$. In the general case of a transition from the nth hole state to the n'^{th} electron state, we can write the initial and final quantum well wave functions as:

$$\psi_i(\mathbf{r}) \equiv |i\rangle = \frac{1}{\sqrt{V}} u_V(\mathbf{r}) \phi_{hn}(z) e^{i\mathbf{k}_{x,y} \cdot \mathbf{r}_{x,y}} \tag{3.44}$$

and

$$\psi_f(\mathbf{r}) \equiv |f\rangle = \frac{1}{\sqrt{V}} u_C(\mathbf{r}) \phi_{en'}(z) e^{i\mathbf{k}'_{x,y} \cdot \mathbf{r}_{x,y}} \tag{3.45}$$

where $u_{V,C}(\mathbf{r})$ denote the envelope functions from the the Bloch theorem for the valence and conduction bands, $\phi_{hn,en'}(z)$ are the band states in the QW in the z-direction and the final $e^{i\mathbf{k}_{x,y} \cdot \mathbf{r}_{x,y}}$ define the plane waves for free motion in the $x - y$-plane. Since the photon momentum is negligibly small in comparison to that of the electrons, we can write: $\mathbf{k}_{x,y} = \mathbf{k}'_{x,y}$. We can now rewrite the matrix elements as two factors:

$$M = M_{CV} M_{nn'} \tag{3.46}$$

where

$$M_{CV} = \langle u_C | x | u_V \rangle = \int u_C^*(\mathbf{r}) x u_V(\mathbf{r}) d^3 \mathbf{r}$$

and

$$M_{nn'} = \langle en' | hn \rangle = \int_{-\infty}^{\infty} \phi_{en'}^*(z) \phi_{hn}(z) dz$$

From $M_{nn'}$, we see that the matrix elements for optical transitions in the QW will be proportional to the overlap of electron and hole states. This will allow us to evaluate some straightforward selection rules; $\Delta n = n' - n$.

For the simplest case of an infinite quantum well, see Section 2.5.1, we can express the $\phi_n(z)$ functions as simple sine functions, giving:

$$M_{nn'} = \frac{z}{d} \int_{-d/2}^{d/2} \sin\left(k_n z + \frac{n\pi}{2}\right) \sin\left(k_{n'} z + \frac{n\pi}{2}\right) dz \qquad (3.47)$$

This gives unity for $n = n'$ and is zero otherwise. Hence we obtain the following selection rule for an infinite quantum well:

$$\Delta n = 0 \qquad (3.48)$$

For finite quantum wells, the electron and hole wave functions with different quantum numbers are not necessarily orthogonal to each other because of differing decay constants in the barrier regions. This means that there are small departures from the above selection rules. However, these $\Delta n \neq 0$ transitions are usually weak and are strictly forbidden if Δn is an odd number, because the overlap of states with opposite parties is zero.

3.2.1 Absorption in Quantum Well Structures

The absorption spectra for quantum wells can be understood from the application of the selection rules, as stated above. We saw that for low photon energies, no absorption can be expected until a specific threshold is reached for the excitation of the electron from the valence band ($n = 1$, h level) to the conduction band ($n' = 1$, e$^-$ level). This corresponds to a $\Delta n = 0$ transition and is hence allowed. The energy threshold can be expressed as:

$$\hbar \omega_0 = \Delta + e_{h1} + e_{e1} \qquad (3.49)$$

This demonstrates that the optical absorption edge will be shifted with respect to the bulk case (where $\hbar \omega_0 = \Delta$) by an amount, which depends on the ground states for electrons and holes in the QW, which is directly related to the thickness of the QW, d. This means that, by a suitable choice of well thickness, we can tailor the absorption threshold energy. Figure 3.7 illustrates the threshold for absorption, expressed in Eq. (3.49), using the energy diagrams for quantum well states.

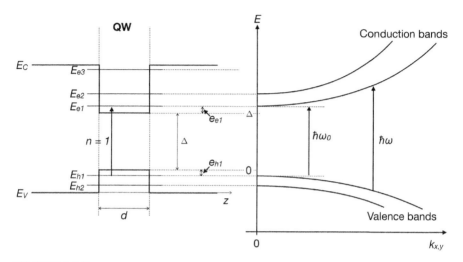

FIGURE 3.7 Schematic illustration of a quantum well structure and the energy band diagram. The absorption edge for the excitation of an electron from the valence to the conduction band is shown as $\hbar\omega_0$.

Since the bands are parabolic and from the conservation energy, a general transition between bands ($n = 1$ states) can be expressed by:

$$\hbar\omega = \Delta + \left(e_{e1} + \frac{\hbar^2 k_{x,y}^2}{2m_e^*} \right) + \left(e_{h1} + \frac{\hbar^2 k_{x,y}^2}{2m_{hh}^*} \right) \tag{3.50}$$

This can be alternatively expressed as:

$$\hbar\omega = \Delta + e_{e1} + e_{h1} + \frac{\hbar^2 k_{x,y}^2}{2\mu} = \hbar\omega_0 + \frac{\hbar^2 k_{x,y}^2}{2\mu} \tag{3.51}$$

Here, we have considered the case, where the valence state corresponds to a heavy hole band with corresponding reduced mass μ. This is also illustrated in Figure 3.7. We can compare this with the bulk case, as expressed in Eq. (3.37). An important difference is that the wave vector for the quantum well spans only the x and y coordinates. This will have a direct consequence for the joint density of states, which is now independent of energy:

$$g(E) = \frac{\mu}{\pi\hbar^2} \tag{3.52}$$

This expression was derived in the previous chapter, see Section 2.6. This considers only the $n = 1$ transition. For increasing energies, further transitions will come into play, and we require the Heaviside function to express the full density of states. The absorption curve will be zero until the threshold energy is reached and then will consist of a series of steps as the photon energy increases. This is schematically illustrated in Figure 3.8. Once the energy for the $n = 2$ transition is reached, a step will be observed in the absorption spectrum, where the step energy occurs at: $E_{n=2} = \Delta + e_{e2} + e_{h2}$.

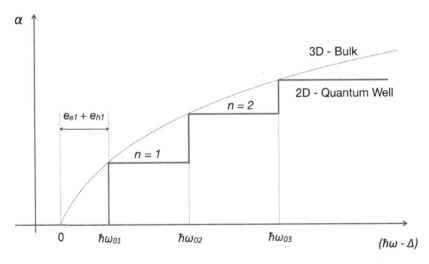

FIGURE 3.8　Absorption coefficient as a function of photon energy for a quantum well structure.

We can now generalize for the threshold energy of the nth level as:

$$\hbar\omega_{0n} = \Delta + \frac{\hbar^2\pi^2 n^2}{2m_e^* d^2} + \frac{\hbar^2\pi^2 n^2}{2m_{hh}^* d^2} = \Delta + \frac{\hbar^2\pi^2 n^2}{2\mu d^2} \qquad (3.53)$$

where we have preserved the selection rule of $\Delta n = 0$. In Figure 3.9, we show experimental data measured on InAs membranes of thickness $\sim$ 3–19 nm on CaF$_2$ support substrates. There are several points worthy of note. Firstly, the absorption edge exhibits a strong shift in energy, particularly for the thinner membranes. Secondly, we can see that the step height in the absorptance, A_Q, is independent of the thickness. In some of the steps, we observe a pre-edge, which shows a double peak feature. This is most clearly seen in the 6 and 9 nm samples. This is due to light and heavy hole transitions being close to one another. This is indicated in the inset of Figure 3.9.

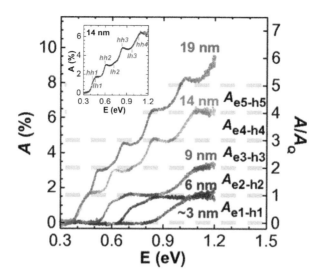

FIGURE 3.9 Absorptance spectra of InAs QMs with thickness of ~ 3 nm, 6 nm, 9 nm, 14 nm, and 19 nm at room temperature, showing universal absorptance steps of $A_Q \sim 1.6\%$ in magnitude. Dashed lines indicate each 1.6% of absorptance. The A_{en-hn} represent the absorptance from the corresponding inter-band transitions, $e_n - hh_n$ and $e_n - lh_n$. Inset shows the data for the 14 nm InAs membrane with light hole (lh) and heavy hole (hh) transitions indicated (Reprinted from Fang et al., 2013. Open access.).

We note here that the quantum membrane (QM) acts very much like a quantum well, where the InAs QMs can be effectively treated as infinitely deep potential wells, because they are confined by air on one side and by a wide band-gap CaF$_2$ substrate on the other side.

Frequently in such experiments, quantum wells exhibit sharp peaks at the step edges, which correspond to the production of the electron–hole coupled states, or *excitons*. These electron–hole pairs are bound by a Coulomb interaction between the charge carriers and enhance the absorption rate at the step edge energy, which of course corresponds to the excitation of further electron hole pairs, and results in the resonant absorption effect. In the case of confined systems, the electron-hole pairs are forced to be closer together than would normally be the case in a bulk solid, which further increases the attractive potential between the charge carriers. This can lead to excitonic states, which are stable at room temperature.

3.2.2   *Optical Emission in Quantum Wells*

The application of quantum well structures in electroluminescent devices is one of the principal commercial applications of these systems. The insertion of QWs into the active regions of a device leads to a significant improvement of the quantum efficiency, since the device will be tuned to the wavelength of interest using the intrinsic properties of the materials used as well as the controlled thickness to produce the desired output energy (wavelength).

In general, light is generated, when electrons in the conduction band recombine with hole states in the valence band. The resulting emission of light, or *luminescence*, typically consists of a peak at the bandgap energy, the width of which depends on the carrier density and temperature. The physical process, which results in photon emission in nanostructured devices, is essentially the same as that for bulk semiconductors. Charge carriers (electrons and holes) are injected via a bias potential or by optical means. These will then rapidly relax to the extremities of the bands (i.e., the bottom of the conduction band for electrons and the top of the valence band for holes) and then direct radiative recombination of the charge carriers will result in the emission of photons, with an energy corresponding to the energy difference between the electron and hole states. For the QW structure, the lowest energy states that are available will correspond to the electrons and holes in the $n = 1$ confined states of the conduction and valence bands, respectively. The energy of the emitted photons will correspond to:

$$h\nu = \Delta + e_{e1} + e_{h1} \qquad (3.54)$$

which will again show that the energy shift associated with the photons with respect to the bulk semiconductor will have an energy corresponding to the gap energy, Δ. The transition is energetically controllable with the width of the active region being set in the fabrication process. By increasing the overlap of the electron and hole wave functions, the emission probability can also be increased for the quantum well. In effect, this will shorten the radiative lifetime, τ_r, making it more efficient and improving the light intensity for say LED devices. In a typical QW structure, the active region will be of the order of 10 nm in thickness. This is well below the critical thickness for misfit dislocations in non-lattice matched epitaxial structures. It is common that such structures are incorporated into the active region at the junction between p and n layers in a semiconducting diode. Such devices are operated in the forward bias mode with the light intensity being controlled by the current in the device and the recombination efficiency.

3.2.3 *Inter-Subband Transitions*

Inter-subband transitions are those that are excited for holes or electrons with the valence or conduction bands, respectively. For example, for the electrons in the QW region this may correspond to a transition between the $n = 1$ and the $n = 2$ band states, as illustrated in Figure 3.10. This contrasts with the inter-band transitions we have been considering in the previous sections, with electrons being transferred between the conduction and valence bands.

Since the energies between subband levels are much closer than those between interband states, the photon energies will be correspondingly lower. For example, in a GaAs/AlGaAs QW structure with a 10 nm well thickness, the energy between the $n = 1$ and the $n = 2$ states will be of the order of 100 meV. This will correspond to radiation in the infrared range (with a wavelength of around 12 m). Such quantum well structures can be exploited in detectors and emitters of infrared radiation.

The excitation of inter-subband transitions occurs via the absorption of linearly (z) polarized light. This will lead to matrix elements for the transition between n and n' states of the form: $\langle n|z|n' \rangle$. The selection rule for such transitions will be $\Delta n = n - n'$, which must be odd due to the parity of the wave functions. The requirement of z-polarized light for exciting the inter-subband transitions means that normal incident light is not sufficient and frequently a metallic grating is incorporated into the device to produce the necessary polarized light.

3.2.4 *Quantum Dots*

Quantum confinement, as we saw in the previous chapter, can be extended to the 1, 2 and 3 dimensions of space. In the latter case of 3D confinement,

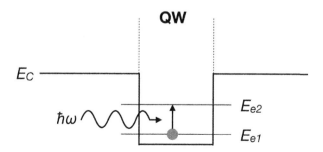

FIGURE 3.10 Inter-subband transition between e1 ($n = 1$) and e2 ($n = 2$) states in the conduction band of a n-doped semiconductor.

the quantum dot is regarded as a 0D structure, since the charge carriers are confined in all directions and there will be no spatial degrees of freedom. This situation was discussed in Section 2.5.3 of the previous chapter. For a rectangular dot with dimensions l_x, l_y and l_z, the energy levels of the system, assuming the well potentials are infinite, can be expressed in the form:

$$E(n_x, n_y, n_z) = \frac{\hbar^2 \pi^2}{2m^*} \left[\left(\frac{n_x}{l_x} \right)^2 + \left(\frac{n_y}{l_y} \right)^2 + \left(\frac{n_z}{l_z} \right)^2 \right] \qquad (3.55)$$

The three quantum numbers are analogous to the first three quantum numbers in an atom (the fourth being due to the spin of the electron) and such structures are referred to as *artificial atoms*. In this case, the spectrum of states (via the quantum numbers) can be tuned by setting the different dimensions in the three spatial directions.

As we noted above, the quantum confinement of the carriers in quantum well structures improves the quantum efficiency for a range of optical effects thus enhancing optoelectronic properties. Indeed, for further confinement in other directions, we can expect increased quantum efficiency due to the increased overlap in the electron and hole wave functions. Furthermore, the discrete nature of the density of states for the quantum dot also reduces the thermal spread of the carriers within their energy bands. Both of these aspects can be expected to reduce the threshold current of laser diodes and improve their quantum efficiency.

3.3   Photonic Band Gap Materials

Nanophotonics is one of the many sub-disciplines of both modern optics and nanotechnologies. An important area of this field is the so-called photonic bandgap (PBG) materials or photonic crystals, which are used to manipulate the optical properties of composites systems and allow the possibility of developing new applications based on the principle of periodically modulating the relative permittivity of a medium. PBG materials are known for the precise control of the electromagnetic properties of materials, which includes: the electromagnetic density of states, phase, and group velocities, field confinement and field polarization. The characteristic length scales for such periodic structures is of the order of magnitude of the wavelength of interest, which will generally mean that the periodicity is not as small as the necessary length scales encountered in quantum confinement, typically lying in the range of 100 nm to 1 m.

Much progress has been made over the last two decades in the area of PBG materials, in terms of both experimental and theoretical studies. As we will see shortly, there is much that is familiar to the student of solid-state physics when we consider the propagation of light in the periodic structures of PBG composites. The analogy holds well for the comparison of the propagation of light in the periodic structure to that of electrons in the periodic potential of a crystal. In solid-state physics, we note that the complex electronic band structures arise from the periodic potential of the atoms in the crystal and are intimately related to the symmetry of the crystal lattice. This leads to band structures and the forbidden energy gap experienced by electrons in crystals. For photonic bandgap materials, we are essentially faced with the same physical problem, where the electron is replaced by the photon and the periodic potential by the periodic dielectric constant of the medium. Of course, while crystalline structures tend to organize under the correct condensing conditions, PBG structures are generally fabricated artificially. In order for the particle (electron/photon) to interact with the periodic structure, its associated wavelength should be comparable to the periodicity of the medium in which it will propagate.

To evaluate the optical modes in a photonic crystal, it is necessary to apply the Maxwell equations in the periodic dielectric medium. This is a rather complex problem and is usually performed using computational methods as analytical solutions are not possible in three dimensions. There are a plethora of possible PBG structures, which can be 1D (multilayers), 2D (bundles of fibers) and of course 3D (PMMA beads for example). In each case, the periodic structure must be inserted into the Maxwell equations and solved using computational methods.

The Maxwell equations for zero electric current and vanishing charge density (since the PBG is insulating) can be expressed in the form:

$$\nabla \cdot \mathbf{D} = 0$$
$$\nabla \cdot \mathbf{B} = 0$$
$$\nabla \times \mathbf{H} = \frac{\partial \mathbf{D}}{\partial t} \qquad (3.56)$$
$$\nabla \times \mathbf{E} = -\frac{\partial \mathbf{B}}{\partial t}$$

We now write the dielectric function for the composite medium as $\varepsilon = \varepsilon(\mathbf{r})$ and for a non-magnetic material, we can write:

$$\mathbf{D}(\mathbf{r}) = \varepsilon(\mathbf{r})\varepsilon_0\mathbf{E}(\mathbf{r}) \tag{3.57}$$

and

$$\mathbf{B} = \mu_0\mathbf{H} \tag{3.58}$$

Inserting this into the Maxwell equations, Eqs. (3.56), yields:

$$\varepsilon_0\nabla \cdot [\varepsilon(\mathbf{r})\mathbf{E}(\mathbf{r},t)] = 0$$
$$\mu_0\nabla \cdot \mathbf{H}(\mathbf{r},t) = 0$$
$$\nabla \times \mathbf{H}(\mathbf{r},t) = \varepsilon(\mathbf{r})\varepsilon_0\frac{\partial\mathbf{E}(\mathbf{r},t)}{\partial t} \tag{3.59}$$
$$\nabla \times \mathbf{E} = -\mu_0\frac{\partial\mathbf{H}(\mathbf{r},t)}{\partial t}$$

We now apply the temporal dependence of the electric and magnetic fields of the electromagnetic radiation: $\mathbf{H}(\mathbf{r},t) = \mathbf{H}(\mathbf{r})e^{i\omega t}$ and $\mathbf{E}(\mathbf{r},t) = \mathbf{E}(\mathbf{r})e^{i\omega t}$. This leads to the full set of coupled equations:

$$\nabla \cdot \varepsilon(\mathbf{r})\mathbf{E}(\mathbf{r}) = 0$$
$$\nabla \cdot \mathbf{H}(\mathbf{r}) = 0$$
$$\nabla \times \mathbf{H}(\mathbf{r},t) = i\omega\varepsilon_0\varepsilon(\mathbf{r})\mathbf{E}(\mathbf{r}) \tag{3.60}$$
$$\nabla \times \mathbf{E}(\mathbf{r},t) = -i\omega\mu_0\mathbf{H}(\mathbf{r})$$

If we take the following operation: $\nabla \times [\nabla \times \mathbf{H}(\mathbf{r})/\varepsilon(\mathbf{r})] = i\omega\varepsilon_0\nabla \times \mathbf{E}(\mathbf{r})$, we can write:

$$\nabla \times \left[\frac{\nabla \times \mathbf{H}(\mathbf{r})}{\varepsilon(\mathbf{r})}\right] = \left(\frac{\omega}{c}\right)^2\mathbf{H}(\mathbf{r}) \tag{3.61}$$

where we have used the last two equations of (3.60) and the relation: $c = 1/\sqrt{\varepsilon_0\mu_0}$. The equation above has the form of an eigenvalue problem, where the term on the LHS shows the operator, and on the RHS, we have the eigenvalue, expressed as ω^2/c^2, and the eigenstate, expressed as $\mathbf{H}(\mathbf{r})$. This master equation determines the possible modes of $\mathbf{H}(\mathbf{r})$. (We note that we can express the electric field in a similar fashion to express the $\mathbf{E}(\mathbf{r})$ modes.) The solutions of this equation can be obtained by a number of methods, for which numerical calculations must be made. In the above, we have used $\mu(\mathbf{r}) = 1$, which is valid for a non-magnetic material. We can now express the real part of the refractive index as:

$$n^2(\mathbf{r}) = \varepsilon(\mathbf{r}) \tag{3.62}$$

Considering Eq. (3.61), we note that this has the form of the generalized wave equation in three dimensions. To demonstrate the similarity of the periodic nature of this problem and its relation to the electron in a crystal, we will consider the 1D photonics structure, which consists of two alternating dielectric layers, with dielectric constants ε_1 and ε_2, with respective thicknesses d_1 and d_2. The crystalline structure will then have an effective lattice parameter of $D = d_1 + d_2$, which we will consider to be in the z - direction, as illustrated in Figure 3.11. The periodicity of the medium imposes the following condition: $\varepsilon(z) = \varepsilon(z+D)$ and $\mu(z) = \mu(z+D)$.

In a similar manner to the confinement effects in quantum well like structures, we can reduce the problem to one dimension, where two types of modes can be distinguished: TE (transverse electric) modes, in which the electric field component is always parallel to the boundaries between the layers and TM (transverse magnetic) modes, in which the magnetic field vector is always parallel to the boundaries. The complex fields obtained from the solution of Maxwell's equations yield:

$$\mathbf{E}(\mathbf{r}) = E(z)e^{i(k_x x + k_y y)}\hat{n}_x; \qquad \text{TE modes} \tag{3.63}$$

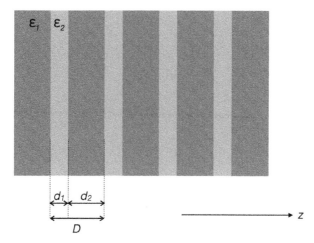

FIGURE 3.11   Periodic dielectric medium consisting of layers of dielectrics with dielectric constants ε_1 and ε_2.

and

$$\mathbf{H}(\mathbf{r}) = H(z)e^{i(k_x x + k_y y)}\hat{n}_y; \qquad \text{TM modes} \qquad (3.64)$$

In each layer of the multilayer stack, $E(z)$ and $H(z)$ will be a superposition of forward $(a_{n,j})$ and backward $(b_{n,j})$ propagating waves. We will simplify the procedure by setting $d_1 = d_2 = d$. These can be represented in the following manner.

$$E_{n,j}(z) = a_{n,j}e^{ik_{zj}(z-nd)} + b_{n,j}e^{-ik_{zj}(z-nd)}; \qquad \text{TE modes} \qquad (3.65)$$

and

$$H_{n,j}(z) = a_{n,j}e^{ik_{zj}(z-nd)} + b_{n,j}e^{-ik_{zj}(z-nd)}; \qquad \text{TM modes} \qquad (3.66)$$

The constants $a_{n,j}$ and $b_{n,j}$ depend on the layer number, n, and the medium, ε_j. The wave vector is expressed as:

$$k_{zj} = \sqrt{\frac{\omega^2}{c^2}\varepsilon_j - k_{||}^2} \qquad (3.67)$$

where $k_{||} = \sqrt{k_x^2 + k_y^2}$. To deduce the $a_{n,j}$ and $b_{n,j}$ constants, we have to apply the boundary conditions the interfaces and using $z = z_n = nd$. These can be expressed in a generic form, where we consider the interface between the nth and $n+1^{th}$ layers, as:

$$E_{n,1}(z_n) = E_{n+1,2}(z_n); \qquad \text{TE modes} \qquad (3.68)$$

$$\frac{d}{dz}[E_{n,1}(z_n)] = \frac{d}{dz}[E_{n+1,2}(z_n)]; \quad \text{TE modes} \qquad (3.69)$$

and

$$H_{n,1}(z_n) = H_{n+1,2}(z_n); \qquad \text{TM modes} \qquad (3.70)$$

$$\frac{d}{dz}[H_{n,1}(z_n)] = \frac{d}{dz}[H_{n+1,2}(z_n)]; \quad \text{TM modes} \qquad (3.71)$$

Applying these boundary conditions along with the expressions for $E_{n,j}(z)$ and $H_{n,j}(z)$ yields the following relations:

$$a_{n,1} + b_{n,1} = a_{n+1,2}e^{-ik_{z2}d} + b_{n+1,2}e^{ik_{z2}d} \qquad (3.72)$$

$$a_{n,1} - b_{n,1} = p_m[a_{n+1,2}e^{-ik_{z2}d} - b_{n+1,2}e^{ik_{z2}d}] \qquad (3.73)$$

where

$$p_m = \frac{k_{z2}}{k_{z1}}; \quad \text{TE modes} \tag{3.74}$$

$$p_m = \frac{k_{z2}\varepsilon_1}{k_{z1}\varepsilon_2}; \quad \text{TM modes} \tag{3.75}$$

Since we still have four unknowns and only two equations we need to establish another two expressions, which we can do by considering the interface between the $(n\text{-}1)^{th}$ and n^{th} layers, where $z = z_{n-1} = (n-1)d$. From this, we obtain:

$$a_{n-1,2} + b_{n-1,2} = a_{n,1}e^{-ik_{z1}d} + b_{n,1}e^{ik_{z1}d} \tag{3.76}$$

$$a_{n-1,2} - b_{n-1,2} = \frac{1}{p_m}[a_{n,1}e^{-ik_{z1}d} - b_{n,1}e^{ik_{z1}d}] \tag{3.77}$$

In establishing these relations, we introduce a further two unknowns and we need to apply the Flouquet–Bloch theorem to rectify this problem. This is analogous to the Bloch theorem we are accustomed to using when dealing with electrons in a periodic potential, and can be stated as follows: If E is a field in a periodic medium, with periodicity $D = 2d$ (where we are using the condition $d_1 = d_2 = d$), then the field must satisfy the following condition:

$$E(z+2d) = e^{2ikd}E(z) \tag{3.78}$$

where k is an (as yet) unidentified wave vector or Bloch wave vector. An analogous expression will exist for the magnetic field component $H(z)$. We now proceed by applying these conditions to the above Eqs. (3.72), (3.73) and (3.76), (3.77). It is worth noting that the problem we are aiming to solve is analogous to the Kronig–Penney model for electrons in a 1D periodic square wave potential, where the Bloch theorem must also be applied in order to obtain the dispersion relation and band diagram. In the current situation, we use the Flouquet–Bloch theorem in conjunction with the boundary condition:

$$E_{n+1,2}(z+2d) = e^{2ikd}E_{n-1,2}(z) \tag{3.79}$$

This allows us to write:

$$a_{n+1,2}e^{ik_{z2}[z+2d-(n+1)d]} + b_{n+1,2}e^{-ik_{z2}[z+2d-(n+1)d]}$$
$$= e^{2ikd}[a_{n-1,2}e^{ik_{z2}[z-(n-1)d]} + b_{n-1,2}e^{-ik_{z2}[z-(n-1)d]}] \tag{3.80}$$

$$a_{n+1,2}e^{ik_{z2}[z-(n-1)d]} + b_{n+1,2}e^{-ik_{z2}[z-(n-1)d]}$$
$$= e^{2ikd}[a_{n-1,2}e^{ik_{z2}[z-(n-1)d]} + b_{n-1,2}e^{-ik_{z2}[z-(n-1)d]}] \tag{3.81}$$

which we can now write as:

$$a_{n+1,2} + b_{n+1,2}e^{-i2k_{z2}[z-(n-1)d]} = e^{2ikd}[a_{n-1,2} + b_{n-1,2}e^{-2ik_{z2}[z-(n-1)d]}] \quad (3.82)$$

Since this equation must hold for all z positive, we can write:

$$a_{n+1,2} = a_{n-1,2}e^{2ikd} \quad (3.83)$$

and

$$b_{n+1,2} = b_{n-1,2}e^{2ikd} \quad (3.84)$$

This allows us to reduce the number of unknown variables from six to four, thus permitting us to solve our system of homogeneous equations. These can now be expressed explicitly as:

$$a_{n,1} + b_{n,1} - e^{-ik_{z2}d}e^{2ikd}a_{n-1,2} - e^{ik_{z2}d}e^{2ikd}b_{n-1,2} = 0 \quad (3.85)$$

$$a_{n,1} - b_{n,1} - e^{-ik_{z2}d}e^{2ikd}p_m a_{n-1,2} + e^{ik_{z2}d}e^{2ikd}p_m b_{n-1,2} = 0 \quad (3.86)$$

$$e^{-ik_{z1}d}a_{n,1} + e^{ik_{z1}d}b_{n,1} - a_{n-1,2} - b_{n-1,2} = 0 \quad (3.87)$$

$$\frac{1}{p_m}e^{-ik_{z1}d}a_{n,1} - \frac{1}{p_m}e^{ik_{z1}d}b_{n,1} - a_{n-1,2} + b_{n-1,2} = 0 \quad (3.88)$$

These equations can be expressed in matrix form as:

$$\begin{pmatrix} 1 & 1 & -e^{-ik_{z2}d}e^{2ikd} & -e^{ik_{z2}d}e^{2ikd} \\ 1 & -1 & -p_m e^{-ik_{z2}d}e^{2ikd} & p_m e^{ik_{z2}d}e^{2ikd} \\ e^{-ik_{z1}d} & e^{ik_{z1}d} & -1 & -1 \\ \frac{1}{p_m}e^{-ik_{z1}d} & \frac{-1}{p_m}e^{ik_{z1}d} & -1 & 1 \end{pmatrix} \begin{pmatrix} a_{n,1} \\ b_{n,1} \\ a_{n-1,2} \\ b_{n-1,2} \end{pmatrix} = 0 \quad (3.89)$$

The solution to this characteristic equation is a simple matter that can be shown to give:

$$\cos(2kd) = \cos(k_{z2}d)\cos(k_{z1}d) - \frac{1}{2}\left(p_m + \frac{1}{p_m}\right)\sin(k_{z2}d)\sin(k_{z1}d) \quad (3.90)$$

As is to be expected, this solution has many characteristics in common with the Kronig–Penney model. Since $\cos(2kd)$ must always lie in the rage [-1, 1], solutions cannot exist, when the absolute of the right hand side is outside this value range. This absence of solutions corresponds to the prohibited values associated with the band gap region. For example, a wave at normal incidence (with $k_{z1} = \sqrt{\varepsilon_1}\omega/c; k_{z2} = \sqrt{\varepsilon_2}\omega/c$) to the photonic crystal with $\varepsilon_1 = 2.25$ and $\varepsilon_2 = 9$ can propagate for $\lambda = 12d$, but not for $\lambda = 9d$.

For each Bloch wave vector k, a particular dispersion relation $\omega(k_\parallel)$ will exist. If we plot all possible dispersion relations on the same graph, we will obtain the complete band diagram, an example is illustrated in Figure 3.12, for a 1D multilayer structure. The dark shaded regions are the allowed bands, for which propagation of the electromagnetic waves through the crystal is possible.

It is important to note that propagating modes can exist even if one of the longitudinal mode numbers, (k_{zj}), is imaginary. The Bloch wave vector at the band edge is determined from $kd = n\pi/2$. For a given direction propagation, characterized by $k_{||}$, we find frequency regions, for which propagation through the crystal is possible, and other regions, in which it is not possible. However, for a 1D crystal, there is no complete bandgap. By this, we mean that there are no frequencies, for which propagation is completely inhibited in all directions. For a wave propagating in vacuum directed into the photonic crystal, only modes with $k_{||} < k = \omega/c$ can be excited, as illustrated by the solid lines in Figure 3.12. In this case, we now observe compete frequency

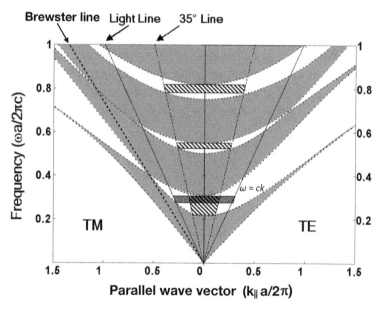

FIGURE 3.12 Photonic band diagram for a one-dimensional photonic crystal having a periodic refractive index alternating between 2.8 and 1.55. Gray regions represent propagating modes within the structure, and white regions represent evanescent modes. Hatched regions represent photonic band gaps, where high reflectivity can be expected for external EM waves over an angular range extending from normal to 35° incidence. The shaded trapezoid represents a region of external omnidirectional reflection. TM and TE represent transverse magnetic and transverse electric polarized modes, respectively.

band gaps $(k_\| < k)$. For these cases, the photonic crystal acts as a perfect mirror, with all incident light being reflected.

Complete band gaps are possible in 3D photonic crystals. It will also be more favorable to have complete band gaps, when the dielectric materials have strongly different dielectric constants. As in the case with semiconductors, we can distinguish between regions above and below the bandgap. In the photonic crystal, the region below the bandgap is called the dielectric band, while the region above it is referred to as the air gap. In the dielectric band, the optical energy is confined in the material with the highest dielectric constant, while in the air gap, the energy will be found in the material with the lowest dielectric constant. Therefore, excitation from one band to another promotes the optical energy from the high dielectric to the low dielectric constant material. In Figure 3.13, we illustrate an example of a 3D fcc-like photonic crystal structure. The range of structures is vast, with filled ball-like and woodpile-like structures being the most common. It should also be noted that the bandgap regions depend on whether the light is polarized for the electric or the magnetic field, as indicated in Figure 3.12.

Another important aspect and application of photonic crystals is the purposely introduced defect in the periodic array. For example, a single positional defect can be used to localize the electromagnetic radiation as in a cavity mode or resonance. A line of defects can be used to guide light. If we

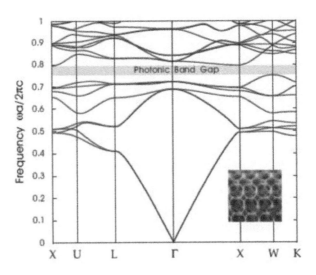

FIGURE 3.13   The photonic crystal band diagram of a closed packed fcc arrangement of spheres. Here the complete photonic band gap region is illustrated.

consider the case of a photon with an energy corresponding to the bandgap region, the light cannot propagate through the PBG material and will be confined to the defect regions. As such, the line of defects will effectively open up a waveguide, so light with a specific frequency within the bandgap can only propagate along a channel of defects, since it will be perfectly reflected or repelled from the bulk of the crystal. Waveguides in photonic structures can transport light around corners almost without loss. Photonic crystals, therefore, have great practical value in miniaturized optoelectronic circuits. Another application of photonic structures is in optical fibers, where the photonic region (2D array) is used as an ideal cladding, since internal reflections will be ensured by the bandgap. The point defect, as we mentioned above, can act as a photonic crystal cavity, and extremely high Q values can be obtained. The experiment has shown Q values in the region of 10^{10}, and theoretically, it is possible to reach values that can be better than 10^{20}. To demonstrate some of these effects, in Figure 3.14(a) we illustrate two types of defects, one in which the localized state emerges from the air band and one which derives from the dielectric band. The simulation of these defects is shown in Figure 3.14(b), showing the z-component of the electric field.

The use of line defects is illustrated in Figure 3.15(a), where we see a simulation of the propagation of an electromagnetic wave along the line defined by the missing holes (line defect). At the T-junction, the wave is split into two equal parts and propagates without loss. The line defect can couple to a cavity in close proximity to the end of a waveguide, as shown in the structure in Figure 3.15(b). In this case, the guided wave mode appears as a band within the bandgap region, as illustrated in Figure 3.15(c).

It is interesting to note that periodic photonic structures can also exist in the natural world. In Figure 3.16, we show an example of the fine structure of a butterfly wing as observed in an SEM micrograph. The 3D periodic structure acts as a photonic material and shows how the coloration of the butterfly wing originates as a photonic effect. In this case, corresponding to the green area of the wing of the parades serositis. The range of colors observed in butterfly wings and other insects covers a broad range and is related to the periodicity of the photonic crystals that occur in nature.

3.4 Plasmons in Nanostructures

In the introduction to this chapter, we discussed the optical properties of conductive or metallic media. Here, we used the Drude model of conductivity to derive the plasma frequency for bulk plasmons, which depends essentially on

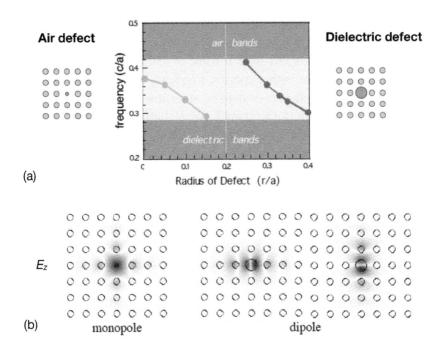

(a)

(b)   monopole   dipole

FIGURE 3.14 (a) The photonic crystal with a point defect. The emergence of defect states within the bandgap region occurs depending on the relative dielectric constant and therefore the states can emerge from either the air bands or the dielectric bands, as shown. (b) Discrete resonance modes, which are localized at the defect in the photonic crystal.

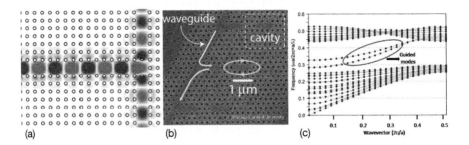

(a)   (b)   (c)

FIGURE 3.15 (a) Illustration of the guided wave modes for line defects. A T-junction at the end of the line shows that this can act as a beam splitter. (b) SEM image of a 2D photonic structure indicating both line and point defects, which act as a waveguide and a cavity, respectively. (c) Photonic band diagram illustrating the guided modes that appear within the bandgap region due to a line defect in a 2D periodic structure.

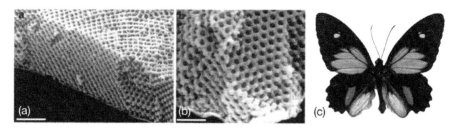

FIGURE 3.16 (a), (b), SEM images of the exposed photonic crystal of the *Parides sesostris* (green area of the wings, see (c)). Bars, (a) 1.2 m; (b) 750 m. Reprinted by permission from P. Vukusic & J. R. Sambles, (2003). *Nature, 424,* 852-855, ©(2003). Nature.

the charge carrier density in the metal, as given in Eq. (3.17). The resulting relative permittivity was also obtained for the Drude model, as expressed in Eq. (3.16). In fact, the two main optical properties of a metal result from its permittivity. One arises from the so-called *skin depth,* and gives rise to the opacity of metals at optical frequencies, the second is the existence of electromagnetic surface waves, which are referred to as a *surface plasmon polariton* (SPP). The skin depth can be expressed as:

$$\delta = \sqrt{\frac{2}{\omega\mu\sigma}} \tag{3.91}$$

which can be derived from Maxwell's equations. Here, μ represents the permeability and σ the conductivity of the metal. For conductors at optical frequencies, the skin depth will be of the order of tens of nm. The bulk plasma frequency will depend on the density, n, of free charge carriers in the metal and can be expressed as:

$$\omega_p = \sqrt{\frac{ne^2}{\varepsilon_0 m}} \tag{3.92}$$

3.4.1   Surface Plasmon Polaritons

To consider the nature of the SPP, we need to look at the interface between the metal, with permittivity ε_m and a dielectric, of permittivity ε_d. Maxwell's equations should be resolved for homogeneous solutions at the interface, which decays into the two media exponentially with the distance from the interface. This is schematically illustrated in Figure 3.17. The dispersion relation of the SPP can be obtained from the boundary conditions at the in-

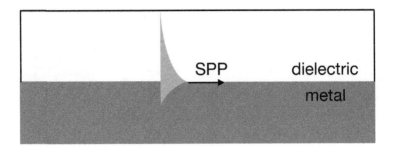

FIGURE 3.17   Propagating surface plasmon polariton (SPP) at a planar metal–dielectric interface.

terface, which yields the in-plane wave vector for the SPP as a function of the light wave vector and the dielectric constants:

$$k_{SPP} = k_0 \sqrt{\frac{\varepsilon_m \varepsilon_d}{\varepsilon_m + \varepsilon_d}} \qquad (3.93)$$

The wavelength of the SPP can be considerably shorter than the wavelength of light in free space, a fact that can enable optical confinement well beyond the *diffraction limit*. The dispersion relation is valid for both real and complex ε_m, i.e., for metals with and without attenuation. The dispersion relation for both bulk and surface plasmons is illustrated in Figure 3.18 along with the light line in the air ($\varepsilon_d = 1$). The dispersion relation shows the case for a metal/air interface, with the dielectric function of the metal being given by the Drude model, as given in Eq. (3.18). We note that the SPP curve sits to the right of the light line, $\omega_0 = ck_0$. For low values of frequency and wave vector, the SPP mode behaves like a photon, however, for larger values of the wave vector, the SPP frequency plateaus to an asymptotic value referred to as the *surface plasmon frequency*, given by:

$$\omega_{sp} = \frac{\omega_p}{\sqrt{1 + \varepsilon_d}} \qquad (3.94)$$

For the case of an air/metal interface, this becomes: $\omega_{sp} = \omega_p/\sqrt{2}$. In the above, we have considered the case of an idealized metal with no imaginary component. In real metals, however, the excitation of the conduction electrons will experience damping due to free electron and inter-band transitions. This means that ε_m will become complex as will the resulting wave vector. The propagating SPP will thus be damped with losses due to absorption within the metal expressed in terms of a characteristic propagation

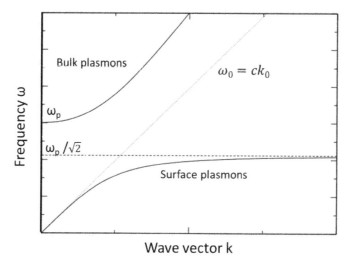

FIGURE 3.18 Dispersion relation for bulk and surface plasmons. Also shown is the light line in the air.

length, $L_{SPP} = 1/2\Im\{k_{SPP}\}$, which has values in the tens to hundreds of microns in the visible spectrum, depending on the particular metal/dielectric configuration. The SPP is characterized by its localization at the interface of the metal and dielectric media. For low frequencies, the decay of the SPP into the metal is given by the skin depth, Eq. (3.91). The field profile in the dielectric extends to a significantly longer distance than that in the metal. The confinement in the direction perpendicular to the interface can be expressed in terms of the two dielectric constants and the wavelength as:

$$z_i = \frac{\lambda}{2\pi}\sqrt{\frac{\varepsilon_d + |\Re\{\varepsilon_m\}|}{\varepsilon_i^2}} \tag{3.95}$$

where $i = d, m$. SPPs are very sensitive to changes in the dielectric constant and are the basis for many different types of sensors. It is interesting to note that at optical frequencies, $\varepsilon_m < 0$ and large, $|\varepsilon_m| \gg \varepsilon_d$, such that from Eq. (3.93), we can write $k_{SPP} > k_0\sqrt{\varepsilon_d}$. This will mean that the plasmon wavelength will differ from that of the free photon wavelength, with the consequence that the induced surface plasmon cannot radiate to the dielectric and becomes trapped at the surface, see Figure 3.19. A thick metallic film can support two independent plasmon waves, one at each surface, with wavelengths being dependent on the permittivities of the dielectrics adjacent

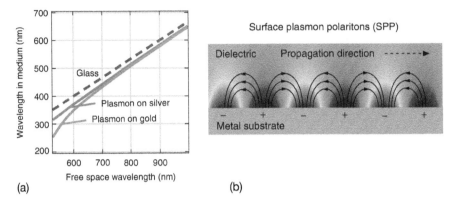

(a)   (b)

FIGURE 3.19 Plasmons propagating on surfaces: (a) The wavelength of an SPP is shorter than the free space wavelength of light or the wavelength within the dielectric, which prevents the plasmon from radiating; (b) the electric field associated with an SPP propagating over the surface of a thick metal film. Reprinted from T. J. Davis, D. E. Gómez & A. Roberts, (2017). *Nanophotonics, 6*, 543, under the Creative Commons Attribution 4.0 International License.

to each surface. As the film reduces in thickness to below the skin depth, the plasmon modes become coupled and mode splitting can occur.

Early work on the generation of surface plasmons was performed by Powell and Swan, who used electron energy loss spectroscopy (EELS) to observe the surface plasmons in Al and Mg foils. A more common approach for the study of SPPs is via photon excitation. However, a simple direct approach via free-space photons is not possible since $k_0 < k_{SPP}$. A projection of the light along the interface will only reduce the momentum by a factor of $\sin \theta$, i.e. $k_\parallel = k_0 \sin \theta$. This fact is evident from the fact that the surface plasmon dispersion curve lies to the right of the free space curve in Figure 3.18. This can be overcome by coupling the photons through a medium such as a prism, which can match the photon and surface plasmon wave vectors. This leads to the condition referred to as *attenuated total reflection* (ATR), as demonstrated by Otto and independently by Kretschmann and Raether in 1968. These geometries are illustrated in Figure 3.20. The coupling via the prism, with permittivity ε_p, allows for an increase in the wave vector parallel to the interface and will take the form:

$$k_\parallel = k_{SPP} = k_0 \sqrt{\varepsilon_p} \sin \theta \qquad (3.96)$$

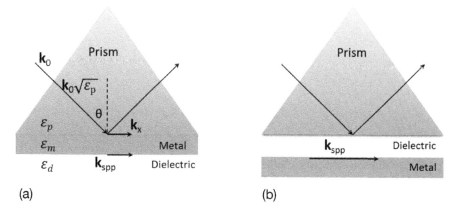

FIGURE 3.20 Attenuated total reflection using (a) the Kretschmann and (b) the Otto geometries to excite surface plasmons along with the metal/dielectric interface.

The Kretschmann approach is one of the most commonly used experimental methods for the excitation of SPPs in thin film geometries. An alternative approach for the coupling of light to surface plasmons is via a periodically patterned surface, with a wave vector designed to match that of the incoming photon:

$$k_{SPP} = k_0 \sin \theta + \frac{2\pi N}{\Lambda} \qquad (3.97)$$

where N is the diffraction order (± 1, ± 2, ...) and Λ is the period of the surface grating.

3.4.2 Localized Surface Plasmon Resonances

As we saw above, SPPs are propagating and dispersive electromagnetic waves coupled to the electron plasma of a conductor at the interface with a dielectric medium. localized surface plasmon resonances (LSPR), however, are non-propagating excitations, which arise due to the excitation of electron density oscillations by incident electromagnetic radiation in confined metallic nanostructures. This is a global effect within the nanoparticle, where the electron gas is collectively displaced under the effect of the electric field. The LSPRs appear as strong absorption peaks in optical spectra and are strongly dependent on the size, shape, composition, and environment of the metallic nanostructures. In Figure 3.21, we show a schematic illustration of the effect of the electric field on delocalized electrons in a nanoparticle in the collective

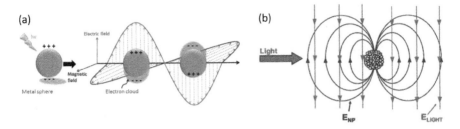

FIGURE 3.21   (a) Illustration of a localized surface plasmon resonance (LSPR) resulting from the collective oscillations of delocalized electrons in response to an external electric field. (b) Illustration of the electric field distribution of incident light and that created by the electron oscillations near the metal NP at resonance. Reproduced from S. Peiris, J. McMurtrie & H. Zhu, (2016). *Catal. Sci. Technol.,* 6, 320, DOI: 10.1039/C5CY02048D with permission from The Royal Society of Chemistry.

excitation of a localized surface plasmon resonance and the calculated averaged electric field distribution around the nanoparticle due to a fundamental mode plasmonic resonance.

   The optical properties of metallic nanoparticles will differ significantly from those of the same elements in bulk form. When the nanoparticle is illuminated with plane-polarized light, the conduction electrons are driven into oscillatory motion by the electric component of the electromagnetic field. The fixed ions of the nanoparticle provide a restoring force, which is related to the curved surface of the nanoparticle. The restoring force allows conditions for the resonant motion of the delocalized electrons in the nanoparticle, but remain localized in the particle itself. An additional consequence of the particle shape is that plasmon resonances can be directly excited by incident light, in contrast to the situation at a flat surface for SPP modes. To consider the principal resonance, we will assume that the particle diameter, $d = 2a$, is much smaller than the wavelength of the incident radiation, which means that we can ignore the effects of phase over the particle, i.e., the electric field is assumed to be constant over the particle. For simplification, we will also assume that the particle is an isotropic sphere surrounded by a homogeneous non-absorbing dielectric of constant ε_d. The metallic sphere is described by a generic complex dielectric function, $\varepsilon(\omega) = \varepsilon_r(\omega) + i\varepsilon_i(\omega)$. To progress, we consider the electrostatic potential, as described by the Laplace equation, $\nabla^2\Phi = 0$, from which the electric field can be expressed as: $\mathbf{E} = -\nabla\Phi$. Considering the spherical symmetry of the problem, where the electric field aligns with the z axis, the general solution of the electric potential takes the form, (Jackson, 1999):

$$\Phi(r,\theta) = \sum_{l=0}^{\infty} [A_l r^l + B_l r^{-(l+1)}] P_l(\cos\theta) \tag{3.98}$$

where $P_l(\cos\theta)$ are the Legendre Polynomials of order l, and θ the angle between the position vector r at point Q and the z-axis, as illustrated in Figure 3.22.

Since the potential must remain finite at the origin, the solution for the potential inside and outside the sphere can be expressed in the form:

$$\Phi_{in}(r,\theta) = \sum_{l=0}^{\infty} A_l r^l P_l(\cos\theta) \tag{3.99}$$

and

$$\Phi_{out}(r,\theta) = \sum_{l=0}^{\infty} [B_l r^l + C_l r^{-(l+1)}] P_l(\cos\theta) \tag{3.100}$$

where the coefficients A_l, B_l and C_l are determined from a consideration of the boundary conditions at $r \to \infty$ and at the surface of the sphere, $r = a$. The condition that $\Phi_{out} \to -E_0 z$ as $r \to \infty$, produces $B_l = -E_0$ and $B_l = 0$ for $l \neq 1$. The other coefficients are evaluated from the boundary conditions at $r = a$, where the tangential and normal components of the electric field

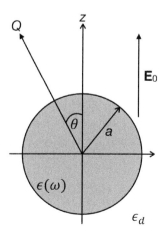

FIGURE 3.22   Coordinate system for a homogeneous sphere, $\varepsilon(\omega)$, immersed in a medium, ε_d, subject to an electrostatic field.

must be equal at either side of the interface. It is found that $A_l = C_l = 0$ for $l \neq 1$, leaving just the coefficients A_1 and C_1. The final solution is given as:

$$\Phi_{in}(r, \theta) = -\frac{3\varepsilon_d}{\varepsilon + 2\varepsilon_d} E_0 r \cos\theta \qquad (3.101)$$

$$\Phi_{out}(r, \theta) = -E_0 r \cos\theta + \frac{\varepsilon - \varepsilon_d}{\varepsilon + 2\varepsilon_d} E_0 a^3 \frac{\cos\theta}{r^2} \qquad (3.102)$$

We note that Eq. (3.102) is the sum of the applied field and the dipole located at the center of the particle. We can rewrite this equation and thus define the dipole moment as follows:

$$\Phi_{out}(r, \theta) = -E_0 r \cos\theta + \frac{\mathbf{p} \cdot \mathbf{r}}{4\pi\varepsilon_0 \varepsilon_d r^3} \qquad (3.103)$$

which thus gives:

$$\mathbf{p} = 4\pi\varepsilon_0 \varepsilon_d a^3 \frac{\varepsilon - \varepsilon_d}{\varepsilon + 2\varepsilon_d} \mathbf{E}_0 \qquad (3.104)$$

This shows that the applied electric field induces a dipole moment inside the sphere that is proportional to the amplitude of the electric field. Introducing the polarizability, α, defined with $\mathbf{p} = \varepsilon_0 \varepsilon_d \alpha \mathbf{E}_0$, we express this as:

$$\alpha = 4\pi a^3 \frac{\varepsilon - \varepsilon_d}{\varepsilon + 2\varepsilon_d} \qquad (3.105)$$

This expression is of the same form as the Clausius–Mossotti relation. From this relation, the polarizability undergoes a resonant enhancement for the condition that $|\varepsilon + 2\varepsilon_d|$ is a minimum, which for small or slow varying $\Im\{\varepsilon\}$, occurs for:

$$\Re\{\varepsilon(\omega)\} = -2\varepsilon_d \qquad (3.106)$$

This relation is referred to as the Fröhlich condition and the associated mode in oscillating fields is the dipole surface plasmon for a metallic nanoparticle. The resonance red-shifts as ε_d increases. The electric fields inside and outside the metal sphere can be evaluated from $\mathbf{E} = -\nabla\Phi$ using Eqs. (3.101) and (3.102), which we express as:

$$\mathbf{E}_{in} = \frac{3\varepsilon_d}{\varepsilon + 2\varepsilon_d} \mathbf{E}_0 \qquad (3.107)$$

$$\mathbf{E}_{out} = \mathbf{E}_0 + \frac{3\hat{n}(\hat{n} \cdot \mathbf{p}) - \mathbf{p}}{4\pi\varepsilon_0 \varepsilon_d} \frac{1}{r^3} \qquad (3.108)$$

Clearly, a resonance in the polarizability will also imply a resonant enhancement in the electric fields. This is the plasmon resonance related to the metallic nanoparticle. This means that the electric field enhancement in this case is due to that of the polarizability.

In the above, we have essentially just considered the situation from the electrostatic point of view. For a small sphere with $a \ll \lambda$, we can present the particle as an idealized dipole in the quasi-static regime. In this case, we allow for time-varying fields, but neglect spatial retardation effects over the volume of the sphere. Under plane wave illumination with $\mathbf{E}(\mathbf{r},t) = \mathbf{E}_0 e^{i\omega t}$, the dipole moment will follow the electric field, where we have $\mathbf{p}(t) = \varepsilon_0 \varepsilon_d \alpha \mathbf{E}_0 e^{i\omega t}$, where α is expressed by Eq. (3.105). The radiation of such a dipole produces a scattering of the plane wave from the sphere, which is represented as the radiation from a point dipole.

The electromagnetic fields associated with the oscillating dipole, can be expressed explicitly as:

$$\mathbf{H} = \frac{ck^2}{4\pi} (\hat{\mathbf{n}} \times \mathbf{p}) \frac{e^{ikr}}{r} \left(1 - \frac{1}{ikr} \right) \tag{3.109}$$

$$\mathbf{H} = \frac{1}{4\pi\varepsilon_0\varepsilon_d} \left\{ k^2 (\hat{\mathbf{n}} \times \mathbf{p}) \times \hat{\mathbf{n}} \frac{e^{ikr}}{r} + [3\hat{\mathbf{n}}(\hat{\mathbf{n}} \cdot \mathbf{p}) - \mathbf{p}] \left(\frac{1}{r^3} - \frac{ik}{r^2} \right) e^{ikr} \right\} \tag{3.110}$$

where the wave vector is given by $k = 2\pi/\lambda$ and $\hat{\mathbf{n}}$ is the unit vector pointing along the direction of the point of interest, Q. We note that the conditions; $\mathbf{H}(t) = \mathbf{H} e^{i\omega t}$ and $\mathbf{E}(t) = \mathbf{E} e^{i\omega t}$ are implicitly maintained. From the optical point of view, what is interesting to note is the consequences of the resonant enhancement of the polarizability on the light scattering and absorption. These can be determined from the scattering and absorption cross-sections from the Poynting vector using Eqs. (3.109) and (3.110), (Bohren and Huffman, 1983):

$$\sigma_{sca} = \frac{k^4}{6\pi} |\alpha|^2 = \frac{8\pi}{3} k^4 a^6 \left| \frac{\varepsilon - \varepsilon_d}{\varepsilon + 2\varepsilon_d} \right|^2 \tag{3.111}$$

$$\sigma_{abs} = k\Im\{\alpha\} = 4\pi k a^3 \Im \left[\frac{\varepsilon - \varepsilon_d}{\varepsilon + 2\varepsilon_d} \right] \tag{3.112}$$

For the case of small particles $(a \ll \lambda)$, the absorption, which scales with a^3, will dominate scattering, which scales as a^6. This means that it is intrinsically difficult to observe scattering from very small objects. Once again both absorption and scattering will be resonantly enhanced due to the Fröhlich condition, Eq. (3.106). For a sphere of radius a, and dielectric function $\varepsilon = \varepsilon_1 + i\varepsilon_2$ in the quasi-static limit, the extinction cross-section can be expressed as:

$$\sigma_{ext} = \sigma_{abs} + \sigma_{sca} = 24\pi^2 \frac{a^3}{\lambda} \varepsilon_d^{3/2} \frac{\varepsilon_2}{(\varepsilon_1 + 2\varepsilon_d)^2 + \varepsilon_2^2} \qquad (3.113)$$

If we consider a more general shaped nanoparticle, described as an ellipsoid with semi-axes $a_1 \leq a_2 \leq a_3$, as specified by $x^2/a_1 + y^2/a_2 + z^2/a_3 = 1$, the polarizability will take on different values along each of the principal axes $(i = 1, 2, 3)$ as given by (Bohren and Huffman, 1983):

$$\alpha_i = \frac{4\pi^2 a_1 a_2 a_3}{3} \frac{\varepsilon(\omega) - \varepsilon_d}{\varepsilon_d + L_i[\varepsilon(\omega) - \varepsilon_d]} \qquad (3.114)$$

where L_i is a geometrical shape factor expressed as:

$$L_i = \frac{a_1 a_2 a_3}{2} \int_0^\infty \frac{dq}{(a_i^2 + q)f(q)} \qquad (3.115)$$

Here, we have $f(q) = \sqrt{(q + a_1^2)(q + a_2^2)(q + a_3^2)}$ and the function L_i is subject to the condition $\sum_i L_i = 1$. We note that for a sphere $L_1 = L_2 = L_3 = 1/3$. It can be further shown that the corresponding extinction cross-section can be expressed as:

$$\sigma_{ext} = \sigma_{abs} + \sigma_{sca} = 24\pi^2 \frac{a^3}{\lambda} \varepsilon_d^{3/2} \frac{\varepsilon_2}{[\varepsilon_1 + (1 - L_i)\varepsilon_d]^2 + \varepsilon_2^2} \qquad (3.116)$$

Many studies have been made of particles of various shapes and sizes (for LSPR within the visible to infrared frequencies). In Figure 3.23, we show some examples of different shapes. The number of resonance peaks is determined by the number of modes in which a particularly shaped nanoparticle can be polarized. For non-spherical particles, this means that multiple red-shifted peaks are expected as compared to these of similar-sized spherical particles. For example, for the case of nanorods, polarization can be achieved along two distinct axes, i.e., the longitudinal and transverse. The particle size will influence the relative intensity of the absorption and scattering cross-sections. The increase of size generally increases the scattering cross-section, while for particles below 20 nm, absorption is the dominant process. The effect of size can be explicitly seen in Figure 3.23(f), where for a rod-shaped particle the longitudinal absorption peak red-shifts with increasing length. Since the width is not changed, the transversal peaks remain unaltered.

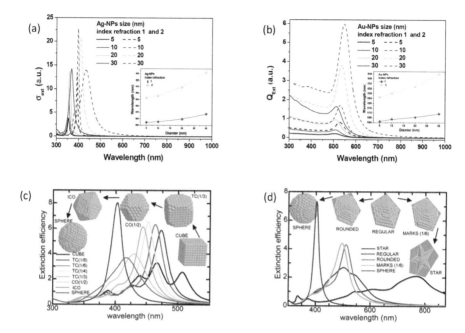

FIGURE 3.23 Extinction cross-section from Mie theory for (a) silver and (b) gold NPs. Both in the function of size with refraction index 1 (line curves) and 2 (dot lines). The inset figure shows a red-shift with the increment of size NPs and refraction index. (c) Extinction efficiencies as a function of the wavelength of the incident light of a silver cube, different truncated cubes, and a spherical nanoparticle. (d) Extinction efficiency as a function of the wavelength of the incident light for the regular decahedron and its truncated morphologies for parallel light polarization. (a) and (b) Reprinted from V.A.G. Rivera, F.A. Ferri, and E. Marega Jr., in *Plasmonics—Principles and Applications*, Ki-Young Kim (Ed.), IntechOpen, (2012). DOI: 10.5772/50753, under the Creative Commons Attribution 4.0 International License. (c) and (d) Reprinted with permission from C. Noguez, (2007). *J. Phys. Chem., 111*, 3806. ©(2007) American Chemical Society.

Another very demonstrative example is shown in Figure 3.24, where a series of metallic particles are illuminated with light and re-emit at a characteristic wavelength depending on their size and shape.

The above description of the dipole particle is strictly only valid for the case of vanishingly small particles. It turns out, however, that calculations are in reasonable agreement with experiment for spherical and ellipsoidal particles below 100 nm, irradiated with visible or near-infrared radiation. In the case of the larger particle, the appearance of phase shifts across the volume of the particle with respect to the driving field means that the above

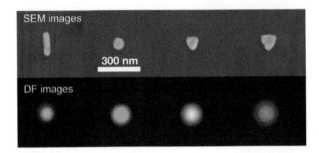

FIGURE 3.24 Gold nanoparticles (upper, imaged with scanning electron microscopy) are each about 30 nm wide and carved by electron-beam lithography into distinct shapes (two triangles, a disc, and a rod) on a silica substrate covered with 20 nm of indium tin oxide. The distinctly colored light emission from each (lower, imaged with dark-field optical microscopy) illustrates that the metal resonance frequency can be tuned simply by varying its size and shape. Reprinted from W. A. Murray & W. L. Barnes, (2007). *Adv. Mater., 19*, 3771–3782, under the Creative Commons Attribution 4.0 International License (Wiley Open Access Articles).

approach becomes invalid. This occurs, when the size of the particle becomes comparable to the wavelength of the radiation, and is a result of the finite velocity of light. The Mie theory, which was originally developed in 1908, considered the scattering and absorption of electromagnetic radiation by a sphere. The Mie approach is one of the few analytical approaches, which allows a solution of the Maxwell equations and describes the electromagnetic field in the form of expansions of normal modes (or harmonics). The quasi-static results valid for sub-wavelength spheres can be expressed as a power series expansion of the absorption and scattering coefficients and retaining the first terms only. In this manner, the cross-section for scattering can be expressed as:

$$\sigma_{sca} = \frac{2\pi}{k^2} \sum_{n=1}^{\infty} (2n+1) \left(|a_1^2| + |b_1^2| \right) \qquad (3.117)$$

while the extinction cross-section takes the form:

$$\sigma_{ext} = \frac{2\pi}{k^2} \sum_{n=1}^{\infty} (2n+1) \Re \left[a_1(d) + b_1(d) \right] \qquad (3.118)$$

Here a_1 and b_1 are the Mie coefficients, which depend on the particle size, refractive index, the surrounding medium, and the Riccati-Bessel functions. For a dilute mono-disperse assembly of spherical nanoparticles, the interaction between them can be neglected allowing an effective extinction co-efficient of the composite medium (nanoparticles plus matrix) with nanoparticle volume fraction f to be evaluated as (Y. Battie et al., 2015):

$$\alpha_{\text{eff}}(a) = \frac{3f\sigma_{ext}(a)}{4\pi a^3} \tag{3.119}$$

where a signifies the nanoparticle radius.

3.4.3 Plasmonic Response of Nanoparticle Assemblies

The above discussion has allowed us to evaluate the localized plasmon resonance for isolated metallic particles in a dielectric medium. When many particles are assembled together, we can consider the effective response of the particle and the medium, in which they are embedded. Such an approach will provide an effective medium, which can be considered to be homogeneous. The problem of particle ensembles as effective media was considered by Maxwell Garnett in 1904. For simplicity, we will consider the case of an assembly of $N \gg 1$ spherical particles of radius a, each of polarizability α, embedded in a matrix of dielectric constant ε_d, enclosed in a spatial region $\mathbb{V}$ of volume, V. If we now place this system in a constant external field $\mathbf{E}_{ext}$, neglecting the electromagnetic interaction between particles, we can say that each particle has a dipole moment of $\mathbf{p} = \alpha \mathbf{E}_{ext}$. The total dipole moment of the assembly being:

$$\mathbf{p}_{Tot} = N\mathbf{p} = N\alpha \mathbf{E}_{ext} \tag{3.120}$$

We can also assign the sample a macroscopic permittivity ε_m and polarization $\mathbf{P} = [(\varepsilon_m - 1)/4\pi]\mathbf{E}$, where $\mathbf{E}$ defines the macroscopic electric field inside the medium. We can therefore describe the total dipole moment as:

$$\mathbf{p}_{Tot} = V\mathbf{P} = V\frac{\varepsilon_m - 1}{4\pi}\mathbf{E} \tag{3.121}$$

The relation between the two electric fields is expressed as:

$$\mathbf{E} = \mathbf{E}_{ext} + \left\langle \sum_n \mathbf{E}_n(\mathbf{r}) \right\rangle; \qquad \mathbf{r} \in \mathbb{V} \tag{3.122}$$

Here, $\mathbf{E}_n(\mathbf{r})$ is the field produced by the nth dipole and $\langle ... \rangle$ denotes the average over the sample. This average can be evaluated as $\langle \sum_n \mathbf{E}_n(\mathbf{r}) \rangle \simeq -(4\pi/3V)\mathbf{p}$, (Markel, 2016). The field can now be expressed as:

$$\mathbf{E} = \mathbf{E}_{ext} - N\frac{4\pi}{3}\frac{\mathbf{p}}{V} = \left(1 - \frac{4\pi}{3}\frac{\alpha}{v}\right)\mathbf{E}_{ext} \qquad (3.123)$$

where $v = V/N$ denotes the specific volume of a particle. We can now substitute Eq. (3.123) into Eq. (3.121), under the assumption that Eqs. (3.120) and (3.121) must give the same total dipole moment. We can thus establish the relation:

$$\frac{\alpha}{v} = \frac{\varepsilon_m - 1}{4\pi}\left(1 - \frac{4\pi}{3}\frac{\alpha}{v}\right) \qquad (3.124)$$

It is now possible to solve for ε_m:

$$\varepsilon_m = \frac{1 + (8\pi/3)(\alpha/v)}{1 - (4\pi/3)(\alpha/v)} \qquad (3.125)$$

This expression is known as the Lorentz formula for the permittivity of a non-polar molecular gas. The denominator takes into account the local field correction. In dilute systems, this denominator is almost unity. We could alternatively isolate the term α/v in Eq. (3.124), which yields:

$$\frac{\alpha}{v} = \frac{3}{4\pi}\frac{\varepsilon_m - 1}{\varepsilon_m + 2} \qquad (3.126)$$

which is just another form of the Clausius–Mossotti relation.

Using this formula, Purcell and Pennypacker (1973), developed an approach based on the solution of N-coupled dipole equations, known as the *discrete dipole approximation* (DDA). The model was developed to study the scattering of light by interstellar dust, with particles ranging in size and shape. The starting point is the discretization of the macroscopic Maxwell equations written in integral form. The dipole field and the Clausius–Mossotti relation must be modified to account for retardation effects and other corrections associated with polarizability and finite frequency. Interestingly, the results for an assembly of N particles of polarizability α as given in Eq. (3.126) are equal to the polarizability of a large sphere, and can be shown to be in agreement with the N-coupled dipole equations. In the context of the DDA, the Lorentz local field correction is rather important to obtain the correct results.

We will now consider the Maxwell Garnett approach to the problem. We take the particle to be of radius a and the permittivity ε_m that is distributed in vacuum either on a lattice or randomly, but uniformly on average over space. The specific volume is $v = V/N$ and the volume fraction of the particles is expressed as: $f = (4\pi/3)a^3/v$. The effective permittivity of this ensemble being given by Eq. (3.125). Using the expression for the polarizability of the particle as: $\alpha = a^3(\varepsilon - 1)/(\varepsilon + 2)$, along with the volume fraction, we obtain:

$$\varepsilon_{MG} = \frac{1 + 2f(\varepsilon - 1)/(\varepsilon + 2)}{1 - f(\varepsilon - 1)/(\varepsilon + 2)} = \frac{1 + (1 + 2f)(\varepsilon - 1)/3}{1 + (1 - f)(\varepsilon - 1)/3} \quad (3.127)$$

This is the Maxwell Garnett mixing formula for small particles in a vacuum and corresponds to the permittivity of a composite and not the material of the particles themselves. If we now place the particles in a medium of permittivity ε_d with the particles having permittivity ε_p, we make the following substitutions; $\varepsilon_{MG} \to \varepsilon_{MG}/\varepsilon_d$ and $\varepsilon \to \varepsilon_p/\varepsilon_d$, in the above formula, which now takes the form

$$\varepsilon_{MG} = \varepsilon_d \frac{1 + 2f(\varepsilon_p - \varepsilon_d)/(\varepsilon_p + 2\varepsilon_d)}{1 - f(\varepsilon_p - \varepsilon_d)/(\varepsilon_p + 2\varepsilon_d)} = \varepsilon_d \frac{1 + (1 + 2f)(\varepsilon_p - \varepsilon_d)/3}{1 + (1 - f)(\varepsilon_p - \varepsilon_d)/3} \quad (3.128)$$

This equation was derived under the assumption that the particles have spherical shape. However, upon inspection, there is no information in the equation regarding the particle shape itself, containing only the permittivities of the particles and host matrix. It turns out that Eq. (3.128) is valid for any particle shape as long as the medium is spatially uniform and isotropic on average.

Rearranging Eq. (3.128), we can write:

$$\frac{\varepsilon_{MG} - \varepsilon_d}{\varepsilon_{MG} + 2\varepsilon_d} = f\frac{\varepsilon_p - \varepsilon_d}{\varepsilon_p + 2\varepsilon_d} \quad (3.129)$$

If the medium now contains particles of different materials, with permittivities $\varepsilon_{p,n}, (n = 1, 2, ..., N)$, Eq. (3.129) can be generalized in the form:

$$\frac{\varepsilon_{MG} - \varepsilon_d}{\varepsilon_{MG} + 2\varepsilon_d} = \sum_{n=1}^{N} f_n \frac{\varepsilon_{p,n} - \varepsilon_d}{\varepsilon_{p,n} + 2\varepsilon_d} \quad (3.130)$$

where f_n is the volume fraction of the nth component. The Maxwell Garnet formula can be expressed in a general form that resembles the Clausius–Mossotti relation, in which we write:

$$\alpha_{MG}(a) = a^3 \frac{\varepsilon_p - \varepsilon_d}{\varepsilon_p + 2\varepsilon_d} = \frac{a^3}{f} \frac{\varepsilon_{MG} - \varepsilon_d}{\varepsilon_{MG} + 2\varepsilon_d} \quad (3.131)$$

There have been several modifications and extensions to the Maxwell Garnett (MG) model. This includes taking into account the size dependence of the nanoparticles using the Mie theory. This leads to the Maxwell Garnett–Mie electric dipole polarizability, which can be expressed as:

$$\alpha_{MGM}(a) = \frac{3i\lambda^3}{16\pi^3\varepsilon_d^{3/2}}a_1(a) \tag{3.132}$$

with a_1 being the first electric Mie coefficient.

A modified MG model includes the effect of a size distribution of the nanoparticles, where the average polarizability can be expressed as:

$$\langle\alpha\rangle = \int_0^\infty \alpha P(a)\mathrm{d}a \tag{3.133}$$

where the polarizability is weighted by the size distribution function, $P(a)$. This leads to the modified Maxwell Garnett (MMG), which can be expressed as (Keita and En Naciri, 2011):

$$\frac{\varepsilon_{MG}(\omega) - \varepsilon_d}{\varepsilon_{MG}(\omega) + 2\varepsilon_d} = f\int_{R_{min}}^{R_{max}} \left(\frac{a}{\langle a\rangle}\right)^3 P(a)\frac{\varepsilon_p(\omega, a) - \varepsilon_d(\omega)}{\varepsilon_p(\omega, a) + 2\varepsilon_d(\omega)}\mathrm{d}a \tag{3.134}$$

Strictly speaking, the dielectric functions should be expressed as a complex quantities. The upper and lower bounds of the integral are taken as R_{max} and R_{min}, respectively, which characterize the size distribution.

A further extension can take both the above corrections into account, the so-called modified Maxwell Garnett–Mie (MMGM) theory. The general expression is expressed as:

$$\frac{\varepsilon_{MG}(\omega) - \varepsilon_d}{\varepsilon_{MG}(\omega) + 2\varepsilon_d} = \frac{3i\lambda^3}{16\pi^3\varepsilon_d^{3/2}}\frac{f}{\langle a\rangle^3}\int_{R_{min}}^{R_{max}} P(a)a_1(a)\mathrm{d}a \tag{3.135}$$

By way of demonstration, in Figure 3.25 we show a comparison of the experimental data and calculations of the MGM and MMGM theories applied to different solutions of Au nanoparticles (SAu). The particle size distributions were evaluated using TEM micrographs. From these size distributions, $P(a)$, the calculations were made based on the MGM and MMGM theories. While the agreement between experiment and theory is rather good, the MMGM approach shows a superior fit.

3.4.4 Applications of Plasmonic Systems

Applications of plasmonic structures have grown steadily since the 1990s, with a number of important experimental advances. Indeed, applications of plasmonic systems are quite diverse, ranging from sensing and bio-sensing to plasmonic nanoscopy and photovoltaics. An exhaustive review is beyond

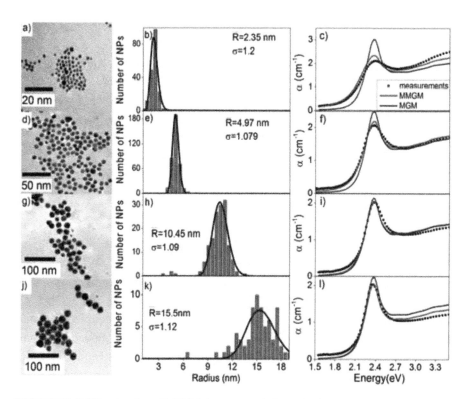

FIGURE 3.25 (a, d, g, j) TEM images and (b, e, h, k) radius distributions of SAu1, SAu2, SAu3 and SAu4 NP colloids, respectively. (c, f, i, l) Comparison between the measured extinction spectra of the corresponding colloidal Au NP solutions and the calculated ones from MGM and MMGM theories. Reprinted from Y. Battie, A. Resano-Garcia, N. Chaoui, & A. En Naciri, (2015). *Phys. Stat. Sol. c, 12*, 142, under the Creative Commons Attribution 4.0 International License (Wiley Open Access Articles).

the scope of this work and our aim is to give a brief overview of some of the main topics in this area.

One of the remarkable effects that can be observed due to the proximity of two particles is the generation of localized regions of extremely large electric fields, which are often referred to as *hot spots*. An illustration is shown in Figure 3.26, where reports indicate electric field enhancements in the region of several orders of magnitude for the longitudinal configuration, Figure 3.26(a). This effect has been widely exploited and in particular, this effect plays an important role in surface-enhanced Raman spectroscopy (SERS), where enhancement factors of over 10^8 have been reported. With such massive enhancement of the signal, SERS is potentially bright enough to observe single molecules.

Many geometries of nanostructures have been fabricated by a variety of methods, often referred to as plasmonic nanostructures, and their optical properties extensively studied. These range from arrays of spherical or non-spherical particles to core-shell systems consisting of composites of metallic and non-metallic structures. The interaction between the various components plays a crucial role in defining the optical properties of the assembly. Plasmon hybridization has been found, in which it is possible to tune the plasmon resonance frequency. Certain plasmonic nanostructures can convert the energy of free propagating radiation to localized energy and vice versa. Such systems can be considered to act as optical antennae.

A number of applications of SPPs, which are traveling waves located or bound at the surfaces of metallic films and which propagate along with the interface, have been found, in which they are used as waveguides. We can consider various types of geometry, such as insulator-metal-insulator (IMI) or metal-insulator-metal (MIM), in which SPP modes can be coupled for specific film thicknesses. Both symmetric and antisymmetric modes can couple, as illustrated in Figure 3.27. Propagating SPPs are bound and propagate at the metal–dielectric interface. In this way, plasmon waves can be con-

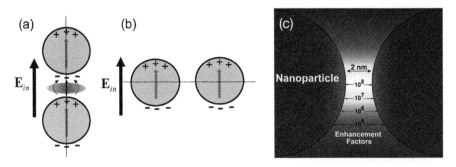

FIGURE 3.26   Longitudinal (a) and transverse (b) modes for a dimer of particles. When the longitudinal mode is excited, the gap between the particles becomes a hot spot. (c) Enhancement factors in SERS in the vicinity of two nanoparticles due to hot spots. (a) Reprinted from S. Hayashi, & T. Okamoto, (2012). *J. Phys. D: Appl. Phys., 45*, 433001; Journal of Physics D: Applied Physics by Institute of Physics and the Physical Society; Institute of Physics (Great Britain) Reproduced with permission of IOP Publishing in the format Book via Copyright Clearance Center.; (b) Reprinted from E. Petryayeva & U. J. Krull, (2011). *Anal. Chim. Acta, 706*, 8, ©(2011), with permission from Elsevier.

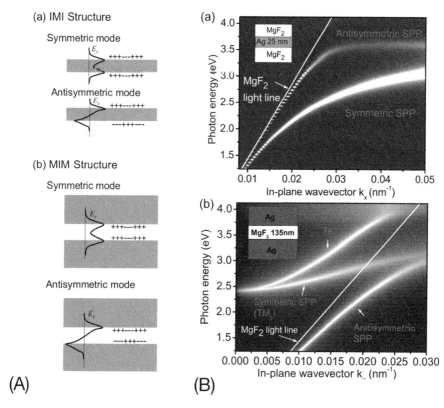

FIGURE 3.27 (A) Symmetrically and antisymmetrically coupled SPP modes in IMI (a) and MIM (b) structures. Dark regions represent metals. (B) Dispersion curves of symmetrically and antisymmetrically coupled SPP modes in IMI (a) and MIM (b) structures. Reprinted from S. Hayashi & T. Okamoto, (2012). *J. Phys. D: Appl. Phys., 45,* 433001; *Journal of Physics D: Applied Physics by Institute of Physics and the Physical Society*; Institute of Physics (Great Britain) Reproduced with permission of IOP Publishing in the format Book via Copyright Clearance Center.

fined in thin metallic strips, providing a method for light propagation in optical waveguides. Other possible structures include V-grooves, Λ-wedges and nanowires, see for example Davis et al. (2017).

The use of plasmonics in improving the efficiencies of photovoltaic devices is another interesting application. In conventional solar cells, photovoltaic absorbers must be optically thick in order to allow near complete light absorption and photocurrent carrier collection. However, high-efficiency solar cells also require minority carrier diffusion lengths several times the material thickness for all photo carriers to be collected, i.e., they naturally require thin layers. These opposing requirements can limit the efficiency of conventional cells. Plasmonic structures can overcome these obstacles by reducing the physical thickness of the photovoltaic absorber. This can be achieved by using metallic nanoparticles as sub-wavelength scattering elements to the couple and trap plane waves from the sun into the absorbing semiconducting layer, Figure 3.28(a). A second option would be to use the metallic nanoparticles as sub-wavelength antennae, in which the plasmonic near-field is coupled to the semiconductor, increasing the effective absorption cross-section, Figure 3.28(b). A third possibility is to use a corrugated metallic film on the rear surface of the photovoltaic layer, which can couple sunlight into SPP modes supported at the metal/semiconductor interface, as well as guided modes in the semiconductor, in which light is converted to photo carriers in the semiconductor, Figure 3.28(c), see Atwater and Polman, (2010).

Perhaps one of the most commonly exploited effects in plasmonics is that due to the shift of the plasmon resonance, typically to longer wavelengths (redshift), upon the aggregation of metallic nanoparticles. Such an effect is used in the widely used nanoplasmonic sensor; the home pregnancy test. The test assay is designed to detect the human chorionic gonadotropin hormone (hCG), which is produced in the uterus from the initial stages of pregnancy. Primary antibodies are immobilized on a sensing strip and specifically tar-

FIGURE 3.28   Plasmonic light-trapping geometries for thin-film solar cells. (a) Light trapping by scattering from metal nanoparticles at the surface of the solar cell. Light is preferentially scattered and trapped into the semiconductor thin film by multiple and high-angle scattering, causing an increase in the effective optical path length in the cell. (b) Light trapping by the excitation of localized surface plasmons in metal nanoparticles embedded in the semiconductor. The excited particle's near-field causes the creation of electron-hole pairs in the semiconductor. (c) Light trapping by the excitation of surface plasmon polaritons at the metal/semiconductor interface. A corrugated metal back surface couples light to surface plasmon polariton or photonic modes that propagate in the plane of the semiconductor layer. Reprinted by permission from Nature: H. A. Atwater & A. Polman, (2010). *Nat. Mater.*, 9, 205, ©(2010).

get hCG. In the pregnancy test, if hCG is present in the urine, then it will be bound to the antibodies on the sensing strip. This is not enough for the sensing and for this, a suspension of gold nanospheres, which have been functionalized to chemically link to secondary antibodies, is added. This means they will link to the hCG on the sensing strip, forming a dense monolayer of gold nanospheres at the surface. The close proximity of the Au nanoparticles produces a red-shift in their emission and hence indicates the positive test. In a test region, the particles have a normally green color, while in close proximity, they appear red. This procedure can be used for sensing other medical conditions such as prostate cancer and HIV-AIDS.

Another form of sensing can be achieved by the alteration of the dielectric constant of the surrounding host material. Again, red-shifted spectra of LSPR are detected in response to covering a layer of metallic particles with analyte molecules. A shift of the dielectric constant by an amount $\Delta\varepsilon_d$ creates a plasmonic frequency shift of $\Delta\omega_n$ relative to its spectral width γ_n, according to the formula:

$$\frac{\Delta\omega_n}{\gamma_n} = -Q\frac{\Delta\varepsilon_d}{\varepsilon_d} \qquad (3.136)$$

This basic principle has been applied to many sensing applications and notably in the area of biology and biomedical applications. For a more in-depth discussion of the various methods used in biosensing and biomedical applications, the review by X. Huang et al. (2009) is a good starting point.

In addition to these biomedical applications, metallic nanoparticles have attracted much interest due to their photocatalytic properties, such as for example in green processing for the synthesis of organic compounds. By using the energy absorbed by metallic nanoparticles and the intense electric fields that are created in proximity to these particles when exciting localized surface plasmon resonances, certain catalytic processes can be performed at room temperatures. This can be used to avoid undesired by-products which are formed at the elevated temperatures normally required (Lang et al., 2014). Indeed there are many examples where catalyzed reactions have been successfully performed using pure metallic nanoparticles and their alloys being driven by the light illumination at room or moderate temperatures. In many cases, efficient photocatalysis can be performed using the conversion of solar energy to chemical energy. Early work on photocatalysis stemmed from the discovery of the splitting of water in TiO_2 by Fujishima and Honda in 1972. While showing early promise, the semiconductor photocatalysis method suffered from the drawback of requiring energies corresponding to the bandgap and in cases such as TiO_2 require ultraviolet radiation and limit their practical use. The photon energies required in metallic photocatalysis are significantly reduced and typically occur in the visible spectrum. Many examples are observed in the literature, such as the oxidation of compounds such as alcohols, amines, and aldehyde as well as the selective reduction of various compounds. The recent review by Peiris et al. (2016) outlines a number of examples.

The technique of scanning near-field optical microscopy (SNOM) is based on the scanning of nanometric surface objects using the hot spot at the end of a pointed probe, which is illuminated by light. As with many modern probe techniques, SNOM comes in a number of different types. For example, hollow tapered tips covered by a metallic layer, such as Al, are often used to image biological cell structures. Apertureless tips are also widely used in medical imaging and use a sharp metal tip, which is directly excited by a focused laser. Another option is to use a plasmonic nanosphere at the end of a tapered dielectric tip. The role of the nanosphere is to act as an antenna, delivering optical energy at the nanoscale and enhancing light intensity scattered or emitted by a nano-object. In all cases, the tip is scanned over the surface, much like that in a typical AFM or scanning probe experiment, thus mapping the surface of the sample. The resolution of this technique is far superior to far-field optical methods, which are wavelength limited. SNOM resolutions can be as low as tens of nm.

In this section, we have considered the plasmonic behavior due to small metallic bodies. However, localized plasmonic responses can also be found from small dielectric inclusions or voids in an otherwise homogeneous metallic body. The dipole moment for such a void is very similar to that of a metallic inclusion in a dielectric medium. For the case of a single spherical nanovoid, we can obtain this by making the transformation: $\varepsilon(\omega) \to \varepsilon_d$ and $\varepsilon_d \to \varepsilon(\omega)$ in Eq. (3.105). The polarizability in this case becomes:

$$\alpha = 4\pi a^3 \frac{\varepsilon_d - \varepsilon(\omega)}{\varepsilon_d + 2\varepsilon(\omega)} \tag{3.137}$$

where a is the radius of the spherical void. In this case, the induced dipole moment will be antiparallel to the applied field. The corresponding resonance or Fröhlich condition is expressed as:

$$\Re[\varepsilon(\omega)] = -\frac{1}{2}\varepsilon_d \tag{3.138}$$

An example of such a void could be a core-shell nanoparticle with a dielectric core and a metallic shell.

A periodic array of such voids can produce a surface plasmon polaritonic crystal (SPC). Such a structure can be considered as an analogue to the photonic crystals we discussed earlier in Section 3.3. In this case, the SPC can be considered to be a periodically modulated dielectric, which generates Bragg reflected SPP modes and folds the smooth interface SPP dispersion curve by an integral number of the Bragg momentum; $k_{Bragg} = 2\pi/d$, where d defines the periodicity in the direction of $\mathbf{k}_{Bragg}$. In this approximation, and not accounting for the gap formation, the momentum of the SPP Bloch modes can be expressed as

$$\mathbf{k}_{SPP-BM} = \mathbf{k}_{SPP} + \mathbf{k}_{Bragg} = \frac{2\pi}{\lambda}\sqrt{\frac{\varepsilon(\omega)\varepsilon_d}{\varepsilon(\omega)+\varepsilon_d}}\hat{\mathbf{u}} + l\frac{2\pi}{d_x}\hat{\mathbf{x}} + m\frac{2\pi}{d_y}\hat{\mathbf{y}} \quad (3.139)$$

where k_{SPP} and k_{SPP-BM} are the SPP wave vectors before and after Bragg scattering in the periodic lattice, and k_{Bragg} is the momentum contributed during the scattering by the crystal. The period of the crystal along the x- and y-directions is $d_{x,y}$, (l,m) are integer numbers, $\hat{\mathbf{u}}$ is the unit vector describing the SPP propagation direction before Bragg scattering, and $\hat{\mathbf{x}}$ and $\hat{\mathbf{y}}$ are the unit reciprocal lattice vectors of the periodic structure.

In Figure 3.29, we show some examples of various SPCs and electric field-induced modifications of their transmission via their SPP-BM dispersion. These structures comprise both 1D and 2D arrays of sub-wavelength apertures having the same period (550 nm). In the 1D case, the array consists of slits having a width of approximately 100 nm. For the 2D structures, square arrays have been created using either a circular hole (diameter 180 nm) or a rectangle (200 nm length, 100 nm width) as the lattice basis.

Plasmonic enhancement of diffraction has been observed in gratings formed by nanoparticles, as described in Section 8.3 of Volume 1. The nanoparticle assembly is formed by photo-induced precipitation in areas illuminated by an optical interference pattern. A measurement of diffracted light as a function of angle and wavelength is shown in Figure 3.30(a) and (b). The intensity of the line profiles, shown in the lower portion of the panels, indicate the same variation as the extinction spectra, Figure 3.30(c), showing that the diffraction enhancement is due to the plasmonic response of the Au nanoparticles.

3.5   Negative Refractive Index Metamaterials

The advent of negative refractive index materials, or *metamaterials*, is a story of intelligent design and materials engineering. While it has only relatively recently been realized, the groundwork was laid in the late 1960s, with the theoretical predictions of V. G. Veselago (1968). In fact, even this

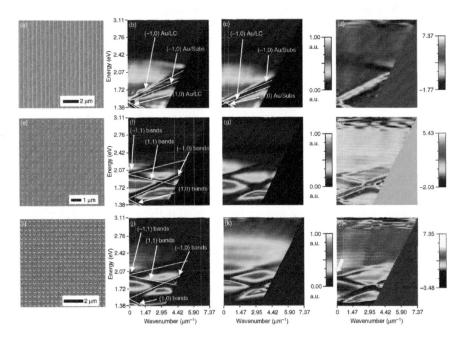

FIGURE 3.29   SEM images and optical transmission dispersion plots of SPC made in a 200 nm-thick Au film. SEM images for a periodic array of (a) slits (60 nm width, 550 nm period), (e) rectangles (200 nm × 50 nm, 550 nm period) and (i) holes (150 nm diameter, 550 nm period). Optical transmission dispersion plots for the structures in LC without and with an applied electric field (12.55 kV cm^{-1}) are shown in (b) and (c) for the slits; (f) and (g) for the rectangles and (j) and (k) for the holes, respectively. (d), (h), and (l) show the change in frequency dispersion associated with the application of the applied field for the slits, rectangles, and holes, respectively. The yellow arrow in (l) indicates where the dynamic response of the transmission was studied. The white lines are the dispersion calculated with Eq. (3.139). It should be noted that, in contrast to the slit array, the SPP bands on the opposite interface in the case of circular and rectangular basis overlap and cannot be distinguished in the plots. Reprinted from G. A. Wurtz, R. J. Pollard, W. Dickson, & A. V. Zayats, (2016). *Optics of Metallic Nanostructures*, in *Reference Module in Materials Science and Materials Engineering, 4*, S. Hashmi (Ed.), Elsevier, 67, ©(2016), with permission from Elsevier.

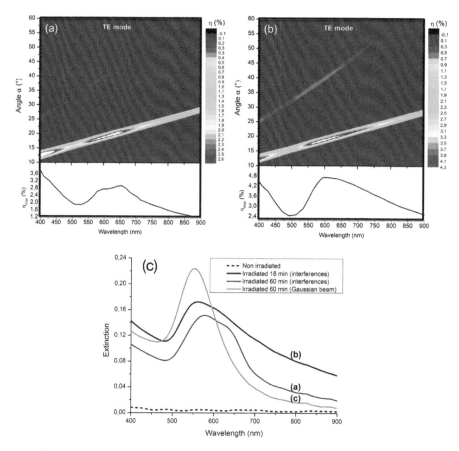

FIGURE 3.30 Diffraction efficiency maps showing the spectral dependence of gratings (a) and (b) irradiated, respectively, for 18 min and 60 min with an interference pattern. The measurement has been performed for the 1st and 2nd order-diffracted beams. The graphs below the diffraction efficiency maps show the corresponding maximum of diffraction efficiency as a function of wavelength. (c) Extinction spectra for a range of Au nanoparticle gratings. The spectra (a) and (b) compare with the gratings of the corresponding diffraction efficiency profiles in the preceding figures (a) and (b). Republished with permission of The Royal Society of Chemistry, from E. Nadal, N. Barros, H. Glénat, J. Laverdant, D. S. Schmool, & H. Kachkachi, (2017). *J. Mater. Chem. C, 5*, 3553; permission conveyed through Copyright Clearance Center, Inc.

was pre-dated by postulations on negative refraction by L. I. Mandelshtam. Despite these early considerations, contemporary technological limitations meant that experimental developments were not possible. In fact, it was not until the later 1990s that Pendry and co-workers provided further impetus in the area with a series of papers regarding the extraordinary properties of metamaterials. This was followed closely by the work of Smith et al. (2000), who demonstrated the negative refractive properties of metamaterials.

Negative-index materials (NIMs) have a negative refractive index, so that electromagnetic waves in such media propagate in a direction opposed to the flow of energy, which is unusual and counterintuitive. There are no known naturally-occurring NIMs. However, artificially designed materials (metamaterials) can act as NIMs. In general, the work on NIMs is related to artificial structures formed by repeated elements, whose physical dimension and spacing are less than the wavelengths of interest, and can be treated as homogeneous to a good approximation. Despite this, metamaterials are generally structured and can be ordered (periodic, quasi-periodic, aperiodic, fractal) or disordered (random).

In order to understand metamaterials, it is necessary to understand the optical response of a material to electromagnetic waves. This has been broadly discussed at the beginning of this chapter. As we saw, the electromagnetic response of a homogeneous medium will in general be governed by two parameters, related to electric and magnetic properties; i.e., the permittivity, $\varepsilon(\omega)$, and the permeability, $\mu(\omega)$. Both of these quantities are dependent on the frequency of the electromagnetic radiation and are complex:

$$\varepsilon(\omega) = \varepsilon_1(\omega) + i\varepsilon_2(\omega) \qquad (3.140)$$

$$\mu(\omega) = \mu_1(\omega) + i\mu_2(\omega) \qquad (3.141)$$

There are many physical parameters of solids that depend on one or both of these quantities, for example the electrical conductivity, the optical absorption, etc. One of the parameters we met at the beginning of the chapter was the refractive index of a material. It was defined as the ratio of the velocity of light in vacuum to that in the medium of interest, but this also gave rise to the relation:

$$n^2(\omega) = \mu(\omega)\varepsilon(\omega) \qquad (3.142)$$

The refractive index will also provide information on the deflection of a light beam as it traverses an interface between two different media, as expressed by the law of Snell–Descartes:

$$n_1 \sin \theta_1 = n_2 \sin \theta_2 \qquad (3.143)$$

where the refractive indices of the two media are n_1, n_2 and θ_1, θ_2 the angles of the light beam with respect to the surface normal at the interface, respectively. Veselago (1968) hypothesized the case, where simultaneously $\varepsilon(\omega) < 0$, or $\sqrt{\varepsilon(\omega)} = i\sqrt{|\varepsilon(\omega)|}$ and $\mu(\omega) < 0$, or $\sqrt{\mu(\omega)} = i\sqrt{|\mu(\omega)|}$, in which case, according to Eq. (3.142) we have $n(\omega) = -\sqrt{|\varepsilon(\omega)||\mu(\omega)|} < 0$. While in nature it is possible for materials to display either $\varepsilon(\omega) < 0$ or $\mu(\omega) < 0$, no such natural materials have been found in which they are simultaneously negative. However, artificially fabricated metamaterials allow for this condition.

It is interesting to note the consequence of the negative refractive index on the deflection of light. This is illustrated in Figure 3.31 according to the Snell–Descartes law. We see that while for the normal refractive index medium, the deflection of the light ray is quite small, for the negative refractive index material, the deflection is much larger.

From Maxwell's equations, we can establish the following relations in terms of the wave vectors for a propagating electromagnetic wave (Dumelow, 2016):

$$\mathbf{k} \cdot \varepsilon \mathbf{E} = 0 \tag{3.144}$$

$$\mathbf{k} \cdot \mu \mathbf{H} = 0 \tag{3.145}$$

$$\mathbf{k} \times \mathbf{E} = \omega \mu_0 \mu \mathbf{H} \tag{3.146}$$

$$\mathbf{k} \times \mathbf{H} = -\omega \varepsilon_0 \varepsilon \mathbf{E} \tag{3.147}$$

Combining Eqs. (3.146) and (3.147), we obtain:

$$\mathbf{k} \times (\mathbf{k} \times \mathbf{E}) = -\omega^2 \varepsilon_0 \varepsilon \mu_0 \mu \mathbf{E} = -k_0^2 \varepsilon \mu \mathbf{E} \tag{3.148}$$

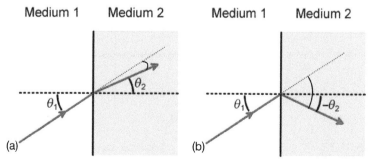

FIGURE 3.31 Ray diagram showing (a) positive refraction and (a) negative refraction, according to Eq. (3.143). Reprinted from T. Dumelow, (2016). *Negative Refraction and Imaging from Natural Crystals with Hyperbolic Dispersion*, in Solid State Physics, Volume 67, R. L. Stamps and R. Camley (Eds.), pp. 103–192, Elsevier, ©(2016), with permission from Elsevier.

where $k_0 = \omega\sqrt{\varepsilon_0\mu_0} = \omega/c$. Using the identity: $\mathbf{k} \times (\mathbf{k} \times \mathbf{E}) = (\mathbf{k} \cdot \mathbf{E})\mathbf{k} - k^2\mathbf{E}$, where for an isotropic medium the first term is zero we find:

$$-k^2\mathbf{E} = -k_0^2\varepsilon\mu\mathbf{E}, \quad\text{or}\quad k^2 = \varepsilon\mu k_0^2 \tag{3.149}$$

which, for the case of a negative refractive index material, yields:

$$k = -\sqrt{\varepsilon\mu}k_0 = -nk_0 \tag{3.150}$$

To understand the refractive behavior, we can compare the power flow (Poynting vector, $\mathbf{S} = \mathbf{E} \times \mathbf{H}$) and the wave vector. For positive values of ε and μ, we find that $\mathbf{k}$ and $\mathbf{S}$ will be parallel, while for negative ε and μ, $\mathbf{k}$ and $\mathbf{S}$ are antiparallel. This can be visualized in Figure 3.32. We note that the wave vector in the negatively refracting medium is in the opposite direction to that of the conventionally regarded ray direction.

A great deal of interest has been attracted by the imaging possibilities that are a consequence of the negative values of ε and μ. This was highlighted in 2000, when Pendry (2000) illustrated the lensing effects of a slab of negative refractive index metamaterial. This is schematically illustrated in Figure 3.33. The construction of this diagram uses the fact that we have $n < 0$ and is made using the Snell law as indicated in Eq. (3.143). Light from an ob-

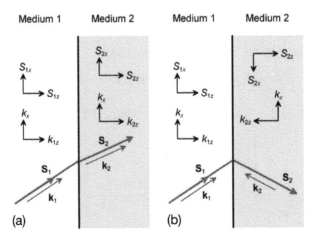

FIGURE 3.32 Wave vector and Poynting vector components for refraction (a) from a medium with positive ε and μ to another medium with positive ε and μ, and (b) from a medium with positive ε and μ to a medium with negative ε and μ. Reprinted from T. Dumelow, (2016). *Negative Refraction and Imaging from Natural Crystals with Hyperbolic Dispersion*, in Solid State Physics, Volume 67, R. L. Stamps and R. Camley (Eds.), pp. 103–192, Elsevier, ©(2016), with permission from Elsevier.

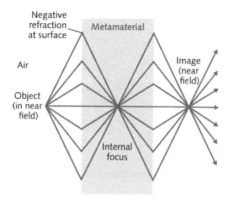

FIGURE 3.33 Ray optics diagram for the focusing effect due to a thin slab of negative refractive index material.

ject placed in front of the slab will be refracted, assuming the slab is not too thin, to a focal point within the slab, and then be refracted by the second surface to provide a further focal point or image outside the slab. This provides an aberration-free image. Pendry (2000) also pointed out that such a lens system would not suffer from the diffraction limit, as would a conventional positive refraction index lens.

In terms of materials, or metamaterials, that exhibit the negative refraction properties outlined above, this refers to artificially fabricated materials or composites, which have distinct properties as compared to the constituent materials, from which they are fabricated. Typically, these are artificial structures, which are globally treated as homogeneous in terms of parameters such as ε and μ. They are made up of an array of sub-wavelength elements, designed independently to respond preferentially to the electric or magnetic component of the electromagnetic wave.

The split-ring resonator (SRR) has been a common element used for response to the magnetic component of the electromagnetic field and was proposed by Pendry et al. (1999). A double SRR is illustrated in Figure 3.34 for circular and rectangular planar elements. Many other geometries have been tested and studied, from planar to 3D elements. These can be fabricated into regular arrays of elements, typically in a square type lattice. In its simplest form, we can represent the SRR as an equivalent LC resonator, where the values of L and C are determined by the specific geometry and dimensions of the object. If we apply a time-varying magnetic field in a direction perpendicular to the plane of the SRR, a circulating current will be induced, according to Faraday's law. Since there is a gap (split) in the SRR, this circulating current

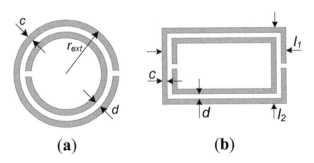

FIGURE 3.34   Double split-ring resonators with (a) circular and (b) rectangular planar geometries.

will cause a buildup of charge across the gap, with the energy being stored capacitively. The LC circuit equivalent will have a resonant response, where the resonance frequency is expressed as $\omega_0 \simeq 1/\sqrt{LC}$. The inductance arises from the current path in the SRR. For $\omega < \omega_0$, the current keeps phase with the driving field and the response is said to be positive. However, as the frequency increases, the currents can no longer keep up with the time-varying magnetic field and begin to lag behind it, which results in an out-of-phase, or negative response.

This result can be expressed in the form of the frequency-dependent effective permeability for the SRR, which can be written as:

$$\mu_{\text{eff}}(\omega) = 1 + \frac{F\omega}{\omega_0^2 - \omega^2 - i\Gamma_m\omega} \tag{3.151}$$

This is a generic form for the SRR, where the factor F is dependent on the SRR geometry. Γ is a resistive damping factor and will determine the linewidth of the resonant response as a function of frequency. The magnetic response, permeability, will have real and imaginary components. These will exhibit a resonance at the value indicated as ω_0. Since the dimensions of the object are chosen parameters in the fabrication process, we can essentially choose the value of resonance desired by adjusting the size of the SRR. A negative magnetic response can be achieved for high-Q resonators.

Electrically negative responses for the permittivity are well-known in metals at frequencies below the plasma frequency. Above this frequency, the metal becomes transparent. Bulk metals are not the only material, which exhibits a negative electric response. Such behavior is also observed in distributed arrays of conductors or even gratings on conductors. The general form of the frequency-dependent permittivity can be expressed as:

$$\varepsilon_{\text{eff}}(\omega) = 1 - \frac{\omega_p^2}{\omega^2 - \omega_0^2 - i\Gamma_e \omega} \qquad (3.152)$$

where ω_p denotes the plasma frequency, as expressed in Eq. (3.17). This means that we can manipulate the properties by adjusting the carrier density and effective mass of the carriers. Changes to geometry can be used to vary these parameters to obtain a negative response in the frequency of interest. A combined negative electric and magnetic response will, therefore, be achievable to obtain the desired outcome of negative effective refractive index over a restricted band of frequencies, as dictated by the geometry of the artificial structures.

Veselago (1968) pointed out that the condition of negative electric and magnetic responses, $\varepsilon < 0$ and $\mu < 0$ leading to $n < 0$, would lead to a number of new phenomena, such as reversed-phase velocity, reversed Doppler shift, radiation pressure becoming a radiation tension, converging lenses becoming diverging and vice versa. Since no naturally occurring negative refractive index materials were known at the time, much of his work was overlooked. However, in recent years, the development of metamaterials has allowed a new playground in materials development to flourish, with many new and exciting innovations. In Figure 3.35, we show a summary of different structures of differing dimensions and functioning over different frequency ranges.

3.6 Summary

The optical properties of materials arise from a number of effects. These can stem from transitions between electron states in the solid or from the free-electron system for conducting materials. Also, the dielectric properties have important effects on the optical properties of insulating materials. The spatial confinement of charge carriers has profound effects on their optical properties. There are further effects that can be exploited in periodic structures with alternating refractive indices, leading to photonic effects that parallel the electronic properties in crystalline materials. It is possible to tailor materials using a mixture of dielectric and metallic components to manufacture specific optical properties to perform specific functions and applications.

The transitions between electronic states have important effects on the optical properties of semiconducting materials. When these are formed into quantum wells, wires or dots, such as in heterostructure, the bound states of the system are profoundly altered. These bound states depend on a number

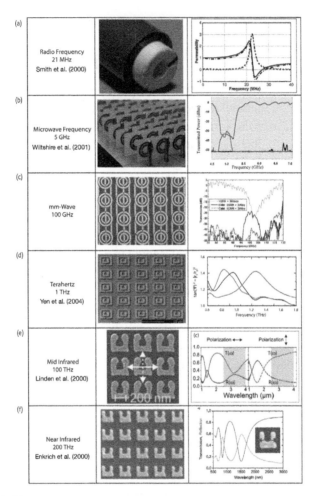

FIGURE 3.35 Summary of metamaterial (MM) results from RF to near optical frequencies. In the left column, we detail the frequency range, in which each MM was demonstrated and note the reference number in this review. The middle column shows a photo of the MM from each publication, and the third column shows some data detailing the MM response. The top row is an investigation of swiss-roll-type magnetic structures to guide magnetic flux in magnetic resonance imaging machines. The second row is the original, work in which NI materials were discovered at microwave frequencies. The third row shows some recent work on MMs at millimeter-wave frequencies. The next column details the first work extending MMs out of the microwave into the terahertz regime. The bottom two columns show the further extension of the SRR magnetic MM medium to MIR and NIR frequencies. Reprinted from W. J. Padilla, D. N. Basov & D. R. Smith, (2006). *Materials Today*, *9*, 28, ©(2006), with permission from Elsevier.

of factors, such as the effective masses of the charge carriers involved, but most importantly on the dimensions of the spatial confinement. This fact has been greatly exploited in the development of semiconductor heterostructures for many device applications. These include lasers, LEDs and optical detectors, to name a few. The transitions can concern both inter-band and inter-subband transitions, with the tailoring of the device dimensions to match the wavelengths of light required for specific applications.

By alternating materials with different dielectric properties in periodic structures, it is possible to manipulate the propagation of light, which can then be channeled and guided. Such systems are known as photonic bandgap materials, and can be ordered in one, two or three dimensions. The propagation of light can be determined from the application of the Maxwell equations, which illustrate that periodic structures naturally lead to band gaps in the dispersion relations for light. The parallel with the electronic band's structures in crystals is a useful analogy since the mathematical treatment is almost identical. For example, for the 1D case, the mathematics is virtually the same as for the Kronig–Penney model, including the Bloch theorem and boundary conditions. The band gaps can be tailored from the periodicity of the structures as well as the differences in the dielectric properties of the components. It is possible to introduce point and line defects in the structures and these can act as cavity states or as guided modes in the structure. It is also interesting to note that the colors of many insects and butterflies are derived from periodic structures which act as photonic crystals.

The optical properties of metallic materials differ greatly from those of dielectrics. This is due to the large population of free electrons in their crystalline lattices. These electrons respond to the electric field of incident electromagnetic radiation. This gives rise to the characteristic metallic, mirror-like appearance of metal surfaces. They tend to reflect a large proportion of the incident light, making them opaque at optical frequencies. In bulk materials, the excitations of the electron gas are referred to as plasmons. These can be bulk or surface localized. The latter being termed surface plasmon polaritons. For reduced dimensional metallic structures, the plasmon becomes localized, or localized surface plasmon resonance (LSPR), since the electrons in the particles will be confined by the ionic potential. This will alter the plasmon frequency, which is both size and shape-dependent. The electric field around such bodies can become locally enhanced, especially for closely spaced objects. This is referred to as a hot spot. There are a number of models available for the treatment of assemblies of nanoparticles, this approach requires a consideration of the interactions between the nanoparticles and can be expressed as an effective medium. The size and shape distributions of nanoparticle assemblies will also have an important influence on their plasmonic response. Plasmonics has found a number of practical applications in diverse areas, such as waveguides, light trapping for the enhancement of photovoltaic conversion, surface-enhanced Raman spectroscopy, and water treatment.

Negative refraction was postulated in the late 1960s and should occur for materials where both the permittivity and the permeability are negative quantities. No known materials are found to possess these two properties simultaneously. This phenomenon remained a curiosity until the late 1990s, when it was finally possible to artificially manufacture such media. These are known as metamaterials. A number of interesting consequences were noted and metamaterials have found a number of applications in optics, such as their use as perfect lenses, cloaking devices and as split-ring resonators.

3.7 Problems

(1) Show that the refractive index, of a non-magnetic material, is related to its dielectric constant via the following expression:

$$n^2 = \varepsilon_r$$

(2) Consider a 1D photonic bandgap material consisting of multilayers of two different media (with different dielectric constants, ε_1 and ε_2). Given that both layers have equal thickness d and that the complex field can be written as:

$$\mathbf{A}(\mathbf{r}) = A(z)e^{i(k_x x + k_y y)}\hat{\mathbf{n}}$$

where for TE modes, $\mathbf{A} = \mathbf{E}$ and $\hat{\mathbf{n}} = \hat{\mathbf{n}}_x$, while for TM modes, $\mathbf{A} = \mathbf{H}$ and $\hat{\mathbf{n}} = \hat{\mathbf{n}}_y$, and we also have:

$$A_{n,j}(z) = a_{n,j}e^{ik_{zj}(z-nd)} + b_{n,j}e^{-ik_{zj}(z-nd)}$$

show, that the characteristic equation defining the band structure has the form:

$$\cos(2kd) = \cos(k_{z1}d)\cos(k_{z2}d) - \frac{1}{2}\left(p_m + \frac{1}{p_m}\right)\sin(k_{z1}d)\sin(k_{z2}d)$$

where $p_m = p_{TE} = k_{z2}/k_{z1}$ and $p_m = p_{TM} = k_{z2}\varepsilon_1/k_{z1}\varepsilon_2$. Interpret this result. Note: You will need to apply the relevant boundary conditions along with the Flouquet â Bloch theorem.

(3) A typical tunnel junction is formed using an oxide layer of thickness 5 nm and dielectric constant $\varepsilon = 5$. Estimate the maximum area of the capacitor plates for Coulomb blockade to be observed at temperatures of

(a) 4.2 K

(b) 300 K

(4) The density of states as a function of energy for a quantum well (QW) has a staircase character, which can be represented as:

$$D_{2D}(E) = \frac{m^* S}{\pi h^2} \sum_{l_z} \Theta(E - e_{l_z})$$

where

$$\Theta(x) = \begin{cases} 1 & ; x > 0 \\ 0 & ; x < 0 \end{cases}$$

and $S = L_x \times L_y$ is the area. Using your knowledge of QW structures draw the density of states per unit area for a GaAs QW of thickness 15 nm. Note you will need to estimate the heights and positions of the steps. NB: $m^* = 0.0067 m_e$.

(5) Describe under what conditions an object will suffer changes to its intrinsic transport properties due to its physical dimensions.

(6) Consider the energy associated with electrons in a quantum dot. Show that the degenerate level (at $k = 0$) in a semiconductor between the heavy and light hole bands is lifted when a quantum dot is formed. (For the purposes of argument use a cubic form of quantum dot.) Show thus that the energy difference in the ground state is given by:

$$\Delta_{VB}^0 = \frac{3h^2\pi^2}{2a^2}\left(\frac{m_{hh} - m_{lh}}{m_{hh}m_{lh}}\right)$$

(7) Show that quantum effects become observable in the condition:

$$\Delta x \leq \frac{h}{\sqrt{mk_BT}}$$

(8) Consider a quantum dot connected to a voltage source V_a. Show that the potential difference across the tunnel junction (1, 2) can be expressed as:

$$V_{12} = \frac{V_a C_{2,1} \pm ne}{C_1 + C_2}$$

(9) Show that the contact resistance across a narrow conductor with M modes can be expressed as follows:

$$G_C^{-1} = \frac{h}{2e^2 M}$$

Prove that this formula will be modified to the following, when finite contacts are used, which have N modes:

$$G_C^{-1} = \frac{h}{2e^2} \left[\frac{1}{M} - \frac{1}{N} \right]$$

(10) A simple and elegant form of considering a multi-terminal device is via the Büttiker formula, which we can express as:

$$I_p = \sum_q [G_{qp} V_p - G_{pq} V_q]$$

where

$$G_{qp} = \frac{2e^2}{h} \bar{T}_{p \leftarrow q}$$

p and q refer to the index of the terminal and the backwards arrow indicates the direction of electron flow. The sum rule condition, that ensures the current is zero when all the potentials are equal, gives:

$$\sum_q G_{qp} = \sum_q G_{pq}$$

Consider a three-terminal device in which the potential on terminal three is zero and show that we can write:

$$\begin{pmatrix} I_1 \\ I_2 \end{pmatrix} = \begin{pmatrix} G_{12} + G_{13} & -G_{12} \\ -G_{21} & G_{21} + G_{23} \end{pmatrix} \begin{pmatrix} V_1 \\ V_2 \end{pmatrix}$$

Obtain the corresponding relation for a four-terminal device.

References and Further Reading

Atwater, H. A., & Polman, A., (2010). *Nat. Mater., 9*, 205.
Battie, Y., Resano-Garcia, A., Chaoui, N., Zhang, Y., & En Naciri, A. (2014). *J. Chem. Phys., 140*, 044705.

Battie, Y., Resano-Garcia, A., Chaoui, N., & En Naciri, A. (2015). *Phys. Stat. Sol. c, 12*, 142.

Bohren, C. F., & Huffman, D. R., (1983). *Absorption and Scattering of Light by Small Particles*. John Wiley and Sons, Inc., New York.

Brongersma, M. L., Hartman, J. W., & Atwater, H. A., (2000). *Phys. Rev. B, 62*, R16356.

Davis, T. J., Gómez, D. E. & Roberts, A., (2017). *Nanophotonics, 6* 543.

Dumelow, T., (2016). *Negative Refraction and Imaging from Natural Crystals with Hyperbolic Dispersion*, in Solid State Physics, Volume 67, Stamps, R. L., & R. Camley (Eds.), pp. 103–192, Elsevier.

C. Enrich et al., (2005). *Phys. Rev. Lett., 95*, 203901.

Fang, H., Bechtel, H. A., Plis, E., Martin, M. C., Krishna, S., Yablonovitch, E., & Javey, A., (2013). Quantum of optical absorption in two-dimensional semiconductors. *PNAS, 110*(29), 11688–11691; https://doi.org/10.1073/pnas.1309563110 (accessed on 30 March 2020).

Ferre Llin, L., & Paul, D. J., (2017). *Thermoelectrics, Photovoltaics and Thermal Photovoltaics for Powering ICT Devices and Systems, ICT – Energy Concepts for Energy Efficiency and Sustainability*, G. Fagas (Ed.), InTech, DOI: 10.5772/65983.

Fox, M., (2001). *Optical Properties of Solids*, Oxford University Press, Oxford.

Fujishima, A., & Honda, K., (1972). *Nature, 238* 37.

Hayashi, S., & Okamoto, T., (2012). *J. Phys. D: Appl. Phys., 45* 433001.

Huang, X., Neretina, S., & El-Sayed, M. A., (2009). *Adv. Mater., 21* 4880.

Jackson, J. D., (1999). *Classical Electrodynamics*. 3rd edition. John Wiley and Sons, Inc., New York, NY.

Joannopoulos, J. D., Villeneuve, P. R., & Fan, S., (1997). *Nature, 386*, p.143–149.

Keita, A.-S., & En Naciri, A. (2011). *Phys. Rev. B 84*, 125436.

Krenn, J. R., Lamprecht, B., Ditlbacher, H., Schider, G., Salerno, M., Leitner, A., & Aussenegg, F. R., (2002). *Europhys. Lett., 60*, 663–669.

Kretschmann, E., & Raether, H., (1968). Zeitschrift für Naturforschung A, *23*, 2135.

Lang, X., Chen, X., & Zhao, J., (2014). Chem. Soc. Rev., *43*, 473.

Linden, S. et al., (2004). *Science, 306*, 1351.

Lu, X., Rycenga, M., Skrabalak, S. E., Wiley, B., & Xia, Y., (2009). Ann. Rev. Phys. Chem., *60*, 167.

Maier, S. A., Brongersma, M. L., Kik, P. G., Meltzer, S., Requicha, A. A. G., & Atwater, H. A., (2001). *Advanced Materials, 13,* 1501.

Maier, S. A., (2007). *Plasmonics: Fundamentals and Applications,* Springer, New York.

Markel, V. A., (2016). *J. Opt. Soc. Am. A, 33* 1244.

Murray, W. A., & Barnes, W. L., (2007). *Adv. Mater., 19* 3771–3782.

Nadal, E., Barros, N., H. Glénat, Laverdant, J., Schmool, D. S., & Kachkachi, H., (2017). *J. Mater. Chem. C, 5,* 3553.

Otto, A., (1968). *Zeitschrift für Physik, 216* 398.

Padilla, W. J., Basov, D. N., & Smith, D. R., (2006). *Materials Today, 9,* 28.

Peiris, S., McMurtie, J., & H.-Zhu, Y., (2016). *Catal. Sci. Technol., 6* 320.

Pendry, J. B., Holden, A. J., Robbins, D. J., Stewart, W. J., (1999). *IEEE Trans. Microwave Theory Tech., 47,* 2075.

Pendry, J. B., (2000). *Phys. Rev. Lett., 85* 3966.

Petryayeva, E., & Krull, U. J., (2011). *Anal. Chim. Acta, 706,* 8.

Powell, C. J., & Swann, J. B., (1959). *Phys. Rev., 115,* 869–875, and *116,* 81–83.

Purcell, E. M., & Pennypacker, C. R., (1973). *Astrophys. J., 186* 705.

Smith, D. R., Padilla, W. J., Vier, D. C., Nemat-Nasser, S. C., & Schultz, S., (2000). *Phys. Rev. Lett., 84* 4184.

Stockman, M. I., (2011). *Opt. Express, 19* 22029–22106.

Stockman, M. I., (2011). *Physics Today, 64,* 39.

Veselago, V. G., (1968). *Sov. Phys. Usp., 10,* 509.

Vukusic, P., & Sambles, J. R., (2003). *Nature, 424,* 852–855.

Wang, B., & Gallais, L., (2013). *Opt. Exp., 21,* 14698.

Webber, M. J., (2003). *Handbook of Optical Materials,* CRC Press, Boca Raton.

Weiner, J., & Nunes, F., (2013). *Light–Matter Interaction: Physics and Engineering at the Nanoscale,* Oxford University Press, Oxford.

Willets, K. A., & van Duyne, R. P. (2007). *Ann. Rev. Phys. Chem., 58,* 267–297.

Wiltshire, M. C. K. et al., (2001). *Science, 291,* 849.

Wurtz, G. A., Pollard, R. J., Dickson, W., Zayats, A. V., (2016). Optics of Metallic Nanostructures, in *Reference Module in Materials Science and Materials Engineering,* S. Hashmi (Ed.), Elsevier, *4* 67.

Yen, T. J. et al., (2004). *Science, 303,* 1494.

Chapter 4

Nanomagnetism

4.1 Introduction: Ferromagnetism and Magnetic Materials

Before we consider the specific aspects of magnetism and magnetic materials at the nanometer scale, it is instructive to summarize some of the principal concepts in magnetism, which will be required for our discussion of nanomagnetism.

We will start with the definition of magnetization. Consider a magnetized body of volume V with a dipole moment given by m, with units $A\,m^2$. The magnetization of the body is expressed as the magnetic moment per unit volume: $M = m/V$ and can be expressed in units of $A\,m^{-1}$. The dipole moment can be examined by placing the magnetized object into an applied magnetic field, H. The energy of interaction between the magnet and the applied field is termed the *Zeeman energy*, and is given by:

$$E = -\mu_0 \mathbf{m} \cdot \mathbf{H} \tag{4.1}$$

Here, we have taken care to express the magnetic moment and the magnetic field in terms of their vector quantities, since the relative directions will have a bearing on the interaction between the two physical quantities. It is fairly common in magnetics to express the energy density, which we write in the form:

$$\mathcal{E} = \frac{E}{V} = -\mu_0 \mathbf{M} \cdot \mathbf{H} \tag{4.2}$$

In the simplest case, we may consider the needle of a compass in a uniform magnetic field. The magnetic energy can be expressed as:

$$E = -\mu_0 m H \cos \theta \tag{4.3}$$

where θ represents the angle between the magnetic field direction and the compass needle, or more correctly the orientation of its magnetization, which should normally lie along the long axis due to magnetostatic energy considerations. The mechanical torque induced from the magnetic field on the compass needle takes the form:

$$\Gamma = -\frac{dE}{d\theta} = -\mu_0 mH \sin\theta = -\mu_0 \mathbf{M} \times \mathbf{H} \tag{4.4}$$

The lowest energy will be obtained for $\theta = 0$, i.e., when $\mathbf{M}$ is aligned along the applied field, $\mathbf{H}$. In this case, there will be no effective torque created on the magnetic compass needle.

For large magnets, the above description should be modified to take into account any inhomogeneities in the magnetic field and magnetization. In this case, we write that energy in the form:

$$E = -\mu_0 \int \mathbf{M}(\mathbf{r}) \cdot \mathbf{H}(\mathbf{r}) dV \tag{4.5}$$

Or we can represent this in terms of the sum over the magnetic moments, $\mathbf{m}_i = \mathbf{m}(\mathbf{r}_i)$;

$$E = -\mu_0 \sum_i \mathbf{m}_i \cdot \mathbf{H}(\mathbf{r}_i) \tag{4.6}$$

The choice of energy expression will depend on the individual system under consideration.

One of the most common problems faced in magnetism is the determination of the magnetization of a sample and its response to an applied magnetic field. From the Maxwell equations, we can express the magnetic flux density (or induction) in terms of the magnetization and the applied magnetic field as:

$$\mathbf{B} = \mu_0(\mathbf{H} + \mathbf{M}) \tag{4.7}$$

For magnetic materials, it is common to consider the magnetic susceptibility, which is defined as:

$$\chi = \frac{M}{H} \tag{4.8}$$

For small values of magnetization, M, the equation of state, $M = M(H)$ can be linearized and a basic classification can be made for paramagnetic ($\chi > 0$) and diamagnetic ($\chi < 0$) materials, which for low applied magnetic fields gives a more or less constant value of the magnetic susceptibility. However, for most cases of interest concerning *magnetic* materials, it is inappropriate to consider χ to be a constant. Using Eq. (4.7), it is possible to provide a first description of the magnetic behavior of a body, where we can incorporate the magnetic susceptibilities:

$$\mathbf{B} = \mu_0 \mathbf{H}(1 + \chi) = \mu_0 \mu_r \mathbf{H} = \mu \mathbf{H} \qquad (4.9)$$

where the relative permeability is $\mu_r = 1 + \chi$. Clearly, for the case where the magnetic signal is weak or negligible, the above relation reduces to $\mathbf{B} = \mu_0 \mathbf{H}$. For a magnetic body, the relationship between the magnetization and the applied magnetic field is non-linear and approaches a finite saturation of magnetization at some elevated field value, $M \rightarrow M_S$. A more precise definition of the magnetic susceptibility can be expressed in the form of the differential magnetic susceptibility at zero applied field:

$$\chi = \left(\frac{dM}{dH} \right)_{H=0} \qquad (4.10)$$

Furthermore, M is not in general a unique function of the applied field. The typical magnetic behavior, as illustrated in Figure 4.1, is described as magnetic hysteresis. The main parameters of interest are shown in the loop; the coercive field, H_C, the remnant magnetization, M_R, the saturation magnetization M_S and the field at saturation, H_{Sat}. These parameters define the *hysteresis loop*. The phenomenon of hysteresis arises from the motion of magnetic domain walls and is intimately related to the magnetic anisotropies in the solid. The specific magnetic state (magnetization and micro magnetic susceptibility, $\partial M / \partial H$) of a sample depends on the sample history. Maxwell's equation do not describe the magnetic hysteresis behavior.

Atomically, the magnetism in solids originates, almost exclusively, from the atomic electrons of the component atoms. The nuclear moments contribute very little to the sample magnetization, but are important for example in resonance imaging and hyperfine interactions. When we talk of saturation, this means that all the available atomic moments are aligned parallel to the applied magnetic field. As a general rule, one electron per atom will contribute an atomic moment of one Bohr magneton, as defined by: $\mu_B = e\hbar / 2m_e = 9.274 \times 10^{-24} \text{A m}^2$, and a magnetization of around one Tesla ($\mu_0 M = 1$ T). Iron, for example, has a magnetization of 2.15 T at room temperature and each atom contributes a moment of about 2.2 μ_B. This means that of the 26 electrons associated with the Fe atom, only about 2 of these contribute to its magnetic moment.

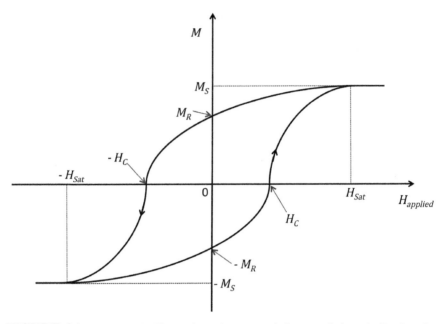

FIGURE 4.1   Schematic illustration of a magnetic hysteresis loop indicating the magnetization as a function of the applied static external field.

To describe the origin of the magnetic moment, we can use a simple model which considers the orbital motion of the electron around the atomic nucleus. Such motion can be considered to be equivalent to a current carrying loop or coil, which can be described by Ampere's law:

$$\oint \mathbf{H} \cdot \mathbf{dr} = NI \tag{4.11}$$

where I is the current in the coil of N windings. Performing the integral around the loop allows us to evaluate the relation between the magnetic field produced by the current in the coil. Considering now the electron orbit as a current in a loop, we can write this current in the form:

$$I = -\frac{e}{T} = -\frac{ev}{2\pi R} \tag{4.12}$$

where v is the velocity of the electron in a circular orbit of radius, R. Since the magnetic moment created by a current carrying coil can be expressed as: $\mathbf{m} = I\mathbf{A}$, where $\mathbf{A}$ defines the vector normal to the plane of the loop of area A, i.e., $\mathbf{A} = A\hat{n}$, where $\hat{n}$ is the unit vector normal to the plane. Thus we can write:

$$m = -\frac{evR}{2}\hat{n} \tag{4.13}$$

The Bohr model provides a wave-like description of the properties of the electron and only specific orbits are permitted. This leads to the quantization of the energies for atomic electrons with a characteristic orbital angular momentum, which is expressed as:

$$L = m_e vR \tag{4.14}$$

where m_e is the rest mass of the electron. Expressing the discrete nature of the orbital angular momentum as $L = n\hbar$, we obtain:

$$m = -\frac{e\hbar}{2m_e}n\hat{n} = \mu_B n\hat{n} \tag{4.15}$$

which demonstrates the derivation of the Bohr magneton, μ_b. Eq. (4.15) shows that for free atoms, the orbital magnetic moment should be quantized with integer values of the Bohr magneton.

One of the shortcomings of the current loop model is that the current loop can be readily destroyed by electrostatic interactions from neighboring atoms in a solid. For example, only about 5% of the 2.2 μ_B moment in Fe is due to such currents. In such cases, the orbital moment is said to be *quenched* by the crystal field. The remaining 95% of the Fe moment is due to the spin angular momentum of the electron. The spin is unrelated to the orbital motion of the electron and survives the crystal field. In magnetism, the moment is frequently referred to as the spin, which is quantized into spin-up (↑) and spin-down (↓) states, with each spin contributing about 1 μ_B. We generally express the orbital and spin magnetic moments in the form:

$$m_l = -g_l\frac{\mu_B}{\hbar}l \tag{4.16}$$

and

$$m_s = -g_s\frac{\mu_B}{\hbar}s \tag{4.17}$$

Here, l and s denote the quantized orbital and spin motion, respectively. The constants g_l and g_s refer to the Landé or g-factor, as discussed below. In general, we take $g_l = 1$ and $g_s = 2$.

An atom will generally contain several interacting electrons. In an isolated atom, these electrons fill the available states, as dictated by *Hund's rules*, in the ground state configuration. This empirical set of rules can be expressed, in order of importance, as:

- Subject to the Pauli exclusion principle, the total spin:

$$S = \sum_i s_i \qquad (4.18)$$

  is maximized.
- Subject to the Pauli exclusion principle and Hund's first rule, the total orbital momentum:

$$L = \sum_i l_i \qquad (4.19)$$

  is maximized.
- Subject to the first two Hund's rules, **L** and **S** couple according to:

$$J = |L + S|; \qquad \text{for electron shells that are more than half-filled} \qquad (4.20)$$

  and

$$J = |L - S|; \qquad \text{for electron shells that are less than half-filled} \qquad (4.21)$$

The total magnetic moment of the atom is given by:

$$\mu_{tot} = -g_J \frac{\mu_B}{\hbar} J \qquad (4.22)$$

where g_J is the Landé factor, which can be written as:

$$g_J = \frac{3}{2} + \frac{S(S+1) - L(L+1)}{2J(J+1)} \qquad (4.23)$$

We note from this equation, for purely orbital motion we obtain $g_J = g_l = 1$, since $J = L$, and for purely spin motion (quenched orbital motion), $g_J = g_s = 2$ and $J = S$.

The Hund's rules presume so-called L–S or *Russel–Saunders* coupling in which we have $J = L + S$. However, for larger atoms with more electrons, a more accurate description is given by jj coupling, where $J = \sum_i = \sum_i (l_i + s_i)$.

While atoms may possess magnetic moments in isolation, in a solid the situation is far from simple. The organization of the magnetic moments in a solid can be very complex and depend on several factors. A large majority of solids display only weak responses to applied magnetic fields, these correspond to diamagnetic and paramagnetic materials. In the former, the constituent atoms of the solid will have filled electron shells, which have no net magnetic moment, and the effect of a magnetic field will be to perturb the orbital motion of the electrons creating a small net magnetic moment with an orientation opposed to that of the applied magnetic field. In a simple classical model, this can be envisaged as a manifestation of Lenz's law. For paramagnetic materials, the electron shells are not completely filled and there will be a net atomic magnetic moment. Since there is no effective orientational preference for the magnetic moments in a paramagnetic material, the net magnetic moment of the solid will be zero. If we apply an external magnetic field to the solid, the individual moments will gradually tend to align in the direction of the applied field. In either, case a small magnetization will be generated which reversibly disappears once the field is removed. Frequently, such materials will be characterized by a small negative magnetic susceptibility for diamagnetic materials and a small positive magnetic susceptibility for paramagnetic solids. For large magnetic fields, the individual moments are fully aligned and saturation occurs. This situation is described by the so-called *Langevin function*. The effects of temperature will essentially be to randomize the moments and are described by the *Curie law*.

When we discuss materials, which we describe as magnetic, we generally mean that the material has a large magnetization, which arises from some form of the internal ordering of the moments, which we describe as an *exchange interaction*. Exchange interactions are of many different origins and will depend on the nature of the constituent atoms and their spatial organization. A full description of such effects is beyond the scope of this brief introduction to magnetism and the interested reader is referred to a textbook on the subject; see References at the end of the chapter for some examples. A simple approach to this problem would be to introduce a spin-dependent Hamiltonian to describe the energy between neighboring spins, such as introduced by Heisenberg and Dirac and can be represented as:

$$\mathcal{H}_{ij} = -2J\mathbf{S}_1 \cdot \mathbf{S}_2 \tag{4.24}$$

Here, the interaction between the two spins, labeled as 1 and 2, is represented by the constant (exchange integral) J. This can be positive or negative. In the former case, the lowest energy state will be obtained for spins which are parallel, as is the case for ferromagnetic coupling, while in the latter case, the ground state is obtained for anti-parallel spins, which describes an anti-ferromagnetic coupling between the spins. Extending this description to a solid, the Heisenberg–Dirac Hamiltonian takes the form:

$$\mathcal{H}_{ij} = -\sum_{ij} J_{ij} \mathbf{S}_i \cdot \mathbf{S}_j \tag{4.25}$$

The exchange integral J_{ij} refers to spins i and j. Some approaches express this equation as:

$$\mathcal{H}_{ij} = -2 \sum_{i>j} J_{ij} \mathbf{S}_i \cdot \mathbf{S}_j \tag{4.26}$$

which avoids counting the interaction twice and thus factor 2 must be introduced. For many cases, only nearest-neighbor interactions are of importance, and the exchange integral can be considered to be a constant for the solid. In this case, we can write:

$$\mathcal{H}_{ij} = -J \sum_{ij} \mathbf{S}_i \cdot \mathbf{S}_j \tag{4.27}$$

The existence of an interaction between spins implicitly indicates that there will be some form of ordering between the spins. In the simplest case, this refers to the alignment of all spins in a full ferromagnetic order. A negative interaction leads to antiferromagnetic ordering, which leads to a zero net magnetization for the solid. This is often considered as dividing the spins into two oppositely oriented magnetic sub-lattices. In cases where there is more than one type of magnetic moment in the solid, a negative exchange interaction can lead to ferrimagnetic ordering. As a result of oppositely aligned sub-lattices, the moment on each lattice will be different, leading to a net magnetization. More complex interactions over several lattice spacings can further lead to canted ferromagnetism and helical structures.

In general, the ordering in spin systems leads to cooperative effects mediated by the exchange interaction. As stated above, the ground state for a ferromagnetic system, at zero temperature, will be that in which all spins are perfectly aligned, in say the z-direction. We can denote this state as $|\Phi\rangle$. In the simplest case for a 1D chain of spins, the spin Hamiltonian, for the Heisenberg model can be expressed as:

$$\mathcal{H}_{ij} = -2J \sum_i \left[\hat{S}_i^z \hat{S}_{i+1}^z + \frac{1}{2}(\hat{S}_i^+ \hat{S}_{i+1}^- + \hat{S}_i^- \hat{S}_{i+1}^+) \right] \tag{4.28}$$

where $\hat{S}^{\pm} = \hat{S}_x \pm i\hat{S}_y$ are the so-called raising and lowering operators. Using this form of the Hamiltonian, we can express the ground state as:

$$\mathcal{H}|\Phi\rangle = -NS^2 J |\Phi\rangle \tag{4.29}$$

where N being the number of spins in the chain. As described above the state $|\Phi\rangle$ can be represented as $|\uparrow\uparrow\uparrow \cdots \uparrow\uparrow \cdots\rangle$. To create an excitation in such a system, we can consider the state where just a single spin is reversed, at say site j, which we represent as: $|j\rangle = |\uparrow\uparrow\uparrow \cdots \uparrow\downarrow\uparrow \cdots\rangle = \hat{S}_j^- |\Phi\rangle$. By flipping the spin at site j, we have effectively changed the total spin of the system by $(1/2 - -1/2) = 1$. The excitation has integer spin, and is thus a boson. Applying the above Hamiltonian, Eq. (4.28), to the state $|j\rangle$, we obtain:

$$\mathscr{H}|j\rangle = 2[(-NS^2J + 2SJ)|j\rangle - SJ|j+1\rangle - SJ|j-1\rangle] \tag{4.30}$$

This is not an eigenstate of the Hamiltonian. To find the first excited state of the system, we must diagonalize the Hamiltonian by looking for solutions of a plane wave of the form:

$$|q\rangle = \frac{1}{\sqrt{N}} \sum_j e^{i\mathbf{q}\cdot\mathbf{R}_j}|j\rangle = \sum_j c_j|j\rangle \tag{4.31}$$

The state $|q\rangle$ is essentially a delocalized flipped spin, which is spread out over all the spins in the system. This will significantly reduce the exchange energy for the excitation. Since $|q\rangle$ is composed of a linear combination of states representing a single flipped spin, the total spin for the state $|q\rangle$ must be $NS - 1$. It is reasonably straight forward to demonstrate that:

$$\mathscr{H}|q\rangle = E(q)|q\rangle \tag{4.32}$$

where

$$E(q) = \frac{-NS}{2} + 2J[1 - \cos(qa)] \tag{4.33}$$

Therefore, the energy of this excitation can be expressed as:

$$\hbar\omega_q = 2J[1 - \cos(qa)] \tag{4.34}$$

The 1D dispersion curve for magnons from Eq. (4.34) is illustrated in Figure 4.2.

We can extend this result to three dimensions, which is given by:

$$\hbar\omega_q = J[z - \sum_m \cos(\mathbf{q}\cdot\mathbf{a}_m)] \tag{4.35}$$

where $\sum_m$ is over nearest neighbor vectors $\mathbf{a}_m$ and z is the coordination number. For the case of small q, we can approximate the above expression to :

$$\hbar\omega_q \simeq J\left[z - \sum_m \left(1 - \frac{(\mathbf{q}\cdot\mathbf{a}_m)^2}{2} + \cdots\right)\right] = \frac{J}{2}a^2q^2 \sum_m \cos^2\theta_m \tag{4.36}$$

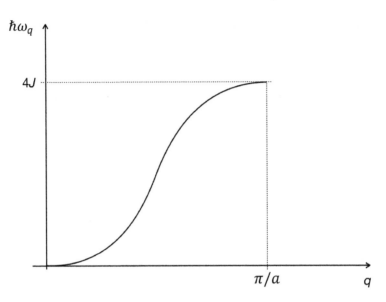

FIGURE 4.2   One-dimensional magnon dispersion curve.

Here θ_m, is the angle between $\mathbf{q}$ and $\mathbf{a}_m$. For the case of small angles, we can further simplify the expression as:

$$\hbar\omega_q \simeq \frac{Ja^2}{2}q^2 = Dq^2 \tag{4.37}$$

For the situation where an external magnetic field is applied, the magnon energy can be expressed in the form:

$$\hbar\omega_q = Dq^2 + g\mu_B\mu_0 H \tag{4.38}$$

The dispersion relation for this equation is illustrated in Figure 4.3 for the case with and without an externally applied magnetic field.

The result above can be used to obtain the temperature variation of the magnetization. This is done first by considering the density of states for the magnons or spin waves, which we can express as:

$$g(q)dq \propto q^2 dq \tag{4.39}$$

and which can also be represented as:

$$g(\omega)d\omega \propto \sqrt{\omega}d\omega \tag{4.40}$$

This has a similar form to that for electrons. The above expression is valid for low temperatures, where only small q and small ω are of importance. The number of magnon modes excited at a temperature T, n_{magn}, can be

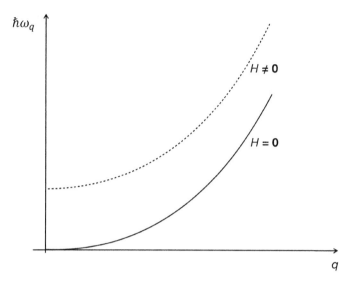

FIGURE 4.3 Three-dimensional magnon dispersion curve with (dashed line) and without (full line) an applied external magnetic field.

evaluated by integrating the magnon density of states over all frequencies after multiplying the Bose-Einstein factor $[e^{\hbar\omega_q/k_BT} - 1]^{-1}$, since as we saw above, magnons are bosons. From this, we find:

$$n_{magn} = \int_0^\infty \frac{g(\omega_q)d\omega_q}{(e^{\hbar\omega_q/k_BT} - 1)} \tag{4.41}$$

Substituting for $x = \hbar\omega_q/k_BT$, we obtain:

$$n_{magn} = \left(\frac{k_BT}{\hbar}\right)^{3/2} \int_0^\infty \frac{\sqrt{x}dx}{e^x - 1} = \left(\frac{k_BT}{\hbar}\right)^{3/2} \frac{\sqrt{\pi}}{2}\zeta(3/2) \tag{4.42}$$

where $\zeta(3/2)$ is the Riemann zeta-function and is independent of temperature and for our purposes a constant. It is thus possible to show that:

$$\frac{M(0) - M(T)}{M(0)} \propto T^{3/2} \tag{4.43}$$

This result is known as the *Bloch $T^{3/2}$ law* and shows a good approximation to experiments at low temperatures. As a rule of thumb, the Bloch $T^{3/2}$ law is generally applicable up to around $T \simeq T_C/2$, with T_C being the order temperature or *Curie temperature*. Above this, the variation of $M(T)$ follows the form $(T_C - T)^\beta$, where β is referred to as the critical exponent, which is usually obtained from a fit to experimental data.

It is worth noting, that the description of ferromagnetic can be based on the band model of metals, for example, the Stoner model of ferromagnetism, where the exchange energy is treated as a mean-field. In this approach, the exchange energy creates a magnetic splitting of the spin-up and spin-down energy bands. This gives rise to a net magnetization, which can be expressed as the difference in the electron densities:

$$M = (n_\uparrow - n_\downarrow)\mu_B \qquad (4.44)$$

Here, $n_\uparrow$ and $n_\downarrow$ are the electron densities for the spin-up ($\uparrow$) and the spin-down ($\downarrow$) bands. This model is frequently used to describe the magnetism in transition metals, such as Fe, Ni, Co, and their alloys. In these metallic systems, the magnetism is referred to as *itinerant*, and arises from the free conduction electrons in the solid.

We noted earlier that the ferromagnetic state implies an alignment between the atomic magnetic moments in the solid. A logical consequence of this is that there must be a preferential orientation of the magnetization with respect to the crystalline axes in the solid. Indeed, this orientation is related to the spin–orbit interaction and has a specific direction, which is strongly correlated to the crystalline structure of a particular atomic species. In short, this orientational preference of the magnetization is referred to as the *magnetocrystalline anisotropy*. Common forms of magneto-crystalline anisotropy are uniaxial (Co) and cubic (Fe and Ni) and are characterized by the strength of the anisotropy constant(s). Other forms of magnetic anisotropy also exist, most common of these is the shape anisotropy, which arises from magnetostatic considerations and originating in a dipolar interaction.

There is one final topic that will also be of use in the following sections and this concerns the existence of magnetic domains. Since the shape of a sample can have a large bearing on the anisotropy of a magnetic sample, the "division" of the magnet into regions of spontaneous magnetization, M_S, with different directions of saturation can reduce the overall energy of the system. This essentially means that the magnetostatic energy of the system can be reduced by the formation of magnetic domains. This is schematically illustrated in Figure 4.4. When the magnetic object, a rectangular shape in this example, is fully magnetized along one axis a stray magnetic field exists, Figure 4.4(a). This stray field creates a magnetic field around the object, which tends to oppose the direction of the magnetization which creates it. This field is called the *demagnetizing field*, H_d. To maintain the magnetization in this state costs magnetic energy. To reduce this energy cost, a two-domain system can be envisaged, where the magnetic body is divided into equivalent magnetic domains, each with a magnetization aligned in opposite directions, Figure 4.4(b). The formation of magnetic domains will, however, have its own energy costs, which arise from the increased exchange energy associated with the transition region between the domains, called *domain walls*. We will discuss this issue below. What is important, is to determine whether the energy of the formation of the domains is less than that of the magnetostatic energy. The system will invariably, in the absence of other external constraints, configure itself in the state with the lowest overall energy.

Frequently, a ferromagnetic sample will exhibit a net zero magnetization in the absence of an applied magnetic field due to the formation of magnetic domains. This is indeed the situation presented in Figure 4.4(b). A further reduction of magnetostatic energy can be achieved for the case of magnetic domain closure, as illustrated in Figure 4.4(c). Once again, we see that $\sum_i M_i = 0$, where i is the domain index, for this domain configuration and the magnetostatic energy will be zero. Offset on this energy "gain" will be the cost of the increased formation of the domain walls for this magnetic configuration. We should also note that the formation of domains and do-

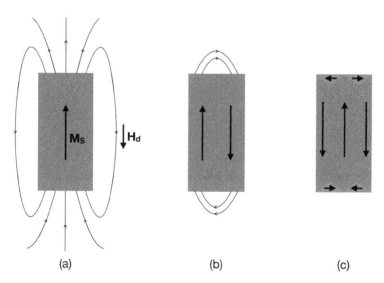

(a)   (b)   (c)

FIGURE 4.4 (a) Illustration of the stray magnetic field created by a permanent magnetic fully magnetized along one direction. (b) The formation of two magnetic domains, with opposing directions of magnetization will significantly reduce the stray field and hence the magnetostatic energy of the magnetic system. (c) When domain closure exists, there is no stray field and hence no additional magnetostatic energy associated with the system.

main walls is intimately related to the magnetocrystalline anisotropy of the ferromagnetic sample. For example, in Figure 4.4(b), the sample must have at least a uniaxial anisotropy, which will permit the sample to be magnetized along with the two opposite directions. For the example illustrated in Figure 4.4(c), the sample would require some form of cubic anisotropy.

The magnetization process can be understood from the consideration of the evolution of the domain state of the sample as a magnetic field is applied. This can be rather complex and will implicitly depend on the energy considerations of the particular sample in question. In general, we can understand the initial magnetization of the sample from the effect of the magnetic field on the sample, which will be generally to increase the size of those magnetic domains with an orientation favorable to that of the applied field direction. Such domains will increase in size at the expense of those with energetically unfavorable directions. This is usually achieved via the displacement of the magnetic domain walls. The effect of the movement of the domain walls will be to increase the magnetization state of the sample. This state will, in fact, be a detailed balance of the various magnetic energies in the sample, including the Zeeman energy associated with the applied magnetic field. Once the field is sufficiently large, only one magnetic domain will remain and if this is not aligned in the direction of the magnetic field, then a further increase of the magnetic field will be required to arrive at full magnetic saturation. If the magnetic field is removed, the magnetic state will relax, but not into the original virgin state with zero magnetization. A remnant magnetic state usually remains and the overall magnetization of the sample will be given by M_R, see Figure 4.1.

In Figure 4.4(c), we indicated a form of domain closure for a rectangular-shaped sample. It will be noted that the domain walls will be of two types, those along the length of the sample, which are 180°, i.e., 180° between the magnetizations of the domains separated by the wall, and those which are 90° domain walls, that we find at the top and bottom of the sample. Further to this, domain walls can also be distinguished as to whether the rotation of the magnetic moments in the wall is parallel or perpendicular to the wall itself, see Figure 4.5.

As we mentioned earlier, there is an energy cost in the exchange energy if one spin is rotated with respect to the other in a ferromagnet. Indeed, this is illustrated from the exchange Hamiltonian, as expressed in Eq. (4.24). For two spins, $\mathbf{S}_1$, $\mathbf{S}_2$, which are at an angle θ with respect to one another, the energy can be expressed as $-2J\mathbf{S}_1 \cdot \mathbf{S}_2 = -2JS^2 \cos\theta$. For $\theta = 0$, this

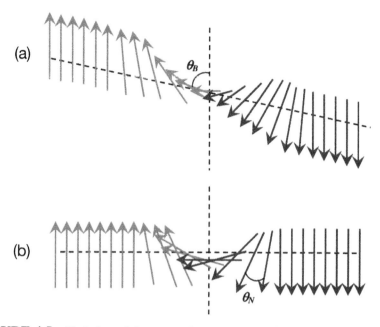

FIGURE 4.5 Variation of the magnetic moment as a function of position along an axis between two magnetic domains. In the case where the moment varies in a direction parallel to the domain wall, we have a Bloch wall (a), while for the case where the moment varies in a direction perpendicular to the wall, we have a Néel wall (b).

energy will be $-2JS^2$. For a small angle θ, $(\theta \ll 1)$, the cost of having $\theta \neq 0$ is roughly $JS^2\theta^2$. For the most common type of domain wall, the Bloch wall, we can consider that the region of transition contains n spins over a certain distance. This is called the domain wall width, with the domain wall energy being shared among the spins across the interface. For a 180° wall, also referred to as a π wall, the energy cost for the wall will be N times $JS^2\theta^2$, where $\theta = \pi/N$, and thus can be expressed as: $JS^2\pi^2/N^2$. Summing over the N spins in the wall width thus gives a total energy of $JS^2\pi^2/N$. In a Bloch wall, we have planes of spins and we are principally interested in the energy per unit area of the wall, σ_{BW}. In a square meter of the wall, there will be $1/a^2$ lines of spins, like the one considered above. Note, a is the lattice parameter. This allows us to determine the wall energy per unit area as:

$$\sigma_{BW} = \frac{JS^2\pi^2}{Na^2} \tag{4.45}$$

This energy will tend to zero as $N \to \infty$. This would seem to indicate that if a domain wall is formed, it will tend to unwind itself and stretch out as far as possible to minimize the energy. To prevent this occurring, the spins need to be anchored in each of the domains. It is, in fact, the magnetic anisotropy which does the anchoring.

Magnetic anisotropy, whether in the form of magnetocrystalline anisotropy or shape anisotropy, will define the orientation of the magnetization vector within the ferromagnet. We usually define an easy axis, or more correctly, a direction of easy magnetization and a hard axis. This means that for a magnetic field applied to the sample, there will be a direction in which it is easier to saturate the sample and a direction in which it is harder to saturate the sample. This refers to the size of the applied magnetic field necessary to bring about saturation. Many experimental techniques are available to measure the magnetic anisotropy, and from this, we can determine the type and strength of the magnetic anisotropy. The more common methods of measurement are standard magnetization measurements such as VSM (vibration sample magnetometry) and SQUID (superconducting quantum interference device) magnetometry, as well as more sophisticated methods such as ferromagnetic resonance (FMR). Once obtained, it is useful to define a functional or phenomenological form of anisotropy energy. This is usually done by considering the directional cosines with respect to the principal axes of the system. For the simplest case of uniaxial anisotropy, we can express this as:

$$E_{Ku} = K_{u1} \sin^2 \theta + K_{u2} \sin^4 \theta + \dots \qquad (4.46)$$

Here, K_{u1} and K_{u2} are the anisotropy constants, which can be positive or negative, and determine how strongly the magnetization is anchored in a particular direction. These constants are usually measured in units of erg cm^{-3} or J m^{-3}. The angle θ refers to that between the magnetization and the easy axis. Frequently, the uniaxial anisotropy is expressed using just the first term of Eq. (4.46). For cubic anisotropy, the energy expression is given as:

$$E_K = \frac{K_1}{4}[\sin^2 2\theta + \sin^4 \theta \sin^2 2\phi] + \frac{K_2}{16} \sin^2 \theta \sin^2 2\theta \sin^2 2\phi \qquad (4.47)$$

In this case, there are two angles, the polar angle, θ, measured from the z-axis and the azimuthal angle, ϕ, measured from the x-axis. Once again, K_1 and K_2 can be positive or negative, with the result that the easy axes will be along the principal axes of body diagonal directions, respectively.

While Bloch type walls are the most common in bulk materials, Néel walls usually occur in thin films, since they reduce the magnetostatic energy by rotating the magnetic moments in the plane of the film.

We now have a picture of the magnetization, which prefers to lie along a particular direction within a magnetic domain, as defined by the magnetic anisotropy of the system. In between the magnetic domains, the magnetic moments must rotate from one orientation to the next, in accordance with the orientation of the magnetic domains. This will mean that there will be components of the magnetization, within the domain walls, that lie along hard axes, and hence will increase the magnetic energy. Since the lower energy configuration will be that in which the magnetization lies along an easy axis, the magnetic anisotropy will act to reduce the width of the domain wall. This is in opposition to what we said with regard to the exchange energy. This, therefore, provides a competition of energies and allows us to determine the wall width with respect to the two energy contributions.

We will consider a simple example, in which we demonstrate the evaluation of the domain wall width using the simplified uniaxial anisotropy; $E_{Ku} = K \sin^2 \theta$, in the formation of a Bloch wall. Using $K > 0$ means that the magnetic moments prefer to lie along $\theta = 0$ or $\theta = \pi$. We sum the contributions from the N spins in the domain wall and then replace this summation with an integral in the so-called continuum limit. This gives an energy contribution of the form:

$$\sum_{i=1}^{N} K \sin^2 \theta_i = \frac{N}{\pi} \int_0^\pi K \sin^2 \theta d\theta = \frac{NK}{2} \tag{4.48}$$

From this, we write the energy per unit area of the wall, which takes the form: $NKa/2$. We now express the total energy per unit area of the domain wall, which includes both anisotropy and exchange contributions, as:

$$\sigma_{BW} = \frac{JS^2 \pi^2}{Na^2} + \frac{NKa}{2} \tag{4.49}$$

This clearly illustrates the competition of the two energy contributions. One term varies as N (so reducing N will reduce the total energy), while the other varies as $1/N$ (so increasing N will reduce the total energy). To obtain a correct balance, we minimize this energy with respect to N. This is performed by taking the equilibrium condition; $d\sigma_{BW}/dN = 0$:

$$\frac{d\sigma_{BW}}{dN} = -\frac{JS^2 \pi^2}{N^2 a^2} + \frac{Ka}{2} = 0 \tag{4.50}$$

It is now a simple matter to determine N:

$$N = S\pi\sqrt{\frac{2J}{Ka^3}} \tag{4.51}$$

This then allows us to determine the domain wall thickness:

$$\delta_{BW} = Na = S\pi\sqrt{\frac{2J}{Ka}} \tag{4.52}$$

The larger the value of J, the thicker the wall will be, while large values of K will reduce the wall thickness, in agreement with our discussion above. Indeed, the energy competition is reduced to the factor $\sqrt{J/K}$ in the above expression. If we back substitute Eq. (4.52) into Eq. (4.49), we obtain:

$$\sigma_{BW} = S\pi\sqrt{\frac{2JK}{a}} \tag{4.53}$$

By introducing the exchange stiffness constant, $A = 2JS^2/a$, for the case of a simple cubic structure, we can simplify the above expressions for the domain wall energy per unit area and the domain wall thickness to:

$$\sigma_{BW} = \pi\sqrt{AK} \tag{4.54}$$

and

$$\delta_{BW} = \pi\sqrt{\frac{A}{K}} \tag{4.55}$$

4.2   Introduction to Nanomagnetism

As we have seen in earlier chapters, the reduction of the physical dimensions of an object can give rise to a number of physical phenomena, and this is also the case when we consider their magnetic properties. The subject of *Nanomagnetism* has undergone enormous advances in recent years and research in this domain is intense. One of the early successes in this area derives from work in the late 1980s on magnetic multilayer systems, and in particular to the discovery of giant magnetoresistance (GMR). This subject was of significant importance to have led to the 2007 Nobel Prize in Physics being awarded to the discoverers, Albert Fert and Peter Grünberg. Magnetic multilayers now form an important class of materials and have some important applications, notably in the magnetic data storage industry. Much of the technology used for the study of magnetic multilayers stems from developments in the 1960s and 1970s, and in particular with respect to the deposition techniques and most importantly to molecular beam epitaxy, which was so successful in the development of semiconductor quantum well systems. Ultrathin magnetic films and structures still play an important role in magnetics research, where fundamental physics can be studied under highly controlled conditions. We will consider in more detail the topic of magnetic multilayers in the following section.

In addition to the 1D size reduction, or planar geometries of thin films and multilayers, size reductions in 2 and 3 dimensions are of enormous interest in magnetics research. For example, much work has been performed on magnetic nanostructures of various forms, such as nanoparticles, nanodots, nanowires, nanorings, etc. In particular, size reduction can be used as a means to manipulate and control magnetic properties. For example, reduced magnetic coordination at a magnetic surface will lead to new forms of magnetic anisotropy or surface anisotropy and can also affect the Curie temperature of a material. Such effects have been studied and exploited in research on magnetic nanoparticles. Furthermore, interactions between magnetic nanoparticles can influence the local effective field of an assembly of nanoparticles.

In addition to the work on low dimensional magnetic structures, much work has been dedicated to the development of experimental techniques used in the study of magnetic nanostructures. A full overview of such work is beyond the scope of this chapter, however, the interested reader may consult the reviews by Schmool and Kachkachi (2015, 2016). In this chapter, we will mainly concentrate on the material aspect of nanomagnetism, outlining the physical phenomena and some of the more important applications of nanostructured magnetic materials.

4.2.1 *Magnetic Energies and Length Scales*

As we have noted, important length scales play a vital role in the study of reduced dimensional structures. We have already seen that when structures reach the critical length scales associated with physical phenomena, the fundamental physical properties are altered. This is, of course, true in magnetism. Magnetism is essentially a quantum mechanical phenomenon, which results from a combination of the Pauli exclusion principle and the repulsive form of the electron–electron potential (Coulomb). A simple manner in which to phenomenologically represent the exchange interaction, as illustrated above, is via the Heisenberg Hamiltonian, which including the Zeeman energy takes the form:

$$\mathcal{H} = -\sum_{i \neq j} J_{ij} \mathbf{S}_i \cdot \mathbf{S}_j - g\mu_0\mu_B \sum_i \mathbf{H}_i \cdot \mathbf{S}_i \qquad (4.56)$$

The first term is as given in Eq. (4.25), while the second term represents the Zeeman energy due to a local field, $\mathbf{H}_i$, acting on spin i. This equation provides the first step in a microscopic description of the magnetic body. This picture can be further improved by adding an anisotropy term, such as a single-site uniaxial term:

$$\mathcal{H} = -\sum_{i \neq j} J_{ij} \mathbf{S}_i \cdot \mathbf{S}_j - g\mu_0\mu_B \sum_i \mathbf{H}_i \cdot \mathbf{S}_i - \sum_i K_i (\mathbf{S}_i \cdot \hat{\mathbf{e}}_i)^2 \qquad (4.57)$$

Here, the anisotropy constant K_i will depend on its position and $\hat{\mathbf{e}}_{ij}$ represents a unit vector in the direction of the easy axis. A more satisfying approach is that introduced by Néel, where the anisotropy will depend on the number of nearest neighbors, meaning that surface or interface spins will have a different anisotropy, leading to the concept of a surface magnetic anisotropy. This can be presented as:

$$\mathcal{H} = -\sum_{i \neq j} J_{ij} \mathbf{S}_i \cdot \mathbf{S}_j - g\mu_0\mu_B \sum_i \mathbf{H}_i \cdot \mathbf{S}_i - \frac{K_S}{2} \sum_i \sum_{j=1}^{z_i} (\mathbf{S}_i \cdot \hat{\mathbf{u}}_{ij})^2 \qquad (4.58)$$

where z_i is the coordination number of site i and $\hat{\mathbf{u}}_{ij} = \mathbf{r}_{ij}/r_{ij}$ is the unit vector connecting the site i to its nearest neighbors. This model is more realistic since the anisotropy at a given site occurs only when the latter loses some of its neighbors, i.e., when it is located at the boundary. The model in Eq. (4.58) is referred to as the *Néel Surface Anisotropy* (NSA) model. We will return to discuss some of the implications of these models for surface anisotropy in magnetic nanoparticles in Section 4.3.4.

While the strength of the exchange interaction between atomic moments (of the order of 0.1 eV/atom) is much larger than the magnetostatic energy (~ 0.1 meV/atom) and the magnetic anisotropy energy (~ 10 eV/atom), these latter two tend to become more important at macroscopic length scales. Such length scales follow from classical expressions for the different energy terms. In the continuum limit, the excess energy due to non-parallel spins, which is a local quantity, can be represented in the form:

$$\varepsilon_{ex} = A(\nabla \hat{\mathbf{u}}_m)^2 \qquad (4.59)$$

where $\hat{u}_m = \mathbf{M}/M$ and A is the exchange stiffness constant. This latter can be expressed in general form as:

$$A = \frac{\eta z S^2 J}{a_{nn}} \tag{4.60}$$

In this expression, we have the coordination number z, which depends on the specific crystalline structure, as does the factor η, which takes the following values: $\eta = 1$ for simple cubic, $\eta = \sqrt{3}$ for bcc and $\eta = 2\sqrt{2}$ for fcc structures.

The principal or characteristic length scales are related to the energy terms we have discussed here and in the previous section. These are:

(i) The *exchange length*:

$$l_{ex} = \sqrt{\frac{A}{2\pi M_S^2}} \tag{4.61}$$

and

(ii) The domain wall width:

$$\delta_{DW} = \sqrt{\frac{A}{K}} \tag{4.62}$$

These expressions give an indication of the length over which the exchange energy dominates over magnetostatic and magnetic anisotropy terms, respectively.

It should be noted that energy considerations will determine the critical size that a ferromagnetic particle maybe, beyond which it is energetically favorable to divide the body into two or more magnetic domains. This will be discussed in more detail below, but will depend on a number of factors, including the sample shape, but also the intrinsic properties of the material. For the present, we can quote the general formula for the radius of a spherical particle:

$$R_{Cr} = \frac{9\pi\sqrt{AK}}{\mu_0 M_S^2} \tag{4.63}$$

We can consider the case of an ultrathin film of say Fe, for which a strong perpendicular anisotropy can occur. From the point of view of the magnetostatic energy, an in-plane magnetic alignment would be favored and therefore we expect the spins to twist or rotate as a function of position across the film. Taking the direction perpendicular to the film to be in the $+z$ direction, and defining θ as the angle between the film normal and the magnetization, we

TABLE 4.1   Magnetic Properties for Selected Materials

Material	Anisotropy	K ($\times 10^6$ erg cm^{-3})	l_{ex} (nm)	δ_{DW} (nm)	R_{Cr} (nm)
bcc Fe	Weak	0.481	3.3	20.3	3.45
fcc Co	Weak	−1.2	4.8	15.8	5.05
hcp Co	Strong	4.12	4.7	8.3	6.85
fcc Ni	Weak	−0.056	7.6	39.2	8.1
fcc Ni$_{80}$Fe$_{20}$	Weak	0.0027	5.1	199	5.4

can analyze the dominant spin configuration from a minimization of the sum of the exchange and magnetostatic energies, which we write as follows:

$$\varepsilon = A \left(\frac{d\theta}{dz} \right)^2 + 2\pi M_S^2 \cos^2 \theta \tag{4.64}$$

This expression is subject to the appropriate boundary conditions. The first derivative of this expression with respect to θ, set to zero, yields:

$$A \frac{d^2\theta}{dz^2} - \pi M_S^2 \cos 2\theta = 0 \tag{4.65}$$

For small deviations, θ, we can obtain an approximate solution of the form: $\theta = \theta_0 e^{-z/l_{ex}}$. The general solution will also depend on any applied magnetic field for which the spin configuration may vary. A rigorous treatment should take into account the surface anisotropy and can show that for a critical thickness, in a thin film with either perpendicular or in-plane anisotropy, all spins are parallel and act as a single macrospin.

In Table 4.1, we list some of the typical values for the critical parameters for selected common ferromagnetic materials. The values are based on bulk parameter values. The single-domain critical size, taken from Eq. (4.63), assumes a spherical particle, since elongated particles can support a single magnetic domain for larger values of length due to shape anisotropy; N.B. spherical particles have no shape anisotropy.

The spin-wave spectrum for a magnetic system depends sensitively on the intrinsic parameters of the system, and is intimately related to physical quantities such as the g-factor, magnetic anisotropy (type and strength), saturation magnetization and the exchange stiffness constant. It can be convenient to define a dynamic length, which characterizes the typical spatial variations in the rf magnetization arising as a consequence of the surface torque exerted

TABLE 4.2   Orbital Magnetic Moment and Magnetic Anisotropy Energy, Based on Calculations for Co Embedded in Pt

	Bulk	Mono-layer	Diatomic wire	Monoatomic wire	Two atoms	Single atom
Orbital moment (μ_B/atom)	0.14	0.31	0.37	0.68	0.78	1.13
Anisotropy energy (meV/atom)	0.04	0.14	0.34	2.0	3.4	9.2

by the surface anisotropy. Such considerations lead to the characteristic dynamic length:

$$l_{dyn} = \sqrt{\frac{At}{K_S}} \qquad (4.66)$$

where t is the film thickness and K_S is the surface anisotropy strength.

4.2.2   Dimensionality and Reduced Coordination Number

In addition to the effects outlined above, i.e., that the reduced coordination number leads to a local variation of the magnetic anisotropy, there are further considerations that must also be taken into account. In general, the electronic structure of the atoms with a smaller coordination number is different from that for the same atoms in the bulk material. As we saw in Chapter 2, the density of states shows that the reduction in coordination number results in a narrowing of the electronic bands. This change in the density of states can alter the imbalance between spin-up and spin-down density of states.

There is an increasing orbital contribution to the magnetic moment with a decrease of dimensionality. This is illustrated in Table 4.2.

Also illustrated in the table is the significant enhancement of the magnetic anisotropy energy resulting from the dimensionality. Additional changes in the magnetic properties of the atoms at surfaces and interfaces can be due to the presence of defects and impurities, strain and alterations in the lattice parameter. Materials in the form of nanoparticles may present crystalline structures, which differ from those of the bulk material. This is indeed the case for cobalt metal, which changes from hcp in the bulk to bcc in particle form for diameters below around 30 nm.

In addition to the considerations above, there are some specific aspects of nanoscale magnets that are crucial for the understanding of their fundamental properties and behavior. This pertains to the interplay of the fundamental properties, such as magnetization and anisotropy as well as the applied magnetic field. Furthermore, when the size of a magnet is in the nanometer regime, their thermal stability poses some important issues and leads to the *superparamagnetic regime*. This, as we shall see, depends on the temperature, magnetic anisotropy and the method of measurement via its characteristic measurement time.

4.3   Magnetic Nanoparticles

4.3.1   Introduction

Magnetic nanoparticles form an important class of magnetic material and research has maintained a high level of interest over the past two decades or more. Much of this is related to the large range of applications of magnetic nanoparticles, some of which we will outline in this section. In addition to the extensive applications of magnetic nanoparticles, the fact that the fabrication of nanoparticles is rather simple and cheap means that this makes them commercially very appealing. In particular, the chemical routes for the preparation of nanoparticles allow for good control of size and importantly good monodisperse distributions of sizes.

In this section, we will provide an in-depth introduction to some of the more important physical models of magnetism in these systems. We will discuss the properties of individual particles and the modification of these properties due to the magnetic interactions between particles in nanoparticle assemblies.

One important consideration for a magnetic body is the critical size of the particle for which the particle is a single magnetic domain. To perform this calculation, we will use a simple approach which considers the energy contributions due to the magnetostatic and domain wall energies. We have already introduced the domain wall energy in the introduction to this chapter. We will therefore start with a consideration of the magnetostatic energy. In its basic form, the magnetostatic energy of a uniformly magnetized body can be expressed as:

$$\varepsilon_{ms} = -\frac{\mu_0}{2} \int \mathbf{M}_S \cdot \mathbf{H}_d dV \tag{4.67}$$

The demagnetizing field can be expressed in terms of the magnetization, $\mathbf{M}_S$, of the body as:

$$\mathbf{H}_d = -\mathcal{N} \mathbf{M}_S \tag{4.68}$$

where $\mathcal{N}$ is the demagnetizing tensor, which will depend on the shape of the magnetic body. Thus we can write

$$\varepsilon_{ms} = \frac{\mu_0}{2} \int \mathbf{M}_S \cdot \mathcal{N} \mathbf{M}_S dV \tag{4.69}$$

In the case of an ellipse of rotation, the demagnetization tensor can be expressed as a linearized matrix for which we can write:

$$\begin{pmatrix} H_{dx} \\ H_{dy} \\ H_{dz} \end{pmatrix} = - \begin{pmatrix} N_x & 0 & 0 \\ 0 & N_y & 0 \\ 0 & 0 & N_z \end{pmatrix} \begin{pmatrix} M_x \\ M_y \\ M_z \end{pmatrix} \tag{4.70}$$

The demagnetization factors, N_x, N_y, N_z define the various components for the particular shape of the sample and are constrained by the relation:

$$N_x + N_y + N_z = 1 \tag{4.71}$$

We can now express the magnetostatic energy in the form:

$$\varepsilon_{ms} = \frac{\mu_0}{2} (N_x M_x^2 + N_y M_y^2 + N_z M_z^2) \tag{4.72}$$

Using the various components of the magnetization, as illustrated in Figure 4.6, we obtain:

$$\varepsilon_{ms} = \frac{\mu_0}{2} M_S^2 (N_x \sin^2 \theta \cos^2 \phi + N_y \sin^2 \theta \sin^2 \phi + N_z \cos^2 \theta) \tag{4.73}$$

For the simple case of a spherical magnetic body, have $N_x = N_y = N_z = 1/3$, from which we find the corresponding magnetostatic energy:

$$\varepsilon_{ms} = \frac{\mu_0}{6} M_S^2 \tag{4.74}$$

The total energy of the single domain spherical particle will be:

$$E_{1Dom} = \varepsilon_{ms} V = \varepsilon_{ms} \frac{4}{3} \pi R^3 = \frac{2}{9} \pi R^3 \mu_0 M_S^2 \tag{4.75}$$

where we have taken into account the volume V of the particle.

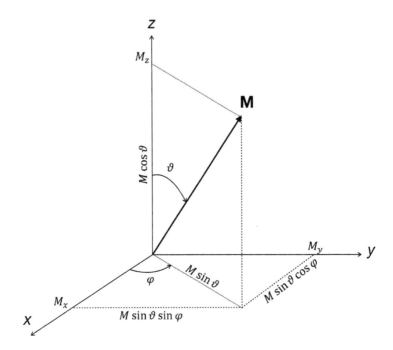

FIGURE 4.6   Coordinate system for the magnetization vector, showing the different components in the various directions of Cartesian and spherical systems.

To see whether it is favorable or not to have a single domain, we will compare this energy to that for a spherical particle which is divided into two equivalent domains, separated by a single 180° domain wall, see Figure 4.7. This domain wall will have a total energy of: $E_{DW} = \pi R^2 \sigma_{BW}$, where R is the particle radius and σ_{BW} the Bloch wall energy given by Eq. (4.54). The total energy of this two-domain particle can now be expressed as:

$$E_{2Dom} = \frac{1}{2} \left(\frac{2}{9} \pi R^3 \mu_0 M_S^2 \right) + \pi R^2 \sigma_{BW} \tag{4.76}$$

For the single domain particle to be energetically favorable, we must satisfy the following condition: $E_{1Dom} < E_{2Dom}$. Substituting for Eqs. (4.75) and (4.76), we obtain:

$$R < \frac{9\sigma_{BW}}{\mu_0 M_S^2} \tag{4.77}$$

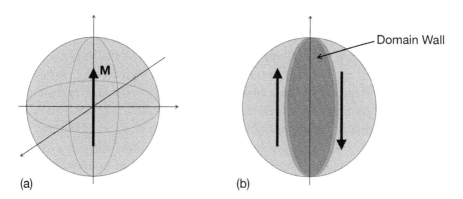

FIGURE 4.7 (a) Single domain spherical magnetic particle. (b) Spherical magnetic particle divided into two equal domains with opposing magnetization directions.

Thus, we find the critical radius for a single domain spherical ferromagnetic magnetic particle to be:

$$R_{Cr} = \frac{9\pi\sqrt{AK}}{\mu_0 M_S^2} \qquad (4.78)$$

where we have substituted, $\sigma_{BW} = \pi\sqrt{AK}$. Values for selected materials are given in Table 4.1. As we mentioned earlier, elongated particles can stabilize into single domain particles for larger sizes due to the effects of shape anisotropy. The specifics can be evaluated in a similar manner, using the energy considerations illustrated in the above example. The effects of magnetocrystalline anisotropy will also have a bearing on the critical size of a single domain particle. Once again, to do this, we would have to consider a specific example of the combination of magnetostatic, magnetocrystalline and domain wall energy contributions.

4.3.2 The Stoner-Wohlfarth Model

The Stoner–Wohlfarth (SW) model dates from 1948 and provides a simple and instructive treatment to magnetic systems that consist of a single magnetic domain. In modern parlance, we would call the particles a macrospin, where all the individual moments act in unison. The SW model is often used as a first introduction to explain magnetic hysteresis. A consequence of the assumption that the magnetization is constant throughout the magnetic particle means that the exchange energy is constant through the particle and that the reversal process will be uniform or coherent. This assumption is generally valid for small, weakly or non-interacting magnetic particles.

In our treatment, we will consider the simplest case where the particle has an ellipsoidal shape (ellipse of rotation), which has a uniform magnetization. This is consistent with a zero contribution from the exchange energy. In the previous section, we introduced the magnetostatic energy and showed that for a spherical particle, this gives a contribution of $\varepsilon_{ms} = \frac{\mu_0}{6} M_S^2$, and no shape anisotropy will exist for the particle. For any other shape of particle, this will not be the case. For example, in a disk shaped body, where the normal to the plane is in the z-direction, the demagnetization factors can be expressed as: $N_x = N_y = 0; N_z = 1$, where we assume an infinite extension in the disk $(x - y)$ plane. In this case, the hard axis will be parallel to the z-axis. Such a geometry is typically used for thin films. In the case of a cylindrical particle, where we take the long axis along the z-direction, we have; $N_x = N_y = 1/2; N_z = 0$. In this case, the z-axis will form the easy-axis. In these previous two examples, the $x-$ and y-directions are identical, as is the case for an ellipse of rotation. Setting $N_x = N_y = N_\perp$ and $N_z = N_\parallel$, we can rewrite the magnetostatic energy in the form:

$$\varepsilon_{ms} = \frac{\mu_0}{2}(N_\perp M_x^2 + N_\perp M_y^2 + N_\parallel M_z^2) = \frac{\mu_0}{2} M_S^2 (N_\perp \sin^2 \theta + N_\parallel \cos^2 \theta) \quad (4.79)$$

We can further simplify the above expression to:

$$\varepsilon_{ms} = \frac{\mu_0}{2} M_S^2 (N_\perp - N_\parallel) \sin^2 \theta + \frac{\mu_0}{2} M_S^2 N_\parallel \quad (4.80)$$

This equation shows that this energy term represents a uniaxial anisotropy, which is contained in the first term on the RHS.

If we consider our magnetic particle to further contain a uniaxial magnetocrystalline component, then we need to take into account the relative orientations of the easy axes for the two contributions. Should they be aligned parallel, we can express the total anisotropy energy in the form:

$$\varepsilon_{anis} = \varepsilon_K + \varepsilon_{ms} = K_u \sin^2 \theta + \frac{\mu_0}{2} M_S^2 (N_\perp - N_\parallel) \sin^2 \theta + \frac{\mu_0}{2} M_S^2 N_\parallel \quad (4.81)$$

We note that factorizing the above allows us to obtain an effective anisotropy for the particle:

$$\varepsilon_{anis} = \left[K_u + \frac{\mu_0}{2} M_S^2 (N_\perp - N_\parallel) \right] \sin^2 \theta + K = K_{eff} \sin^2 \theta + K \quad (4.82)$$

where $K_{eff} = [K_u + \mu_0 M_S^2 (N_\perp - N_\parallel)/2]$. In the above expression, we have taken $K = \mu_0 M_S^2 N_\parallel /2$, which has no angular dependence and can be considered to be a constant. If the two anisotropies are crossed, as for example in the following expression:

$$\varepsilon_{anis} = K_u \cos^2 \theta + \frac{\mu_0}{2} M_S^2 (N_\perp - N_\parallel) \sin^2 \theta + K \tag{4.83}$$

this can be re-arranged and expressed as:

$$\varepsilon_{anis} = \left[\frac{\mu_0}{2} M_S^2 (N_\perp - N_\parallel) - K_u \right] \sin^2 \theta + K' \tag{4.84}$$

Once again, this will reduce to the form of Eq. (4.82), where the effective anisotropy constant will be $K_{eff} = [\mu_0 M_S^2 (N_\perp - N_\parallel)/2 - K_u]$. This illustrates that for crossed anisotropies, the effective anisotropy constant will be the difference between the two anisotropy strengths. As we saw previously, should the anisotropies be parallel, their strengths will be additive in the effective anisotropy constant.

We will now consider the case of a magnetic particle with uniaxial anisotropy in the presence of an applied magnetic field. The energy expression that forms the simplest case of the Stoner-Wohlfarth model can be expressed as:

$$\varepsilon = K \sin^2 \theta - \mu_0 M_S H \cos(\theta - \phi) \tag{4.85}$$

In this expression, the angle θ represents the angle between the easy-axis and the magnetization vector, and the angle ϕ is the angle between the applied magnetic field, H and the easy-axis. The second term in the above expression represents the Zeeman energy. We note, that in this case we consider $K > 0$ and the energy minimum will occur for $\theta = 0$. At equilibrium, the magnetization direction, as expressed by θ, will point along the direction which minimizes the energy, for which we write $\theta = \theta^*$. The energy minimization condition can be expressed as:

$$\left(\frac{\partial \varepsilon}{\partial \theta} \right)_{\theta = \theta^*} = 0, \quad \text{and} \quad \left(\frac{\partial^2 \varepsilon}{\partial \theta^2} \right)_{\theta = \theta^*} \geq 0 \tag{4.86}$$

For the first condition, we obtain:

$$\frac{\partial \varepsilon}{\partial \theta} = 2K \sin \theta \cos \theta + \mu_0 M_S H \sin(\theta - \phi) = 0 \tag{4.87}$$

Writing $H_K = 2K/\mu_0 M_S$, which is known as the anisotropy field, and normalizing with $h = H/H_K$ and $m = M/M_S$, we can express the first minimization condition as:

$$[\sin \theta \cos \theta + h \sin(\theta - \phi)]_{\theta = \theta^*} = 0 \tag{4.88}$$

The second minimization condition, which derives from the second derivation of the energy with respect to θ, yields:

$$[\cos 2\theta + h\cos(\theta - \phi)]_{\theta=\theta^*} \geq 0 \qquad (4.89)$$

For a general ϕ, there will be no analytical solution, these only exist for
$\phi = 0, \pi/4$ and $\pi/2$. The variation of the energy with the angle θ is illustrated in Figure 4.8 for various magnetic field strengths, h, which is applied
at an angle $\phi = 30°$. We note that as the field strength increases, the magnetization is forced along a specific direction, and eventually aligns parallel to
the applied field when it is sufficiently strong to overcome the effects of the
magnetic anisotropy.

We can define the longitudinal magnetization, i.e., the projection of the
magnetization along the direction of the applied field. This can be expressed
in reduced form as: $m_{\parallel} = \cos(\theta - \phi)$. In a similar manner, we define the
transverse magnetization in the form: $m_{\perp} = \sin(\theta - \phi)$. For the analytic
cases of $\phi = 0$ and $\phi = \pi/2$, we can write:

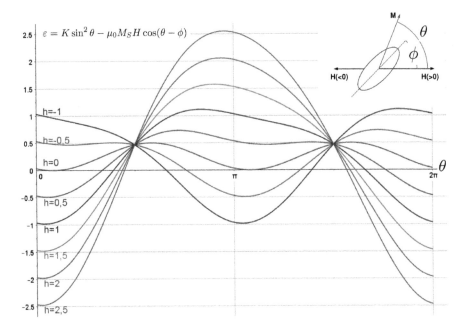

FIGURE 4.8 Variation of the energy landscape with angle θ for different values
of the normalized magnetic field $h = H/H_K$, as indicated. The field orientation is
held fixed and the easy axis is at $\phi = 15°$ with respect to the anisotropy axis. Reproduced from Wikipedia: https://fr.wikipedia.org/wiki/Modéle de Stoner-Wohlfarth.

(1) $\phi = 0$; $\cos \theta^* = -h \Rightarrow \theta^* = \cos^{-1}(-h)$ for $h \leq 1$, otherwise, $\theta^* = 0, \pi$, giving a square hysteresis loop, as illustrated in Figure 4.9.

(2) $\phi = \pi/2$; $\theta^* = \sin^{-1}(h)$ when $h \leq 1$, otherwise, $\theta^* = \pi/2$, which yields $m_{\parallel} = h$, Figure 4.9, and $m_{\parallel} - \perp\sqrt{1 - h^2}$.

Other configurations are illustrated in Figure 4.9. We can consider two types of hysteresis curve: (i) The longitudinal hysteresis curve, with the reduced magnetization $m_{\parallel}$ taken along the direction of the applied field, Figure 4.9, or (ii) The transverse hysteresis curve, with the reduced magnetization $m_{\perp}$ taken along the direction perpendicular to the applied field.

In the model described here, the coercive field is found at $H = H_C = H_K$. The loop is broadest when $\phi = 0$ and becomes narrower as ϕ is increased, and completely collapses to a line for $\phi = \pi/2$. The boundaries of hysteresis can be determined from the simultaneous nulling of the first and second derivatives of the energy. Such a consideration leads to the expression:

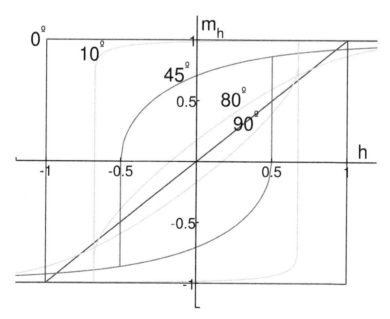

FIGURE 4.9 Longitudinal hysteresis loop for various angles ϕ ($= 0°, 100°, 45°, 80°$, and $90°$) of the field $h = H/H_K$, with the easy-axis. Note the square loop for $\phi = 0°$ and the diagonal line $m_{\parallel} = h$ for $\phi = 90°$. The latter displays a break in the slope at the value of the anisotropy field. Reproduced from Wikipedia: https://fr.wikipedia.org/wiki/Modéle de Stoner-Wohlfarth.

$$H_{SW} = \frac{H_K}{(\sin^{2/3}\phi + \cos^{2/3}\phi)^{3/2}} \tag{4.90}$$

This equation describes the so-called switching field, i.e., the applied field required to invert the direction of the magnetization. This corresponds to a reversal of the magnetization direction and has some important implications for practical applications, such as in the writing process for magnetic storage. In Figure 4.10, we illustrate the form of the magnetic reversal condition as expressed in Eq. (4.90). This curve is commonly known as the Stoner-Wohlfarth astroid. The switching of the magnetization will occur along the solid line of this graph. In the simple case where $\phi = 0$, we have $H_{SW} = H_K = 2K/\mu_0 M_S$.

It is worth pointing out that the uniaxial model that we have considered can be taken as a condition for a metastable state for the magnetization or energy. This corresponds to the case where the initial orientation of the magnetization along $\theta = \pi$ is separated by an energy barrier from another sta-

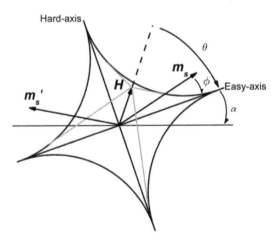

FIGURE 4.10　The Asteroid Rule of the Stoner-Wohlfarth model for a single-domain magnetic particle. The particle is specified by its magnetic moment m_s, anisotropy constant K, and easy-axis direction. ϕ is the angle between the magnetization vector m_s and the easy-axis of the particle, and θ is the angle between the applied field vector H and the easy axis. The equilibrium direction of m_s (or m'_s) is parallel to one of the lines tangent to the asteroid and passing through the tip of the magnetic field vector (shown in gray). Reprinted figure with permission from Physical Review B as follows: Wenzhe Zhang, Gang Xiao, & Matthew J. Carter, (2011). *Physical Review B, 83*, 144416. ©(2011) by the American Physical Society.

ble state for $\theta = 0, 2\pi$. In zero applied, field we have a symmetric energy: $\varepsilon = K \sin^2 \theta$. The energy barrier has a height of KV, the anisotropy constant times the volume of the particle, see Figure 4.11.

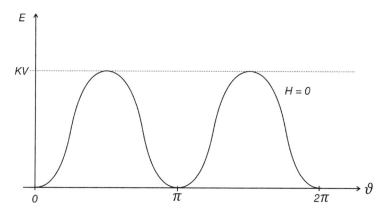

FIGURE 4.11   Angular dependence of the energy for a uniaxial particle in zero applied magnetic field.

4.3.3   *Superparamagnetism and Thermally-Assisted Reversal*

At non-zero temperatures, the presence of thermal activation can assist the magnetization to overcome an energy barrier and to thus reach a stable magnetic state. This problem has been extensively studied, notably by L. Néel as early as 1949, and subsequently by W. F. Brown in the early 1960s, who sought to account for thermal excitations in a theoretical framework.

The *Néel–Brown model* considers the probability of the magnetization of a magnetic particle being reversed via thermal fluctuations after a given time, t. The probability can be expressed as an exponential law of the form:

$$P = e^{-t/\tau} \tag{4.91}$$

The constant τ represents the thermal stability, and in its simplest form can be expressed by the Arrhenius relation:

$$\tau = \tau_0 e^{E_B(H,T)/k_B T} \tag{4.92}$$

The energy barrier height, E_B, can depend on the applied magnetic field and the temperature via the anisotropy constants, $K = K(T)$. The factor $k_B T$, where k_B is the Boltzmann constant, represents the thermal energy and τ_0 is an intrinsic characteristic reversal time, which depends on various parameters, including those intrinsic to the nanoparticle and extrinsic parameters such as the measurement time, τ_m, which will be a characteristic of the method used in a measurement. It is common to use the *attempt frequency*, $v_0 = 1/\tau_0$, to characterize the reversal process.

In Brown's approach, thermal activation is taken into account by means of a random (fluctuation) field, which excites the resonant precessional frequencies of the magnetization around the direction corresponding to the energy minimum. This means that τ_0 will depend on factors such as the form of the magnetic anisotropy, the magnetization, the g-factor as well as the damping parameter, α. We will discuss these parameters in relation to the dynamics of the magnetization in a later section. It is common to take the value of τ_0 to be a constant in the range $10^{-9} - 10^{-11}$ s.

Eq. (4.95) has some rather important implications for nanoparticle systems. For example, if the energy barrier is relatively small, the thermal fluctuations will produce rapid reversal processes between the energy minima. Such behavior mimics atomic paramagnetism and the magnetic particle, which is in a ferromagnetic state, is said to be *superparamagnetic*. This, however, is not the full story. The temporal scale is equally important in defining the superparamagnetic state. This is where the measurement time becomes a crucial factor. For instance, for a magnetic measurement which samples the state of magnetization on a short time scale, the magnetic moment can be thought of as being frozen into one of the minima during the measurement process. In terms of the time scales, this corresponds to $\tau \gg \tau_m$. In this case, we say that the particle exhibits stable ferromagnetism and is said to be in a *blocked state*. The temperature at which this transition occurs, i.e., where $\tau = \tau_m$, is referred to as the *blocking temperature*. From this, we can establish the blocking temperature as:

$$T_B = \frac{E_B}{k_B} \frac{1}{\ln(\tau_m/\tau_0)} \qquad (4.93)$$

We can also express the condition for superparamagnetism in a slightly different form, which states that the system will be superparamagnetic for the following condition:

$$\ln\left(\frac{\tau_m}{\tau_0}\right) > \frac{E_B}{k_B T} \geq 0 \qquad (4.94)$$

What is important to note is the subtle interplay between the temperature, characteristic measurement time and the barrier height. All three parameters play an equally important role in determining the magnetic stability of the magnetic particle. If we take the example of a Co nanoparticle of 3 nm diameter, the variation of its magnetic state as a function of temperature can be schematically illustrated as shown in Figure 4.12. Here, we show that for very low temperatures, the particle displays stable magnetism with hysteresis. Above the blocking temperature, the particle becomes unstable, superparamagnetic and there will be no hysteric behavior, as is the case of paramagnetism. Note we have assumed a fixed measurement time. At very high temperatures above the Curie temperature, the sample becomes truly paramagnetic.

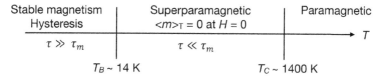

FIGURE 4.12 Schematic illustration of the variation of the magnetic state of a Co nanoparticle of 3 nm as a function of temperature.

As we indicated above, in the simplest case, the energy barrier can be expressed as $E_B = KV$, where there is no applied magnetic field. This shows that the barrier height is directly dependent on the size of the nanoparticle. This is why superparamagnetism is such an important consideration in magnetic nanosystems. Stability in the nanoparticle can, however, be countered by using magnetic materials with a high intrinsic magnetocrystalline anisotropy, such as Co, CoPt, FePt, etc. It is also important to note that interparticle interactions can also aid the stabilization of the magnetic state. Interactions, such as dipole–dipole interactions, leads to a modified form of the Arrhenius law, known as the Vogel–Fulcher law, and can be expressed as:

$$\tau = \tau_0 e^{E_B(H,T)/k_B(T-T_0)} \tag{4.95}$$

Here, T_0 is an effective temperature, which is proportional to H_i^2, where H_i is an effective field created by the interaction, and increases with the interaction strength. The application of an external magnetic field will also have a stabilizing effect and will modify the Arrhenius relation of the form:

$$\tau = \tau_0 e^{E_B(1-H/H_K)/k_B T} \tag{4.96}$$

where H_K is the anisotropy field. The above expressions represent a simplified treatment of the effects of an additional field, whether in the form of an externally applied field or in that due to interactions with other magnetic particles. The principal conclusion is that this will tend to stabilize the magnetization of a particular nanoparticle. We will discuss more explicitly the role of interactions in Section 13.3.5, below.

What the above does not highlight, is the importance of the direction of the additional field contribution with respect to the anisotropy axis of the particle. An idea of the alteration of the energy landscape can be seen for example in Figure 4.8, where a magnetic field is applied at 30° to the easy axis of a uniaxial particle. Here, at zero fields, the energy variation is symmetric around the axis, however, once the field is applied, the curve becomes biased, with one energy minima becoming deeper than the other. This will, therefore, bias the switching behavior, with the deeper minimum being more stable and hence harder to switch. This will be reflected in the switching probabilities for both energy minima. Eventually, as the field becomes stronger, the shallow minimum is no longer stable and the particle magnetization will be fixed in the deep energy minimum. It will be noted that the position or angle of the energy minimum will shift for the general case, and will eventually align along the direction of the applied field. Clearly, the specifics will depend on the orientation and strength of the applied field. There are two special cases that can be considered. In the first case, we can consider the applied field along the direction of the easy axis. This will deepen the energy minimum parallel to the direction of the applied field and reduce that in the opposite direction, i.e., at 180°. As the field increases, the one will grow at the expense of the other until only the minimum along the applied field exists, and the magnetization will be completely stabilized. In the second special case, we consider the field applied at a direction perpendicular to the easy axis, i.e., at $\phi = 90°$ or 270°. This case is displayed in Figure 4.13, where we show a series of curves of the energy landscape as a function of the applied field. Here, we note that as the field increases from $H = 0$, the two energy minima move equally towards the direction of the applied field along $\phi = 270°$. The minima are of equal depth due to symmetry considerations and the magnetization can shuttle between the two at a frequency determined by the switching probability and dependent on the height of the barrier between the two minima. Eventually, at $H = H_K$, the two minima merge into a broad minimum and the magnetization is fully stabilized.

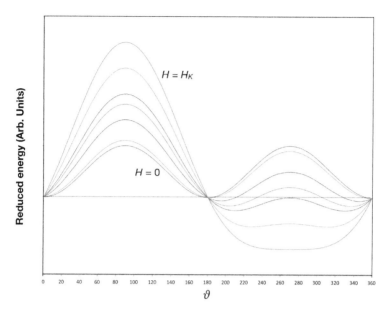

FIGURE 4.13 Energy landscapes for an applied field between $H = 0$ and $H = H_K$, applied along the direction $\phi = 270°$.

4.3.4 Many-Spin Description of a Magnetic Nanoparticle

In the previous sections of this chapter, we have considered that a particle can be described as a coherent magnetic entity in which all the individual magnetic moments or spins are rigidly coupled. Such a description can be termed a *macrospin*. For magnetic nanoparticles of very small dimensions, of say less than 10 nm (for example for cobalt), a large proportion (over 50%) of its atoms will be located on its surface. As for any surface, the surface atoms of the nanoparticle may undergo some form of lattice reconstruction. This can lead to crystal-field symmetry breaking with strong local inhomogeneities, see Figure 4.14.

In order to take into account the effects of a large proportion of surface spins, it is necessary to use a microscopic approach involving the atomic magnetic moments. Such an approach must also account for the local environment within the particle, including microscopic interactions and single-site anisotropy. This means that we need to adopt a many-spin approach to the problem, where we consider each of the N atoms/spins of the nanoparticle. Such considerations lead to the use of a Hamiltonian of the form expressed in Eqs. (4.57) and (4.58). These equations permit the treatment of

FIGURE 4.14   Surface atoms for an icosahedral nanoparticle (left) and a quasi-spherical particle.

the nanoparticle as an assemble of spins constrained by their physical proximity and exchange. In the former, the surface anisotropy value can be set at a different value from that in the core. This corresponds to the so-called transverse surface anisotropy (TSA) and for a spherical particle will lead to a tangential easy axis for $K_i < 0$, while for $K_i > 0$ the surface spins are radially oriented, leading to a hedgehog-like configuration. While for spherical particles, due to their symmetry, the global effect would provide no net anisotropy, for non-spherical particles, such as ellipsoids of rotation, there will be a residual overall effect providing a distinguishable surface anisotropy which should be experimentally measurable. In the case of Eq. (4.58) for the Néel surface anisotropy (NSA), the particular environment of nearest neighbors will define a local anisotropy at a particular spin site i. Importantly, this approach will distinguish spins which are in the interior region of the particle and have full coordination from those in the surface region, where the coordination number varies.

In specific cases, there can be a coincidence between the TSA and NSA models. For example, for the case of a (100) plane with a simple cubic structure, a surface spin will have four nearest neighbors in the surface plane and one in the plane below the surface. Analysing the form of the NSA, we have, for spin $|\mathbf{S}_i| = 1$, the anisotropy term of Eq. (4.58) given by: $\mathscr{H}_{NSA} = 2K_S - K_S S_{i,z}^2$. Therefore, for the case of $K_S > 0$, the easy axis will be perpendicular to the plane, i.e., parallel to $\pm\hat{\mathbf{e}}_z$, while for $K_S < 0$, the (100) surface becomes an easy plane. By dropping the irrelevant constant, we can express the NSA as: $\mathscr{H}_{NSA} = -K_S(\mathbf{S}_i \cdot \hat{\mathbf{e}}_i)^2$, which is identical to the TSA expression. A more in-depth comparison of the NSA and TSA models can be found in (Schmool and Kachkachi, 2015) and references therein.

For a magnetic particle of a specific finite dimension with N spins, we can define formally the magnetization as:

$$\mathbf{M} = \frac{1}{N}\sum_i \mathbf{S}_i \tag{4.97}$$

We can take the thermodynamic average of this quantity, called the *induced magnetization*, which takes the form:

$$\langle \mathbf{M} \rangle = \frac{1}{N}\sum_i \langle \mathbf{S}_i \rangle \tag{4.98}$$

This magnetization will vanish for finite systems in the absence of a stabilizing magnetic field due to the Goldstone mode, which corresponds to the global rotation of the magnetization. At low temperatures, however, the spin system is aligned with respect to one another, leading to a net or *intrinsic magnetization*, which is defined as:

$$M = \sqrt{\langle \mathbf{M}^2 \rangle} = \sqrt{\left\langle \left(\frac{1}{N}\sum_i \mathbf{S}_i \right)^2 \right\rangle} \tag{4.99}$$

At low temperatures, all spins in the particle are coupled more or less rigidly and the magnetization can be considered as a macrospin. When a magnetic field $\mathbf{H}$ is applied, the magnetization $\mathbf{M}$ displays a non-zero average in the direction of the field, leading to a net induced value for the induced magnetization.

The connection between the induced and intrinsic values of the magnetization can be expressed in the form of the so-called *superparamagnetic relation*, written as:

$$\langle \mathbf{M} \rangle = M\mathscr{L}(Mx) \tag{4.100}$$

where $x = NH/k_B T$ and $\mathscr{L}(x) = \coth(x) - 1/x$ is the Langevin function. The validity of this expression has been shown to be limited to cases where the particles are rather large, for which the change in M is small with respect to the change of field, and cannot be used where $\mathscr{L}(x)$ is of the order 1 or less. The field dependence of the magnetization for very small particles is rather complex and requires computational methods to provide a more accurate description for the general case. An outline is given in (Schmool and Kachkachi, 2015) and references therein.

The description of a magnetic nanoparticle as either a single coherent macrospin or an assembly of exchanged coupled individual spins depends on a number of factors, which result from the complex interplay of parameters such as particle size and shape, the prevailing anisotropy and in particular its strength, the ambient temperature and applied field. The macrospin approach, also referred to as the one spin problem (OSP) has the advantage of being simple to apply and to understand and is widely used in the interpretation of experimental measurements for this reason. However, its range of applicability is subject to a number of limitations as described above. Indeed, the multi-spin problem (MSP) is mainly used to study fundamental issues relating to finite size and surface effects, which cannot be ignored. Both approaches are important and indeed should be considered as being complementary. The limits of the macrospin are more thoroughly discussed by Rohart *et al.* (2007). In Figure 4.15, we illustrate the various regimes for the description of a magnetic nanoparticle, where use the example for Co. Here we note that for particles above 45 nm, we need to consider multiple domains, while in the range 10–45 nm, a macrospin model adequately described the magnetic behavior of the Co nanoparticle. Below 10 nm, however, it is necessary to use a multispin approach, where the surface and bulk spins are no longer sufficiently collinear. Indeed, the smaller the particle, the more the surface becomes a dominating influence of the magnetic properties of the particle.

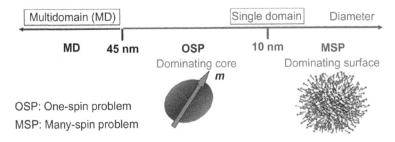

FIGURE 4.15   Magnetic regimes for a spherical Co nanoparticle as a function of its size. Reprinted from D. S. Schmool and H. Kachkachi, (2015). *Single Particle Phenomena in Magnetic Nanostructures*, in Solid State Physics, Volume 66, R. L. Stamps and R. Camley (Eds.), pp 301–423, Elsevier, ©(2015), with permission from Elsevier.

4.3.5 *Magnetic Nanoparticle Assemblies: Interactions*

In real magnetic systems, we are most commonly confronted by an assembly of nanoparticles. In such cases, the interactions between the particles can have an important effect on the global magnetic behavior of the assembly. For dilute assemblies, it is possible to neglect these effects, though there is usually some residual effect which implies a modification of the magnetic properties. In this section, we will consider principally the effect of the interaction between particles via a dipolar mechanism. Other forms of coupling can also play a role, but will generally depend on the nature of the intervening media between the particles. For example, for a conducting medium RKKY, interactions can be important, while for an insulating matrix, superexchange interactions may exist in certain cases. The former depends on the interparticle separation, falling off as $1/r_{ij}^3$, as is the case for dipolar interactions. For the case of superexchange, the nature of the matrix is important. The consideration of interparticle interactions is rather complex and compounded by the possibility of having more than one interaction mechanism at play simultaneously. Furthermore, accounting for the size and shape distributions in nanoparticle assemblies as well as thermal fluctuations leads to a very complex situation.

As we noted earlier, the relaxation time can also be affected by interactions since it can modify the energy barrier, E_B. This will also depend on the form of anisotropy present within the particle. Such effects may lead to a change of regime, where the properties of the nanoparticle are no longer superparamagnetic for example. Interactions between particles can also lead to the presence of collective states between the nanoparticles in an assembly, such as the superferromagnetic state, which will also be temperature-dependent. We will begin our discussion with the consideration of the dipolar interaction.

The basic form of the magnetic dipolar interaction between two magnetic nanoparticles, with magnetic moments $\mathbf{m}_i$ and $\mathbf{m}_j$, can be represented by the relation:

$$E_{ij}^{Dip} = \frac{1}{4\pi r_{ij}^3} \left[\mathbf{m}_i \cdot \mathbf{m}_j - \frac{3(\mathbf{m}_i \cdot \mathbf{r}_{ij})(\mathbf{m}_j \cdot \mathbf{r}_{ij})}{r_{ij}^2} \right] \qquad (4.101)$$

Here, $\mathbf{r}_{ij}$ represents the vector joining the particles i and j. This general expression is valid for any orientation of the two moments with respect to one another and also for any spatial disposition of the particles with respect to their moments. It is possible to express this in terms of the polar and azimuthal angles in a form, $E_{ij}^{Dip} = E_{ij}^{Dip}(\theta_i, \phi_i; \theta_j, \phi_j; \vartheta, \varphi)$, where ϑ, φ define the orientation of the vector r_{ij}. This expression is a little long and cumbersome and not very practical. We can greatly simplify the expression in the case where the two magnetic moments are aligned, as by the action of an applied magnetic field. This is much more practical and a reasonable assumption for many practical applications where an applied field is used in experimental conditions. In this case, we can set $\theta_i = \theta_j = \theta$ and $\phi_i = \phi_j = \phi$. The energy expression can be written as:

$$E_{ij}^{Dip}(\theta, \phi, \vartheta, \varphi) = \frac{m_i m_j}{4\pi r_{ij}^3} \left\{ 1 - 3[\sin\theta \sin\vartheta \cos(\phi - \varphi) - \cos\theta \cos\vartheta]^2 \right\}$$

(4.102)

It is instructive to consider some special cases to illustrate the effect of dipolar coupling for the case of two magnetic moments. Firstly, we can simplify the above expression by setting the position vector between the moments to lie parallel to the y-axis, we obtain:

$$E_{ij}^{Dip}(\theta, \phi) = \frac{m_i m_j}{4\pi r_{ij}^3} (1 - 3\sin^2\theta \sin^2\phi)$$

(4.103)

We clearly note that if the magnetic moments lie in any direction in the $x - z$ plane, then the energy will always take the same value. This is logical, since they are always parallel and in the plane perpendicular to the vector r_{ij}. The same is not true if we consider the variation in the $y - z$ plane. In this case, the variation of the dipolar interaction energy will be given by:

$$E_{ij}^{Dip}(\theta, \phi) = \frac{m_i m_j}{4\pi r_{ij}^3} (1 - 3\sin^2\theta)$$

(4.104)

Here, we note that this represents the variation of the moments from parallel in the direction perpendicular to vector joining them to parallel along the axis joining them. These examples are illustrated in Figure 4.16(a) and (b). In Figure 4.16(c), we illustrate the variation of the energy term expressed in Eq. (4.104).

What is interesting to note is that just considering the interaction between a pair of moments, it is possible to see the origin of a magnetic anisotropy, which is that which is shown in Fig 4.16(c) and has the form of a uniaxial anisotropy. This is an important realization since the behavior of an assembly of nanoparticles will clearly be different from the one of an isolated particle. To account for all interactions in an assembly, we need to sum all of the contributions from all the nanoparticles. This can be performed as:

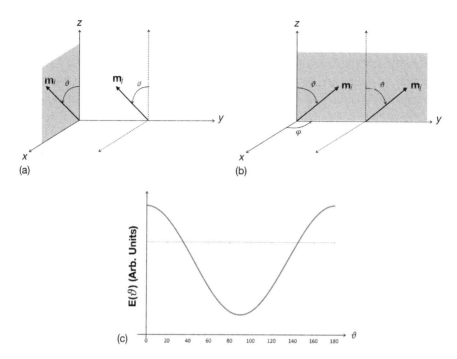

FIGURE 4.16 Variation of the magnetic moments of two nanoparticles (a) in the plane perpendicular to the axis joining them $(x - z$ plane) and (b) from the direction perpendicular to the axis joining them to the direction parallel to the axis $(y - z$ plane). (c) Variation of the dipolar energy for the case illustrated in (b) as given by Eq. (4.104).

$$E_{Tot}^{Dip} = \frac{1}{2}\sum_i \sum_{j \neq i} E_{ij}^{Dip}(\theta, \phi, \vartheta, \varphi) \qquad (4.105)$$

While it is evident that this summation over all the particles in an assembly is complex, a number of approaches and simplifications are possible in practice which renders it possible to account for the overall effect of the interactions. This said, it should also be evident that the global effect will depend on the specifics of the spatial distribution of all the magnetic nanoparticles. Of particular importance will be the particle density and the overall form of the spatial distribution. Variations in particle size and orientation, including their anisotropy axes, will further complicate the treatment and interpretation of experimental results. It is common to use averaging and size distributions to assist in this evaluation. We will consider this in more detail when we discuss ferromagnetic resonance measurements in magnetic nanoparticle assemblies in a later section.

Mössbauer spectroscopy has proven to be a powerful tool in the study of magnetic particles and many studies have been performed on a broad range of magnetic systems, where size, shape, anisotropy, surface effects and interactions have been investigated. For a more detailed introduction, see for example Schmool and Kachkachi (2015, 2016) and Dormann et al. (1997) and references therein. We saw previously that the modified form of the Arrhenius law, as expressed in the Volger-Fulcher model, can be used as a phenomenological description which englobes the overall effect of interactions, Eq. (4.95). The effect of the dipolar interaction field, B_i, on the blocking temperature in the weak field limit can be expressed by:

$$T_B \simeq \frac{KV}{k_B \ln(\tau_m/\tau_0)} \left\{ 1 - \frac{\mu^2 \langle B_i^2 \rangle}{(2KV)^2} \left[\frac{4}{3} \ln\left(\frac{\tau_m}{\tau_0}\right) - 1 \right] \right\} \tag{4.106}$$

Deviations to the model, first presented by Morup and Tronc (1994), are expected for large surface anisotropies for smaller particles where spin canting effects can occur. A more simplified model expresses directly the interaction energy, E_{int}, an adjustable parameter, which takes the form:

$$T_B \simeq \frac{KV + E_{int}}{k_B \ln(\tau_m/\tau_0)} \tag{4.107}$$

The density of nanoparticles plays a vital role in the strength of interactions, and it is important to note the interplay of particle size and separation in this context. This will affect the interparticle separation as well as the effective energy barrier height, which determines the switching properties and hence the superparamagnetic behavior, if any. Frequently, in practical situations, the nanoparticle assembly will have a size distribution, and this will lead to a distribution of the switching field. This will also compound the description of the assembly.

With regard to the possibility of other interaction mechanisms, the treatment is rather complex and depends crucially on the nature of the intervening media between the nanoparticles. For example, the RKKY interaction is in itself a complex mechanism, which occurs for metallic materials and depends on a number of parameters, among which is the form of the Fermi surface of the metallic matrix into which the magnetic nanoparticles are embedded. The case for crystalline solids involves the consideration of the orientation of the crystalline lattice between the particles and the relative orientation of the magnetic moments will have to account for all other magnetic moments in their vicinity. This problem is extremely complex and has not really been treated in a rigorous manner for the case of magnetic nanoparticle assemblies. We will outline some of the principal factors regarding the RKKY interaction in Section 4.4.4. The case has been extensively studied for magnetic multi-layer systems due to the relatively simple geometry, in which we can define the orientations of the crystalline axes for the different layers. In general, we can consider an effective interaction that characterizes the overall effects of the interactions between particles in the form of an effective field, as we have effectively done for the case of the dipolar interaction.

4.3.6 *Applications of Magnetic Nanoparticles*

As we mentioned in the introduction to this section, the interest in magnetic nanoparticles is strongly driven by the broad range of applications of these systems. Some examples of potential and actual applications are: catalysis including nanomaterial-based catalysts, biomedicine, and tissue-specific targeting, magnetically tunable colloidal photonic crystals, microfluidics, magnetic resonance imaging, magnetic particle imaging, data storage, environmental remediation, nanofluids, optical filters, defect sensors, and cation sensors. This list is rather long and a full treatment would be beyond the scope of this book. What is readily clear is the strong bias towards the use of magnetic nanoparticles in biomedical applications, which include labeling, filtering, and imaging applications. This is potentially a huge market. Magnetic data storage also offers huge potential for future device applications. In this section, we will outline some of these principal applications. It is worth noting that the literature in research on magnetic nanoparticles is vast and this is particularly true of the research into biomedical applications. While we will give a brief summary below, a good starting point for the interested reader would be a series of review articles published in J. Phys. D: Appl. Phys in 2003 and updated in 2009, by Pankhurst et al. (2003, 2009), Berry and Curtis (2003), Berry (2009), Tartaj et al. (2003) and Roca et al. (2009).

In the area of biomedical applications of magnetic nanoparticles, there are four principal areas of application that can be defined in the diagnosis and treatment of diseases: (i) Magnetic separation of biological bodies in the development of diagnostics, (ii) Magnetic nano-carriers which can contribute to drug delivery, (iii) Radio-frequency controlled magnetic particles for cancer treatment via hyperthermia, and (iv) Magnetic resonance imaging enhancement. In each of these areas, the magnetic nanoparticles are used in a specific manner, relying on the magnetic properties of the nanoparticles to perform a specific task. In order to be able to perform these tasks, however, it is usually necessary to functionalize them so as to be able to direct them to the specific region where the tasks are to be performed. The layering and coating of magnetic nanoparticles used in biomedical applications also serve to improve their biocompatibility and immunogenicity. The aggregation of magnetic nanoparticles is another factor, which must be dealt with if they are to be of practical use as biological tools in therapy and diagnostics. Once again, the coating of the nanoparticle is necessary to avoid such effects. Indeed, the surface modification of the magnetic particle also requires protection from an aggressive environment such as the human body. Magnetic particles can be absorbed by proteins and phagocytosed by the vascular endothelial system (Guo et al., 2018). The functionalization process thus serves also as a stabilization technique to allow the particle to survive in the body and thus be useful for the various applications indicated above.

There are many options available for the functionalization of magnetic nanoparticles. Some examples include polyethylene glycol (PEG), polyethyleneimine (PEI), folic acid (FA) liposomes, as well as noble metals and inorganic materials. In Table 4.3, we give some examples of commonly used compounds in the functionalization of magnetic nanoparticles. Some of these are illustrated in schematic form in Figure 4.17, with further architectures shown in Figure 4.18. Clearly, there is a vast range of treatments and coatings that can be applied to the surface of the nanoparticles and the choice will largely depend on the nature of the application envisaged, as is indicated in Table 4.3.

Once functionalization has been performed, the nanoparticles must be injected into the body, which generally means into the blood circulation, which serves as its transport mechanism to the region of interest for treatment or diagnostics. Studies have shown that once in the circulatory system, nanoparticles of sizes above 50 nm do not transport in a diffusive manner, aided by pressure gradients from the blood vessels. In fact, they tend to remain in circulation and attach to the walls of the vascular system. This increases the risk of thrombosis and is generally to be avoided. This leads to the notion that only particles of 5–10 nm are appropriate for most forms of therapy (Berry and Curtis, 2003).

The biomedical applications of magnetic nanoparticles are generally classified into the *in vivo* and *in vitro*. In the former, the principal uses are for therapeutic (hyperthermia and drug targeting) and diagnostic applications, which generally means NMR imaging enhancement. For the latter, the main

TABLE 4.3 Commonly Used Compounds for the Surface Modification and Functionalization of Magnetic Nanoparticles [RES: reticuloendothelial system. Adapted from (Guo et al., 2018)]

Compound	Advantages	Applications
PEG	Enhanced water solubility of NPs, reduced RES phagocytosis, increased blood circulation time	MRI, tumour diagnosis and treatment
PEI	Good biocompatibility	Gene and drug vectors
Polyvinyl alcohol (PVA)	Elevated stability and reduced particle aggregation	MRI, vectors, bio-separation
Glucan	Elevated stability and extended in-vivo circulation time	Drug vectors
Chitosan	Good stability and biocompatibility	Vectors, thermotherapy
Liposome	Good biocompatibility	Treatment of tumours, thermotherapy and MRI
FA	Good biocompatibility, essential small molecule vitamin	Treatment of tumours (breast, cervical and ovarian cancers) targeted receptors and diagnosis
Gold	Biocompatible, optical properties for biological applications	Tumour diagnosis and MRI

applications concern the use of magnetic nanoparticles in diagnostic separation, selection, and magnetorelaxometry.

For *in vivo* applications, the role of particle size and surface functionality are fundamental factors. For example, even in the absence of surface functionality, small iron oxide nanoparticles in the superparamagnetic regime have a strong influence on the *in vivo* biodistribution which is dependent on their diameter. Iron oxide particles in the 10-40 nm range can be important for prolonged blood circulation. They can cross capillary walls and are frequently phagocytosed by macrophages which transit to the lymph nodes and bone marrow.

An important *in vivo* application is therapeutic hyperthermia. This is thermotherapy aimed at killing tumor cells with thermal energy at a specific tem-

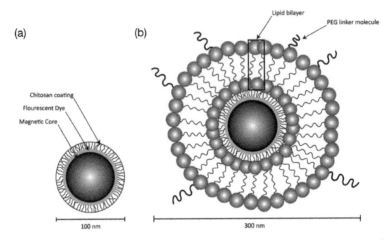

FIGURE 4.17   Principal structure of the two magnetic nanoparticles (MNPs). (a) The chi-MNP consists of a magnetic iron-oxide core covered by a lipophilic green fluorescent dye and a second layer of chitosan coat that prevents aggregation with other MNPs; (b) The lip-MNP is made from the chi-MNP by encapsulation inside a liposome. The liposomes were additionally PEGylated for use in vivo. Reprinted from T. Linemann, L. B. Thomsen, K. G. du Jardin, J. C. Laursen, J. B. Jensen, J. Lichota, & T. Moos, (2013). *Pharmaceutics, 5*, 246, under the Creative Commons Attribution License 3.0.

perature. It can be used alone or as part of a therapy program in conjunction with surgery, radiotherapy or chemotherapy. There have been many studies on the thermal treatment of cancers dated back to the 1950s. For example, $\gamma-Fe_2O_3$ particles with diameters in the range of 20–100 nm exposed to ac magnetic fields at 1.2 MHz were experimented with in early studies. The therapeutic efficiency of paclitaxel has been reported to improve by 10–100 times when maintained at 43°C for 30 minutes. Furthermore, the effectiveness of chemotherapeutics with low cytotoxicity at normal temperature can be doubled after heating. The range of treatments that have been experimented with, including a range of materials as well as field strengths and frequencies, has lead to procedures which involve the dispersion of the magnetic nanoparticles throughout the target tissue and then applying the relevant magnetic field strength and frequency so as to heat the region of interest, wherein broad terms, treatment aims to maintain a temperature of around 42°C for 30 minutes, which is sufficient to destroy cancer cells. The principal issue is that by targeting the cancerous tumor cells, the heating is highly localized and restricts the heating so as to limit damage to healthy tissue. The first clinical trials in human patients were performed in 2007 using aminosilane coated

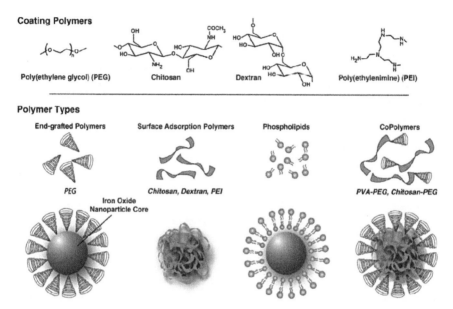

FIGURE 4.18 Illustration depicting different architectures for assemblies of polymers onto the surface of magnetic nanoparticle cores. Reprinted from O. Veiseh, J. W. Gunn, & M. Zhang, (2010). *Adv. Drug Deliv. Rev., 62*, 284, ©(2010), with permission from Elsevier.

iron oxide nanoparticles injected into various sites throughout a glioblastoma multiform tumor, a severe form of brain cancer (Maier-Hauff et al., 2007). Tumor site injections were made based on MRI scans. The superparamagnetic particles, of core size 15 nm, were dispersed in water at a concentration of 112 $mg_{Fe}\,ml^{-1}$. The tumors were injected with 0.1–0.7 ml of the magnetic fluid and then subjected to a magnetic field in the range 3.8–13.5 $kA\,m^{-1}$ at a frequency of 100 kHz. Follow-up CT scans noted that the magnetic deposits were stable over several weeks after injection. Patient tolerance to the treatment was good with promising results.

The mechanism of hyperthermia treatment in tumor tissue is thought to be due to the cancer cells' greater sensitivity to elevated temperatures. Thermotherapy may function via one of the following mechanisms: the reduction of vascular endothelial cell regeneration which destroys the vascular structure, reducing the activity of the enzyme system on tumor cell membranes, destroying mitochondrion which results in energy supply disorder and the inhibition of the activity of DNA polymerase and ligase in tumor cells resulting in DNA and RNA synthesis being disrupted and regulating the expression of apoptosis-related genes thus inducing apoptosis of cancer cells.

The hyperthermia used in the treatment of cancer relies on the local heating produced in the magnetic nanoparticles. This results from the hysteric properties of the magnetic nanoparticles themselves. This can be quantified as the heat generated per unit volume as expressed by the area of the hysteresis loop multiplied by the frequency:

$$P_{FM} = \mu_0 f \oint H \mathrm{d}M \qquad (4.108)$$

Other mechanisms, such as eddy current losses and FMR are excluded since the frequencies involved are too low for any significant effects to be expected. Clearly, the heat generated will depend specifically on the magnetic anisotropies of the particles as well as size-related effects. Above the superparamagnetic limit, there is no implicit frequency dependence of Eq. (4.108) and we can determine this from static magnetization measurements. In principle, we can expect large hysteresis for highly anisotropic materials such as NdFeB or SmCo, however, these are not practical for hyperthermia treatments due to the high magnetic fields that would be required, though minor loops could also yield significant heating. Square loop M-H curves should give the best results and this can be achieved using particles with uniaxial anisotropies which are aligned with the applied field, see Figure 4.9(a).

Another mechanism for generating heat in nanoparticle systems, which can be exploited for hyperthermia applications, is using magnetic fluids in which the particles are in the superparamagnetic state. Such a system is also referred to as a *ferrofluid* and exhibits some interesting properties in itself. The physical basis of the heating produced in SPM particles with an AC magnetic field is understood based on the Debye model and relies on the relaxation mechanisms of the magnetization subject to a changing magnetic field. This can mean that the magnetization of the nanoparticles can lag behind the applied field for frequencies superior to the inverse of the relaxation time. Relaxation in itself will depend on the intrinsic properties of the nanoparticle, via the Néel relaxation, with characteristic time, τ_N, but also on the fluid properties whereby the particles can rotate in the liquid, giving rise to a Brownian rotation that is characterized by a relaxation time, τ_B. Generally, the Brownian rotation losses are maximized at lower frequencies than this due to the Néel relaxation. For small-amplitude fields, the response of the nanoparticle in an AC field can be described by the complex susceptibility, $\chi = \chi' + i\chi''$, where both real and imaginary components are frequency dependent. The out-of-phase χ'' component gives rise to heat generation according to (Rosensweig, 2002):

$$P_{SPM} = \mu_0 \pi f \chi'' H^2 \qquad (4.109)$$

where $\chi'' = \omega \tau \chi_0 / (1 + \omega^2 \tau^2)$, in which the total relaxation time is expressed as $1/\tau = 1/\tau_B + 1/\tau_N$ and χ_0 is the static susceptibility. For the Brownian relaxation, the characteristic time depends on the fluid viscosity, η, and the hydrodynamic volume of the particle, which is larger than its magnetic volume due to the adsorbed surfactant coating, δ, with $V_H = (1 + \delta/R)^3 V_M$, R being the particle radius and $V_M = 4\pi R^3/3$ is the magnetic volume of the nanoparticle. The relaxation time is expressed as:

$$\tau_B = \frac{3\eta V_H}{k_B T} \qquad (4.110)$$

The Néel relaxation time is given by (Brown, 1963):

$$\tau_N = \frac{\sqrt{\pi}}{2} \tau_D \frac{e^\Gamma}{\Gamma^{3/2}} \qquad (4.111)$$

with $\tau_D = \Gamma \tau_0$ and $\Gamma = K V_M / k_B T$, where the anisotropy K can be of magnetocrystalline or shape origin and $\tau_0 \sim K^{-1}$. The interpretation of Eq. (4.109) is that if M lags behind H, there is a positive conversion of magnetic energy into internal energy in the form of heat.

The measurement of heat generation is usually expressed in units of SAR or specific absorption rate ($W\,g^{-1}$). Multiplying SAR by the particle density yields P_{FM} and P_{SPM}. From this, it is found that for most practical magnetic nanoparticles, the applied field strengths should be around 100 $kA\,m^{-1}$ if they are to fully saturate. Minor loops can be used, but will yield lower SARs. SPMs are capable of generating significant levels of heating at lower fields and offer more promising potential in hyperthermia applications.

Drug delivery or targeting has emerged as a key technology in biomedical science and is one in which magnetic nanoparticles can play a vital role. Conventional drug deliveries rely on supplying a sufficient amount of the pharmaceutical substance, such that enough arrives at the desired region of application, and often require flooding the body with the drug via intravenous injection. The drug is then being transported via the circulatory system. This can often lead to deleterious side-effects as the drug will also attack regions of normal healthy tissue. Side-effects are well-known in chemotherapies and other treatments via the general systemic distribution of therapeutic drugs.

Recognition of these problems led researchers to propose alternative methods of delivery and in particular the use of magnetic carriers to target specific sites such as tumors within the body. The aims of this approach are first to reduce the extent of systemic distributions of the cytotoxic drug, to reduce the associated side-effects and secondly to reduce the dosage required by a more effective localized targeting of the drug.

The principal idea is to use the magnetic particle as a vehicle upon which a cytotoxic drug is attached. The nanoparticles will form an assembly of surface-functionalized biocompatible carriers which are injected intravenously into the patient in the form of a ferrofluid. Once in the bloodstream, an external magnetic field is used to create a high gradient field which forces the magnetic particles to concentrate in a specific target site, as illustrated schematically in Figure 4.19. Once the drug has been delivered to the target zone, it can be released from its host via enzymatic activity or changes in physiological conditions, such as pH, osmolality or temperature. The physical principles responsible for directing the magnetic targeting system are

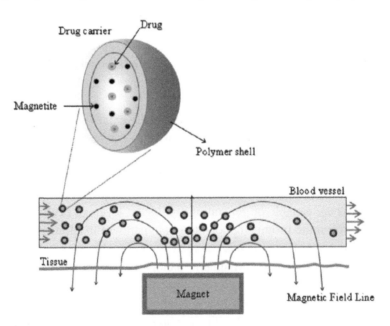

FIGURE 4.19   The schematic representation of the magnetic drug targeting; magnetic drug carriers disintegrate in the target zone and release the drug. Reprinted from J. Yang, H. Lee, W. Hyung, S.-B. Park, & S. Haam, (2006). *J. Microencapsulation, 23*, 203, by permission of the publisher (Taylor & Francis Ltd, http://www.tandfonline.com (accessed on 30 March 2020)).

similar to those used in magnetic separation. We will discuss these aspects shortly in our treatment of the magnetic separation applications of magnetic nanoparticles.

The effectiveness of the delivery system used in magnetic targeting will depend on a number of factors, including the magnetic field strength and gradient of the applied field, the magnetic properties of the MNP system, their concentration and quantity. Added to this, we must also remember that the environment in which the particles are introduced can have an important influence on the effectiveness of delivery. For example, the hydrodynamic conditions in the bloodstream, such as the blood flow rate, infusion route and circulation time, will play a key role. To compound the issue, physiological issues will also determine how effective treatment can be. Factors such as tissue depth to the target site, reversibility, and strength of drug/carrier binding as well as the tumor volume have to be taken into account.

For the purpose of delivery, larger magnetic particles are easier to direct using an external field and can better withstand the flow dynamics in the bloodstream. The field gradients can be supplied by strong permanent magnets such as NdFeB fixed over the region of the target site. The strengths and gradients of fields required will also depend on the type of nanoparticles employed as well as factors related to the target site. Targeting is expected to be more effective when the target site is situated in regions of slower blood flow and close to the source of the magnetic field.

In addition to the targeting for drug delivery, this method can also be applied for the delivery of nucleic acids and proteins. Furthermore, the manipulation and control of cells and sub-cellular structures is possible via magnetic nanoparticle actuation. This can be adapted for examining cellular mechanics, ion channel activation kinetics as well as tissue engineering and regenerative medicine. This signals a new approach to the manipulation and remote control of specific cellular components, see Pankhurst et al. (2009).

In biomedicine, it can be advantageous to separate out specific biological components from their native environment in order to prepare concentrated samples for analysis or other uses. Magnetic separation is one method of achieving this goal and employs magnetic nanoparticles, typically in an *in vitro* environment. Such separation is usually performed in a two-step process; the first is labeling or tagging the biological entity with a magnetic nanoparticle or particles, and then subsequently separating them out via a fluid-based separation device. The tagging process is achieved through the chemical modification of the magnetic particle surface, as we discussed previously. Specific binding sites on the surface of cells are targeted by antibodies or other biological molecules. Since the antibodies specifically target the matching antigen, this provides an accurate process for labeling cells. For cells and larger bodies, agglomerates of particles can be used.

Once a body has been labeled, it is possible to then separate it from the native material by passing the fluid through a region in which an appropriate magnetic field gradient has been set up. For the separation to occur, a magnetic force must act upon the magnetic nanoparticles, which will then drag the magnetic particles attached to the biological entity. The magnetic force can be expressed in the form:

$$F_m = (\mathbf{m} \cdot \nabla)\mathbf{B} \tag{4.112}$$

where **m** is the magnetic moment the nanoparticle and **B** the magnetic induction. This force, therefore, depends on the gradient of the applied magnetic field and can be expressed as (Pankhurst et al. 2003):

$$F_m = V_M \Delta\chi \nabla \left(\frac{1}{2}\mathbf{B} \cdot \mathbf{H}\right) \tag{4.113}$$

Here, V_M represents the magnetic volume of the magnetic particle and $\Delta\chi$ the effective susceptibility. If $\Delta\chi > 0$, the magnetic force acts in the direction of the steepest ascent of the energy density $\mathbf{B} \cdot \mathbf{H}/2$, a scalar field. This explains why, for example, when iron filings are brought near to the pole of a permanent magnet, they are attracted towards that pole. It is also the basis for the motion of magnetic particles used in the biomedical applications discussed above.

For the magnetic particle attached to its tagged molecule/cell in the fluid, the force to which it is subjected to must overcome the hydrodynamics force of the particle in the fluid. This force takes the form:

$$F_d = 6\pi\eta R_M \Delta v \tag{4.114}$$

R_m represents the radius of the magnetic particle and $\Delta v = v_m - v_w$ is the difference in the velocities of the cell and the water, which is used as a reference for the biological fluid. Equating the forces expressed in Eqs. (4.113) and (4.115), we can express the velocity of the particle with respect to the fluid (water) as:

$$\Delta v = \frac{R_M^2 \Delta\chi}{9\mu_0\eta}\nabla(B^2) = \frac{\xi}{\mu_0}\nabla(B^2) \tag{4.115}$$

where ξ is termed the magnetophoretic mobility, a parameter used to characterize the manipulability of a magnetic particle.

In its simplest form, magnetic separation can just involve the application of a permanent magnet to the side of a test-tube to attract the magnetic particles, which are attached to the species we wish to remove from the fluid. While being simple, it isn't necessarily the most efficient means of species removal. It is often preferable to produce regions of high magnetic field gradients to capture the magnetic nanoparticles as they flow through the carrier medium. The magnetic particles can be trapped along with the biomaterials for removal, allowing the fluid and other contents to continue on their path, thus filtering out the unwanted species. Despite this method being more efficient than the first, clogging and obstruction can occur in the column. The application of field gradients from a quadrupole arrangement creates a radial gradient in the outward direction and can more efficiently separate out the magnetically tagged components in the fluid. By moving the magnetic field instead of the fluid, it is possible to achieve fluid flow fractionation. This process splits the fluid outlet into fractions containing the tagged cells or proteins with differing magnetophoretic mobilities.

Magnetic resonance imaging (MRI) is a powerful tool used frequently in medical diagnostics to image biological tissues. Its importance cannot be overestimated in biomedical applications. Physically, the technique relies on a large number of protons present in the biological matter to measure the magnetic precessional resonance of the proton, which has a very small magnetic moment. Since there is a large quantity of water in biological systems, the protons are commonly derived from the hydrogen nuclei in water. The resonance phenomena at play are the nuclear magnetic resonance (NMR) where a static field, $\mathbf{B_0}$, is applied to align the magnetic moments, while a small time-varying field is applied perpendicular to this which drives the resonance. The resonance is analogous to the Larmor precession, having a characteristic resonance frequency, which is expressed as: $\omega = \gamma B_0$, see Section 6.2. We will describe in more detail the derivation of this expression in Section 4.7.2. So given that the gyromagnetic ratio for the proton is $\gamma = 2.67 \times 10^8 \, \mathrm{rad\,s^{-1}\,T^{-1}}$, in a applied field of $B_0 = 1\mathrm{T}$, the Larmor frequency will be in the radio frequency range with a value of $\omega_0/2\pi = 42.57$ MHz. In practice, a pulsed-field is often used with a duration which is short enough to allow a coherent response from the magnetic moment of the protons in the MRI scanner. The response is measured as a function of time using pick-up coils, which detect both the resonance frequency and the relaxation process. The precessional motion can be decomposed into the different Cartesian components, from which two relaxation times are defined:

$$m_z = m(1 - e^{-t/T_1})$$
(4.116)

$$m_{x,y} = m\sin(\omega_0 t + \phi)e^{-t/T_2}$$
(4.117)

These are expressed as T_1 and T_2, the longitudinal (spin-lattice) and transverse (spin-spin) relaxation times, respectively. The former relates to the energy transferred to the lattice from the excitation field, resulting in a heating of the sample. The latter is the relaxation that results in the decoherence of the spin motions and can also be affected by local inhomogeneities in the applied longitudinal field, leading to a shift in the relaxation time, which is expressed in the form:

$$\frac{1}{T_2^*} = \frac{1}{T_2} + \gamma \frac{\Delta B_0}{2}$$
(4.118)

where ΔB_0 is the variation of the field due to variations in the homogeneity of the applied field or in the magnetic susceptibility of the system. The relaxation times, T_1 and T_2^*, can be shortened by the use of a magnetic contrast agent. There are a number of commercially available agents, such as PM Gd ion complexes and SPM nanoparticles, which are commonly iron oxides. The SPM particles are magnetically saturated by the applied fields in the MRI scanner. These provide a strong perturbation to the local field leading to significant local inhomogeneities in the field which will effectively shorten the relaxation time T_2^* as can be seen from Eq. (4.118).

It is well-known, that the inclusion of magnetic particles within tissues enables a much larger signal to be obtained, thus making it significantly more sensitive. What is required for MRI imaging, is for the particles to be selectively taken up by specific cells, which make up the tissue of interest? For example, dextran-coated superparamagnetic iron oxide (SPIO) is biocompatible and was first used as a liver-specific contrast agent. They are selectively adsorbed to the reticuloendothelial system (RES). The improved enhancement of MRI relies on the differential uptake of the nanoparticles by different tissues. The particle size is also important: particles of 30 nm are more rapidly retained by the liver and spleen, while particles of diameter 10 nm or less are not so easily imaged. The smaller particles will thus have a longer duration in the circulatory system and are collected by RES cells throughout the body, including in the lymph nodes and bone marrow. Small particles have also been used as contrast agents for the vascular system and the central nervous system. Since the particle uptake by tumors is rather weak as they do not have an effective RES, contrast agents do not alter their image. This can be used to identify malignant tumors in the lymph nodes as well as in the liver and brain. Indeed, we tend to classify two main groups of particles according to their size; SPIOs, for particles of 50 nm and larger, and USPIOs (ultra-small superparamagnetic iron oxides), which have a diameter of less than 50 nm. As an illustration of the MRI image enhancement, we show a sequence of MRI scans, Figure 4.20, showing the different mapping techniques using the different relaxation times and comparing the effects of the injected nanoparticles before and at different stages after injection for endothelial progenitor cells (EPCs). Research is intense in this field and applications have reached a point where the use of nanoparticles in MRI is standard. Magnetic nanoparticles are also commonly used as an image enhancement tool in positron emission tomography (PET). Here, the high specific activity of radiopharmaceuticals is used to obtain images of high quality, providing a further diagnostic tool with high sensitivity. It is also common for both MRI and PET to be used simultaneously in imaging biological tissues.

Theranostics is the domain of biomedicine, which aims to both identify a disease state and simultaneously deliver therapy as a new approach to personalized medicine. This challenge requires magnetic nanoparticles to be multifunctional. Since magnetic nanoparticles can be used as both a vehicle for targeted drug delivery as well as an agent for image contrast in MRI, they would be an ideal candidate for such applications. One of the most interesting stimuli-responsive NP-based drug carriers is arguably the magnetoliposome. This is a combination of a liposomal drug carrier and magnetic nanoparticles, which also allows a triggered response for releasing the drug payload. Liposomes can be designed to be thermosensitive, undergoing a phase transition from an impermeable gel state to a permeable liquid-crystalline state when a defined temperature barrier is reached. Combining these two independent functionalities yields a versatile nanoplatform, which may provide combined drug delivery and hyperthermia treatment at a specific target site under co-instantaneous tracking via MRI. In Figure 4.21 we illustrate a number of nanoparticle designs that can be used in a range of biomedical applications as discussed above.

While we have outlined some of the main applications of magnetic nanoparticles in biomedicine, this is really just the tip of a very large iceberg. Progress is very fast in this field and it is a superb example of what can be achieved by an interdisciplinary approach such as this, bringing together

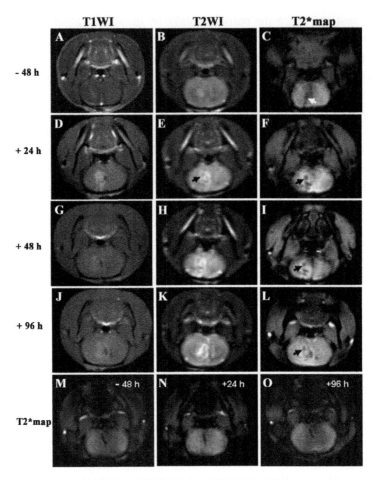

FIGURE 4.20   MRI tracks the USPIO-labeled EPCs in vivo. (A-C) Images collected at 48 h before injection of the labeled cells into the tail vein of the rats. The arrow shows the passage of the needle in the right basal ganglia for the C6 cell injection. (D-F) Images captured at 24 h after injection of the labeled cells. A low signal (arrow) presented at the margin of the tumor under the T2-weighted image (E). The low signal was more significant on the T2* map (F). These low signals were different from the passage of the needle in panel C. The solid tumor appeared as hyperintensity Gd-enhanced under T1-weighted image (D). (G-I) Images captured at 48 h after injection of the labeled cells. (J-L) Images captured at 96 h after injection of the labeled cells. (M-O) T2-weighted images of the control group with non-labeled EPCs at the indicated time points. Reprinted with permission from L. Wang, L. Chen, Q. Wang, L. Wang, H. Wang, Y. Shen, X. Li, Y. Fu, Y. Shen & Y. Yu, (2014). *Oncol. Rep., 32*, 2007. ©(2014) Spandidos Publications.

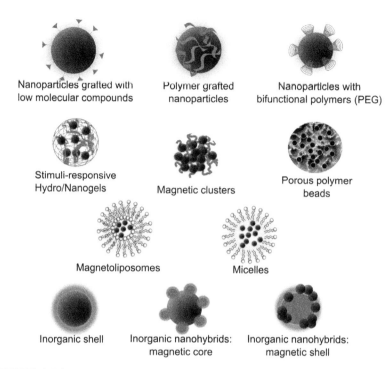

Nanoparticles grafted with low molecular compounds

Polymer grafted nanoparticles

Nanoparticles with bifunctional polymers (PEG)

Stimuli-responsive Hydro/Nanogels

Magnetic clusters

Porous polymer beads

Magnetoliposomes

Micelles

Inorganic shell

Inorganic nanohybrids: magnetic core

Inorganic nanohybrids: magnetic shell

FIGURE 4.21   General types of iron oxide nanoparticle arrangements with polymers, molecules, and inorganic nanoparticles. These architectures can be used in a number of nanoparticles/hybrids that are not depicted to scale. Reprinted from *Biotechnol. Adv., 33*(6), K. Hola, Z. Markova, G. Zoppellaro, J. Tucek, and R. Zboril, Tailored functionalization of iron oxide nanoparticles for MRI, drug delivery, magnetic separation and immobilization of biosubstances, 1162–76, Copyright (2015), with permission from Elsevier.

physicists, chemists, biochemists, biologists, and medical practitioners. This is probably one of the most interdisciplinary areas of current scientific research.

4.4   Magnetic Nanostructures

Magnetic nanostructures can come in a variety of forms and essentially their physical properties will depend on the shape of the magnetic entity, their magnetic anisotropies and on any interactions between the nanostructures, typically via dipolar interactions. At play are the magnetic length scales as discussed at the beginning of this chapter. There has been sustained and varied interest in the various geometries available, via lithographic and other means, to control and tailor magnetic properties. Length scales are important since they will define the size of the entities for which the object is a single domain and also the "preferred" magnetic domain configuration that is observed. Added to this will be the effect of any applied magnetic field, including its direction, and also the frequency for time-varying fields. In some respects, the properties of magnetic nanostructures follow the discussion for magnetic nanoparticles, see Section 4.3, where effects such as superparamagnetism must also be taken into account. The principal difference here is that when we discuss the properties of magnetic nanostructures, we generally mean those pertaining to ordered and well-defined structures. In this section, we will provide a brief overview of some of the more representative properties of magnetic nanostructures, which we classify according to their form. There are some issues pertaining to the domain structures and critical sizes that will also be discussed as well as the dynamics associated with wall and vortex motion in these nanostructures. All other aspects pertaining to the true spin dynamics of nanostructured systems will be dealt with in Section 4.7 of this volume.

4.4.1   Nanodots

Nanodots can come in a variety of shapes, most typically are disks, squares, rectangles, and ellipses, though triangles and other forms have also attracted interest. These are probably the simplest forms of nanostructures that we can consider. As we noted above, the size and particular shape of the nanostructure will play a significant role in determining its magnetic ground state. By this, we mean its magnetic domain structure. In Figure 4.22, we illustrate the effect of the nanodot shape on the structure of magnetic domains, where a thin Fe film has been patterned using focused electron beam induced deposition (FEBID). In these structures, the large axis is maintained at the same length. The magnetic domain pattern is analyzed using magnetic force microscopy (MFM). The figure shows the nanodot topography, MFM image and the simulation of the magnetic domain structure as evaluated using the OOMMF (object-oriented micromagnetic framework) package. The comparison between the experiment and simulation is very satisfying. What is immediately evident is that the domain structure maintains the symmetry of the object shape, where we see clearly the n-fold symmetry maintained for the shapes from the ellipse (uniaxial or two-fold symmetry) to the hexagon, with its six-fold symmetry. The final panel shows a circular nanodot, which

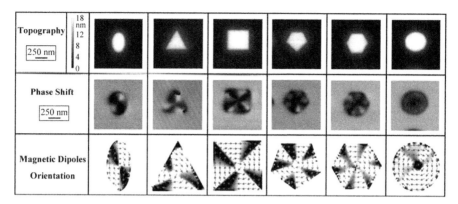

FIGURE 4.22 MFM study of magnetic nanodot structures. The top panel shows the topography for each nanodot, the center panel the corresponding MFM phase shift image, and the lower panel the OOMMF simulated domain structure for each shape. Reprinted from M. Gavagnin, H. D. Wanzenboeck, D. Belic, M. M. Shawrav, A.Persson, K. Gunnarsson, P. Svedlindh & Emmerich Bertagnolli, (2014). *Phys. Stat. Sol. A, 211*, 368, under a Creative Commons Attribution License. Wiley Publishes Open Access.

should have a continuous variation of the local orientation of the magnetization in the plane of the disk. In each case, the zero-field magnetic configuration is characterized by a central point about which the orientation of the local magnetization circulates. This is referred to as a magnetic vortex, which is a magnetic discontinuity or singularity and has a chirality and orientation or polarity, which can point either up or down. The convention for chirality (q) and polarity (p) is illustrated in Figure 4.23. As we discussed at the beginning of this chapter, one of the most fundamental characteristic length scales in magnetism is the exchange length. The energy balance of the exchange with the magnetostatic and anisotropy energies (and the Zeeman energy if there is an applied field) will determine the domain size and the wall width. Should these characteristic distances be of the order of the size of the nanodot, then single domain configurations are the most likely outcome. The nature of the vortex can be potentially exploited for state storage purposes, with the bit state being coded using either the core polarity or vortex chirality. The principal advantage over the conventional storage system being that there are four different possible states, allowing for a more efficient storage scheme.

As we noted above, vortex structures are fairly common in the domain patterns for magnetic nanodots. The application of a magnetic field will dis-

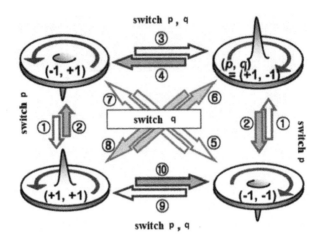

FIGURE 4.23   Illustration of the various chirality, polarity combinations and the switching between the different modes.

place the vortex core, much in the way it displaces a domain wall, i.e., favoring the increase of the magnetization of the nanodot in the direction of the applied field. For the sake of simplicity, we will consider the nanodots of a circular form. In Figure 4.24, we show the form of the hysteresis loop for a magnetic nanodot. The rather unusual shape can be understood as follows: the central region of the cycle shows a reversible behavior, which arises from the reversible displacement of the vortex core, which shifts position as a small magnetic field is applied. The behavior remains reversible until the core is driven out of the nanodot. This is referred to as annihilation. Further application of the magnetic field will gradually take the object to saturation. Upon reducing the field, a vortex core is nucleated. Hysteretic behavior is observed only in the region between the vortex core annihilation and nucleation. The different stages of the hysteresis loop are illustrated in the figure. As we noted earlier, the specific domain structure is intimately linked to the size and shape of the nanodot. The calculated phase diagram for the ground state domain structures is shown in Figure 4.25.   The curve is a universal phase diagram for soft magnetic dots since it has bee normalized to the exchange length. Near the origin, the magnetic state will either lie along the long axis of the nanodot when the magnetostatic energy is sufficiently compensated, or in the plane of the disk. These are essentially single domain states. Vortex states are stabilized for larger structures and occupy the upper region of the phase diagram. The transition between the vortex state and the single domain state is generally characterized by a quasi-uniform *leaf* state

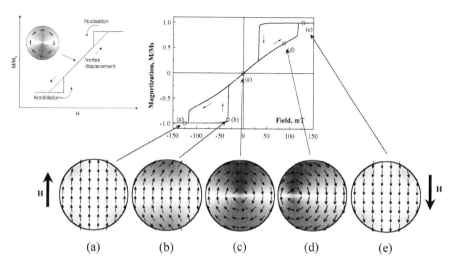

FIGURE 4.24 Hysteresis loop for a circular magnetic nanodot with a magnetic vortex core.

for thin circular dots. The vortex core has a radius which is comparable to the exchange length, $L_{\text{exch}} = L_E = \sqrt{A/M_s^2}$, which is typically on a scale of around 10 nm.

In addition to the vortex structures discussed above, anti-vortex structures are also observed in magnetic nanostructures and are predicted from simple theoretical models based on a Heisenberg Hamiltonian with dipolar interactions or other forms of anisotropy. The forms of the vortex and anti-vortex structures in magnetic nanosystems are illustrated in Figure 4.26. The type I vortex is the structure we discussed above. The other structures result from other conditions related to the sample shape and local anisotropies as well as interactions with vortex structures. Typically the energy for vortex formation is proportional to $\ln R$, where R is the vortex size.

Further to the static behavior of nanodots, there has been intensive work on the study of the dynamic properties of the vortex. In this case, a small alternating magnetic field is applied to the nanodots, which excites a characteristic gyrotropic motion of the vortex core, typically around the center of the dot. This motion exhibits a resonance character, where a large oscillatory response is observed at a given frequency. The motion of the magnetic vortex core is analogous to the Larmor motion of a charged particle and will be treated below. This was illustrated by A. A. Thiele (1973), who proposed a dynamic equation for the steady-state solution of the domain motion in magnetic bubbles. This approach, which was based on the Gilbert equation and

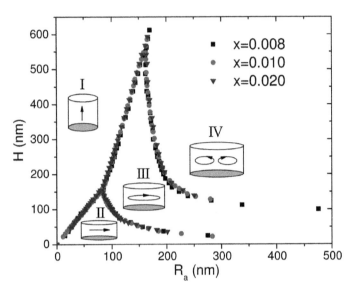

FIGURE 4.25   Scaled phase diagram of an elliptically shaped magnetic nanoparticle $(Ka^3/D = 1$ and $Ja^3/D = 5000)$ as a function of its semi-major axis (R_a) and
height (H) with an aspect ratio 2. Here a denotes the lattice constant of the material
and x is a scaling factor, which aligns the triple point of phases (I), (II), and (III).
The four competing phases are (I) out-of-plane ferromagnetism, (II) in-plane ferromagnetism, (III) single vortex state, and (IV) double vortex state. Reprinted figure
with permission from W. Zhang, R. Singh, N. Bray-Ali, & S. Haas, (2008.) *Phys.
Rev. B, 77*, 144428, ©(2008) by the American Physical Society.

is also related to the Landau–Lifshitz equation of motion, was adopted and is
the standard approach to dealing with the motion of the vortex core. We will
discuss the details of the Landau–Lifshitz equation in the section concerning
spin dynamics (see Section 4.7). The core position is expressed as follows:

$$\mathbf{G} \times \frac{d\mathbf{X}}{dt} + \mathbf{D}\frac{d\mathbf{X}}{dt} - \frac{\partial W(\mathbf{X})}{\partial \mathbf{X}} = 0 \qquad (4.119)$$

where $\mathbf{X} = \mathbf{X}(x,y)$ is the position vector of the vortex core, $\mathbf{G} = G\mathbf{z}$ is the
gyromagnetic vector, with $G = -2\pi qpLM_s/\gamma$ being referred to as the gyro-
magnetic constant. In this expression, p and q are the topological charges, as
indicated in Figure 4.23, and L is the thickness of the layer. The second term
represents the dissipation of the motion of the vortex core and the final term
describes the restoring force acting on the vortex shifted from the equilibrium
position, where $W(\mathbf{X})$ is a magnetic potential energy and describes the interaction of the vortex with in-plane applied fields and other vortices that may
be present. The above equation was explicitly introduced by Huber (1982)

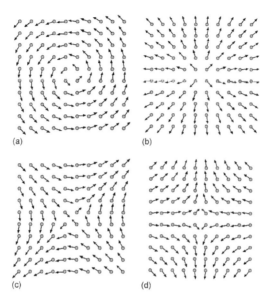

FIGURE 4.26 Schematic illustration of the vortex-like configurations for a square lattice: (a) type I vortex, (b) type I anti-vortex, (c) type II vortex and (d) type II anti-vortex. Reprinted from *J. Magn. Magn. Mater, 324*, B.V. Costa and A.B. Lima, Dynamical behavior of vortices in thin-film magnetic systems, 1999–2005, Copyright (2012), with permission from Elsevier.

and is the basis for many theoretical studies of vortex dynamics. Typically, the vortex gyrotropic mode has an excitation frequency in the hundreds of MHz to low GHz range, and will depend on a number of parameters, such as the dot size and shape as well as the fundamental magnetic properties of the dot material.

In the case where vortex couples to other vortices, interactions will give rise to a frequency (energy) shift and there are several modes which are coupled, as illustrated in Figure 4.27. The energy shift depends on the phase and polarity of the two vortex modes. As with other resonant systems, the variation of the resonance frequency can be related to the effective field or energy of the system and will determine the direction of the variation of the resonant frequency.

A comparison of the dynamics in a vortex and anti-vortex structures shows that the same excitation field produces oppositely moving vortex cores and a phase shift in the displacement of the vortex core. This is intimately related to the local spin configuration around the core region. This is schematically illustrated in Figure 4.28. We have not mentioned the excitation modes

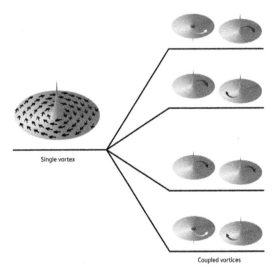

FIGURE 4.27 When two disks are brought close together (right), the magnetic vortices begin to move together. Their motion can be in phase (bottom two levels), or out of phase (top two levels). The vortex cores can also point in the same direction or in opposite directions, leading to four possible types of coupled motion.

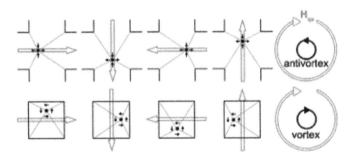

FIGURE 4.28 Coupling of an antivortex (upper sequence) and a vortex (lower sequence) to a quasistatic rotating magnetic field $\mathbf{H}_{qs}$ represented by green arrows. The Zeeman energy is minimized by the illustrated core deflections because of the increase of the volume of the domain which has a parallel magnetization with respect to the field. Both sequences correspond to one circulation (black circles), whereas $\mathbf{H}_{qs}$ only counter rotates in the case of the antivortex. Reprinted figure with permission from T. Kamionka, M. Martens, K. W. Chou, M. Curcic, A. Drews, G. Schütz, T. Tyliszczak, H. Stoll, B. Van Waeyenberge & G. Meier, (2010). *Phys. Rev. Lett.*, *105*, 137204, ©(2010) by the American Physical Society.

which occur at higher frequencies, since these correspond to the case, where the dynamics pertaining to the spin excitation of the system. This is the subject of the following section, where we consider the ferromagnetic and spin-wave resonance excitations of the system. The vortex dynamics discussed here are more akin to the domain wall dynamics of magnetic systems that are not at saturation. The domain wall resonance behavior shows resonant frequencies that also lie in the 100s of MHz range.

4.4.2 Antidots

Magnetic antidots are quite literally the opposite of a magnetic dot. In this case, the dot structure is made up of voids, which are typically lithographically formed in a thin magnetic layer. These will usually be of circular or square shapes in a regular array. The array can also have a structure such as a square or hexagonal form, with size and periodicity being important parameters which can also be used to modify the magnetic properties of the system. The principle parameters are defined in Figure 4.29, where we show the periodicity, p, the dot diameter, d, the effective diameter, d_{eff}, which takes into account the effects of the dot edge which can be effectively non-magnetic, and $\lambda = p - d$ the nearest edge-to-edge separation between the antidots. One of the main efforts at the understanding of the magnetic properties of antidot systems is their static properties and the magnetic configuration. These are typically studied via magnetometry and domain imaging,

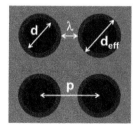 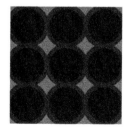

FIGURE 4.29 Left: Definition of the geometrical parameters of the arrays: p, d, d_{eff}, and λ. Orange color represents the magnetic material, while black areas are non-magnetic holes with diameter, d, and gray areas are the holes with effective diameter, d_{eff}. Right: Scheme of the astroid-shaped dots formed by the intersection of the holes in the dot (D) regime, where $p < d_{eff}$. Reprinted figure with permission from C. Castán-Guerrero, J. Herrero-Albillos, J. Bartolomé, F. Bartolomé, L. A. Rodríguez, C. Magén, F. Kronast, P. Gawronski, O. Chubykalo-Fesenko, K. J. Merazzo, P. Vavassori, P. Strichovanec, J. Sesé, & L. M. García, (2014). *Phys. Rev. B, 89*, 144405, ©(2014) by the American Physical Society.

where much of the interpretation is aided by computational simulations via packages such as OOMMF, MuMag, and mumax. In Figure 4.30, the effect of the antidot structuring on the magnetic hysteresis loop is illustrated in A, where the reference film (a) has a fairly isotropic behavior. The magnetic layers are 30 nm thick permalloy films. We note that the coercive field is significantly enhanced by an order of magnitude. The distribution of the antidot lattice is seen to influence the magnetic anisotropy, as seen from Figure 4.30 B, showing the easy (EA) and hard (HA) axes as a function of the orientation of the applied field, θ. The four-fold symmetry for the square lattice and the six-fold symmetry for the honeycomb and rhombus lattices are illustrated to the right in Figure 4.30 B. In this example, the square lattice structure exhibits the largest anisotropy.

The variation of the coercive field as a function of the periodicity is also the subject of many studies. It should be remembered that the variation of the periodicity will alter the effective coverage of the film and will have a critical value since the antidot structure will cease to exist once the period of the structure is of the same value of the diameter of the antidots themselves. In this case, the resulting structure will be the astroid-shaped dots, as illustrated in Figure 4.29. The general variation of the coercive field as a function of periodicity, p, is illustrated in Figure 4.31. The general indicated behavior shows that the coercive field increases with a reduction of the periodicity or surface coverage, $c = 1 - \pi d^2 / 2\sqrt{3} p^2$. This region corresponds to the antidot (AD) regime, where the dots are well spaced. Below this, a region defined as intermediate (INT) can be defined, where the distance between the dots, λ, is small but positive, and a change in the H_C dependence is clearly observed. As the separation between the dot structure is further reduced, the λ values will become negative and we enter the dot (D) regime or island structure. The coercive field in this regime drops off quite rapidly.

The magnetic reversal in antidot films can be severely altered with respect to the unpatterned thin film and domains tend to be aligned along the rows between the dots. The formation of magnetic domains is again dependent on the intimate relation of dot size and periodicity.

4.4.3 *Nanorings*

A magnetic nanoring can be viewed as a magnetic disk with the center removed. As with other nanostructures, the alteration of the magnetic properties arises due to the size and shape constraints imposed on the magnetic

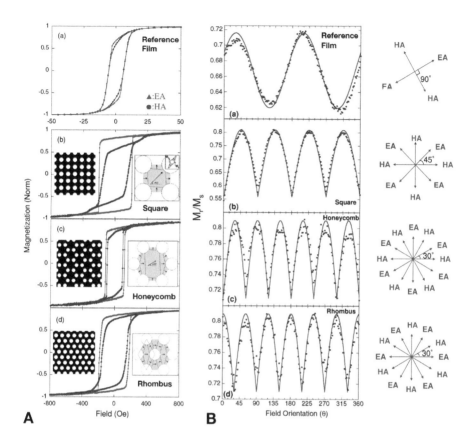

FIGURE 4.30 A: Representative hysteresis loops for field applied along with the EA and HA directions (as marked in the legend) for (a) continuous film, (b) square, (c) honeycomb, and (d) rhomboid lattice geometry. Insets on each M–H loop show the antidot structure (left) and the magnetic domain structure. B: The normalized remanent magnetization (squareness M_r/M_s) as a function of field orientation θ for (a) reference continuous film, (b) square lattice, (c) honeycomb lattice, and (d) rhomboid lattice antidot arrays. The dotted traces are experimental data, and the continuous lines were obtained from curve fitting. To the right of the curves are the easy axis (EA) and the hard axis (HA) distributions of the respective lattice geometry. Reprinted from C. C. Wang, A. O. Adeyeye & N. Singh, (2006). *Nanotechnol., 17*, 1629. Nanotechnology by Institute of Physics (Great Britain); American Institute of Physics Reproduced with permission of IOP Publishing in the format Book via Copyright Clearance Center.

domains supported by the structures. Once again, the size effects will be material dependent and in particular when the size of the structure is of the order of the exchange length. One particularity of the magnetic nanoring structure

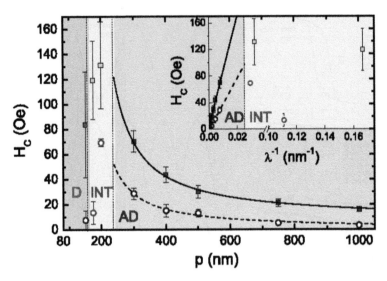

FIGURE 4.31   Coercive field H_C dependence on p of the antidot arrays on Co and Py are plotted as solid squares and open circles, respectively. The blue, green, and red colors represent data belonging to the AD, INT, and D regimes, respectively. In the inset, H_C vs λ^{-1} is plotted, showing a linear trend for p above the crossover. The function $H_C \propto 1/\lambda$, with the obtained value of $d_{\text{eff}} = 169$ nm and 166 nm, is plotted as solid and dashed black lines for cobalt and permalloy, respectively. Reprinted figure with permission from C. Castán-Guerrero, J. Herrero-Albillos, J. Bartolomé, F. Bartolomé, L. A. Rodríguez, C. Magén, F. Kronast, P. Gawronski, O. Chubykalo-Fesenko, K. J. Merazzo, P. Vavassori, P. Strichovanec, J. Sesé, & L. M. García, (2014). *Phys. Rev. B, 89*, 144405, ©(2014) by the American Physical Society.

is the fact, that the ground state often displays a vortex-like structure without a core. This means that near-zero remanence is possible. In Figure 4.32, we show some typical M-H loops for magnetic discs and nanoring structures (Zhang and Haas, 2010). From the saturated state, a reduction of the applied field will lead to a slight decrease of the magnetization to a remanent state, which is referred to as the *onion* state, as illustrated in Figure 4.32. Once a small reverse field is applied, a sharp drop in the magnetization will be observed as the sample switches from the onion state to the vortex state. A further jump will occur in the magnetization as the sample is returned to an onion state in the opposite sense. This allows us to define four stable magnetic states; two onion states in opposing directions and two vortex states with opposite chiralities.

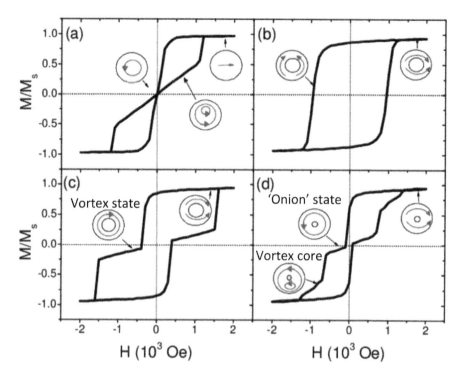

FIGURE 4.32 Magnetic switching processes in nanoring structures and their associated hysteretic behavior. (a) Hysteresis of magnetic disks, (b) one-step O–O switching of rings, (c) double switching O–V–O of rings, and (d) triple switching O–V–VC–O of rings. Reprinted figure with permission from W. Zhang & S. Haas, (2010). *Phys. Rev. B, 81,* 064433, ©(2010) by the American Physical Society.

4.4.4 *Artificial Spin Ice*

Spin ice systems refer to arrays of elongated magnetic dots which are of dimensions that support single domain structures, but are large enough to be athermal. Each magnetic dot can be represented as a single macro or Ising spin. Magnetic interactions between the dots via magnetostatic forces give rise to *frustration* and the ground state will depend on the physical distribution of the magnetic dots. The spin configuration can be altered by the application of a magnetic field. The ability to manufacture such systems is a direct result of the advanced state of fabrication tools, allowing the preparation of high-quality magnetic elements of reliable form integrity on a large scale.

The frustrating interactions can be observed throughout condensed matter physics. One of the early systems to be studied was water ice, where the frustration in the position of protons arises due to the tetragonal geometry of the crystal, and has lent its name to the analogous frustration in magnetic systems. The principal geometries studied have been square and kagome lattices. In either case, the interactions between the elements or macrospins produce frustration in the system, leading to the occurrence of thermodynamically metastable states. As such, artificial spin ice systems provide excellent models for disorder studies in solids.

Starting with a simple square geometry, we illustrate the nature of the interlinked vertices at which four conflicting spins meet, where each has a corresponding magnetic charge at its center. The general lattice is illustrated in Figure 4.33(a), indicating the 2D directions. In Figs. 4.33(b) and 4.33(c), the remanent configurations are illustrated after applying a saturation field in the [10] and [11] directions, respectively. The $2^4 = 16$ possible configurations of the artificial spin ice vertices are shown in Figure 4.33(d). The 2D square symmetry defines four vertex types, indicated as I–IV. Of these, those obeying the so-called *ice rules* with two-in and two-out (type-I and type-II) are energetically split due to the mixing of first and second nearest-neighbor interactions. This gives rise to a two-fold degenerate ground state (GS) with long-range order, which is formed from a checkerboard tiling of alternating type-I vertices (Morgan et al., 2011). The nearest-neighbor ordering imposes an antiferromagnetic second and third nearest-neighbor order, see Figure 4.34. It is noted that for saturation in the [11] direction, the GS configuration at remanence is unambiguous, but the GS after saturation in the [10] direction is not unique. Furthermore, it has been shown that applications of applied fields with fixed amplitudes are unable to drive the system to its GS (Budrikis et al., 2012).

The kagome lattice has a more complex looking lattice structure, being based on a hexagonal array whose vertices have three macrospins. In this case, the spin ice rule dictates a two-in one-out or one-in and two-out vertex configuration. The basic geometry and ground state for this system is shown in Figure 4.35. The ground state of the three-ring kagome lattice has two flux closure rings. The magnetic configurations for one to seven rings are illustrated in Figure 4.36, where the GS and the excited states are shown, with energies given relative to the GS, defined as $\Delta E = 0$. One of the principal findings was that as the number of rings is increased, the ability to achieve low-energy states becomes more and more difficult. This difficulty in achieving the lowest energy states is a direct result of the inherent frustration in these systems. Thermally activated changes have also been extensively con-

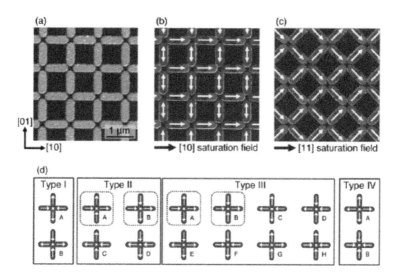

FIGURE 4.33 (a) Atomic force microscopy image of the artificial square spin ice array. The major symmetry axes are indicated in the lower-left corner. (b) Schematic illustration of the remanent magnetic configuration of the spin ice array after applying a saturation field parallel to the [10] direction and (c) parallel to the [11] direction. (d) The 16 possible remanent magnetic configurations for the artificial square spin ice vertices. After applying a saturation field along the [10] direction there is a fourfold degeneracy of the possible remanent vertex configurations. In comparison, after removing the [11] field, there is only one possible remanent vertex configuration. Reprinted from V. Kapaklis, U. B. Arnalds, A. Harman-Clarke, E. Th. Papaioannou, M. Karimipour, P. Korelis, A. Taroni, P. C. W. Holdsworth, S. T. Bramwell & B. Hjörvarsson, (2012). *New J. Phys., 14,* 035009, under Creative Commons CC BY 3.0 license.

sidered in the literature, where thermal energy is used to anneal the system and can permit the system to relax into lower energy spin configurations, though the high frustration tends to limit access to the absolute long-range ordered GS.

A variant on the spin ice geometry is the chiral ice system, consisting of a 2D array of nanomagnets in which the magnetization points in one of two directions along two perpendicular axes, as illustrated in Figure 4.37. The resultant of this structure provides a vertex, which is chiral, i.e., it cannot be superimposed onto its mirror image. Each vertex is associated with four nanomagnets oriented at 90° with respect to each other and has a net magnetization M_V, being the sum of the magnetization of each element at the vertex. Thermally activated relaxation is generally accompanied by the rotation of the net magnetization at individual vertices in a unique direction; state

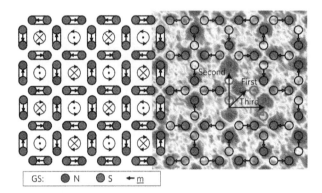

FIGURE 4.34 Left: Schematic illustration of GS ordering, chiral vectors representing alternating flux closure loops. Right: The schematic superimposed on an MFM image. North/south poles are inferred by out-of-plane field contrast, represented using a red/blue color scheme. The green circles mark the first-, second- and third-nearest neighbors. Reprinted by permission from Nature: J. P. Morgan, A. Stein, S. Langridge, & C. H. Marrows, (2011). *Nat. Phys., 7*, 75, ©(2011).

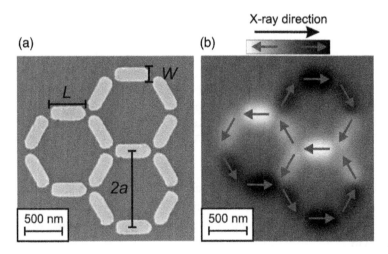

FIGURE 4.35 Three-ring building blocks of artificial kagome spin ice: (a) SEM image of a three-ring structure with nanomagnets of length $L = 470$ nm, width $W = 170$ nm, and thickness $d = 5$ nm patterned within a lattice parameter $a = 500$ nm. (b) Corresponding magnetic contrast image from which the orientation of the nanomagnet moments can be determined using X-ray magnetic circular dichroism (XMCD). Reprinted figure with permission from A. Farhan, A. Kleibert, P. M. Derlet, L. Anghinolfi, A. Balan, R. V. Chopdekar, M. Wyss, S. Gliga, F. Nolting & L. J. Heyderman, (2014). *Phys. Rev. B, 89*, 214405, ©(2014) by the American Physical Society.

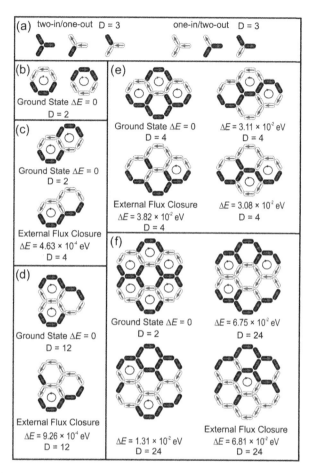

FIGURE 4.36 Three-ring building blocks of artificial kagome spin ice: (a) SEM image of a three-ring structure with nanomagnets of length $L = 470$ nm, width $W = 170$ nm, and thickness $d = 5$ nm patterned within a lattice parameter $a = 500$ nm. (b) Corresponding magnetic contrast image from which the orientation of the nanomagnet moments can be determined using X-ray magnetic circular dichroism (XMCD). Reprinted figure with permission from A. Farhan, A. Kleibert, P. M. Derlet, L. Anghinolfi, A. Balan, R. V. Chopdekar, M. Wyss, S. Gliga, F. Nolting, & L. J. Heyderman, (2014). *Phys. Rev. B, 89,* 214405, ©(2014) by the American Physical Society.

A to state B, as shown in Figure 4.37. The figure shows the favorable pathway for evolution from the A state, which flows to B rather than state D. This system, therefore, can be envisaged as an energy ratchet, in which the energy supplied by the applied field is transformed into the clockwise motion of the net magnetization. This is an example of a chiral system and is manifested in a number of phenomena in nature and in artificial systems. The total reversal

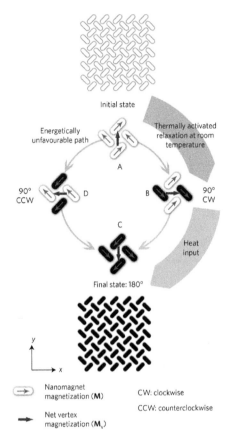

FIGURE 4.37   Schematic representation of the chiral ice and evolution of the net magnetization at individual vertices within the array. The thermally activated evolution after initial saturation along the $+y$ direction (state A) is illustrated, with the white or black color, respectively, indicating the direction of the magnetization M toward the positive or negative y-axis. The thin orange arrows represent the magnetization within the nanomagnets while the net magnetization at each individual vertex, M_v, is indicated by the large brown arrows. The thermal relaxation at room temperature takes place stepwise via a clockwise rotation of the net magnetization by 90° to state B and is indicated by the blue arrows. When the system is heated above room temperature in the presence of a bias field, the average magnetization can locally rotate further (orange arrows), to state C. Considering the evolution from state A, state D statistically occurs with very low probability. The net vertex magnetization therefore consistently rotates clockwise. Reprinted by permission from Nature: S. Gliga, G. Hrkac, C. Donnelly, J. Böchi, A. Kleibert, J. Cui, A. Farhan, E. Kirk, R. V. Chopdekar, Y. Masaki, N. S. Bingham, A. Scholl, R. L. Stamps, and L. J. Heyderman, *Nat. Mat., 16*, 1106 (2017), ©(2017).

of the system from state A to state C will thus occur almost exclusively via the ABC pathway (Gliga et al., 2017).

4.5 Skyrmions

As will be apparent to most readers, the normal ground state of a magnetic thin film is typically an in-plane uniform magnetization. However, in certain materials more exotic ground states can exist due to asymmetric interactions between neighboring spins. One such interaction is the Dzyaloshinskii–Moriya interaction (DMI), which for neighboring spins has an energy of the form:

$$E_{DMI} = \mathscr{D}_{ij} \cdot (\mathbf{S}_i \times \mathbf{S}_j) \tag{4.120}$$

Here, $\mathscr{D}_{ij}$ is the DM exchange, which establishes the strength of the interaction. It has been found, that such interactions can stabilize small perturbations in the magnetization, which are point-like regions of reversed magnetization surrounded by a swirling spin state with a certain chirality. Such a state is known as a *skyrmion*, which is treated as a quasi-particle and can be described as a topological state, defined by a topological integer or *winding number*. The mathematical concept of the skyrmion was introduced by Tony Skyrme in the early 1960s. Skyrme proposed the concept of topological defects in quantum field theory, which describe particles as a wave-like excitation with a finite lifetime. Such defects were considered to be topologically protected by the winding number, which was considered to be constant. This provides the necessary stability for the excitation. While the theory didn't catch on in quantum field theory, it was picked up in condensed matter physics to describe such phenomena as quantum Hall systems, certain liquid-crystal phases, and Bose-Einstein condensates. In the following, we will provide a brief introduction to the physics of skyrmions. A more detailed description is well beyond the scope of the current text. For the interested reader, a number of articles are reviews are available, see, for example, Nagaosa and Tokura (2013).

A magnetic skyrmion can take several forms. They can, for example, be Bloch-like or Néel-like and they are often represented in order parameter space to show the symmetry. In Figure 4.38, we show the forms of the Bloch (a) and Néel (b) magnetic skrymions, both of which have the same order parameter state (c). The distinction between the Bloch or Néel type depends on the specific values of the magnetic anisotropy, K_u, and the DMI exchange

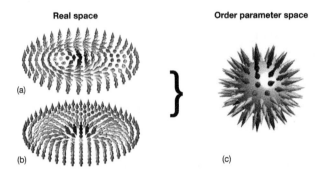

FIGURE 4.38 Schematic illustration of the Bloch-like skyrmion (a), the Néel-like skyrmion (b) and (c) the depiction of these skyrmion states in order-parameter space.

interaction parameter, $\mathscr{D}$. The skyrmion can also be described by the quality factor, expressed as:

$$Q = \frac{2K_u}{\mu_0 M^2} \tag{4.121}$$

and which determines the extent of the skyrmion in real space. This looks a little like the characteristic length scales we discussed in Section 4.2.1. The classification of the skyrmion (winding) number or topological charge is defined as:

$$N = \frac{1}{4\pi} \int \int \left(\frac{\partial \mathbf{m}}{\partial x} \times \frac{\partial \mathbf{m}}{\partial y} \right) \cdot \mathbf{m} \, dx \, dy \tag{4.122}$$

The value of N will be characteristic for different types of excitation, for example, the magnetic vortex state has a value of 0.5, while skyrmions have a value of 1. Some examples of other states are illustrated in Figure 4.39.

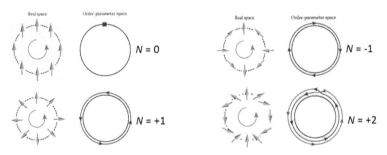

FIGURE 4.39 Characteristic skyrmion numbers for different excitation states.

Skyrmions can exist as a regular array, as illustrated in Figure 4.40, where we show the schematic illustration (a) and the visualization of such a state using LTEM (b). In confined magnetic systems, the skyrmion can also exist in isolation and can be made to travel as a soliton using a magnetic field or a current. This is shown schematically in Figure 4.40(c) and visually in Figure 4.40(d). Such systems have attracted attention for potential applications in data storage.

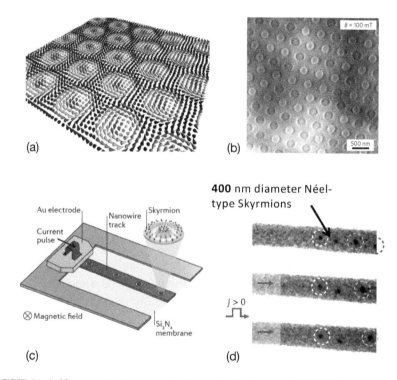

FIGURE 4.40 (a) Depiction of a skyrmion lattice and (b) LTEM image of the regular skyrmion array: the triangular skyrmion lattice in a 100 mT field. Bright and dark colors correspond to the two helicities ($\pm\pi/2$) of skyrmions, which are randomly distributed even though the positions of skyrmions are ordered. (Yu et al., 2012). (c) The current pulse-driven motion of skyrmions in a magnetic nanowire with (d) images obtained from scanning transmission X-ray microscopy (STXM). Reprinted by permission from Nature: [S. Woo, K. Litzius, B. Krüger, M.-Y. Im, L. Caretta, K. Richter, M. Mann, A. Krone, R. M. Reeve, M. Weigand, P. Agrawal, I. Lemesh, M.-A. Mawass, P. Fischer, M. Kläui & G. S. D. Beach, (2016). *Nat. Mater.* *15*, 501, ©(2016).

In addition to the static states of skyrmions, much interest has been addressed to the dynamics of these states. Indeed, as we saw in the above, the skyrmion state can be readily coaxed to move under the action of an applied field and an electrical current. The excitation spectrum for Bloch and Néel skyrmions has been modeled using the LLG equation with a DMI energy density of the form (Mruczkiewicz et al., 2017):

$$\varepsilon_{DMI} = \mathscr{D}[m_z(\nabla \cdot \hat{m}) - (\hat{m} \cdot \nabla)m_z] \qquad (4.123)$$

where $\hat{m} = \mathbf{M}/M$ is the reduced magnetization vector. This energy density is added to the other relevant terms in the system to simulate the resonant excitations of the isolated vortex and skyrmion states. A stability diagram of the skyrmion number as a function of the state (from vortex to Bloch and Néel skyrmions) is shown in Figure 4.41(A). Here, the variation of the states is illustrated from simulations of the ground state as a function of the uniaxial anisotropy, K_u, and the DMI exchange interaction parameter, $\mathscr{D}$. In Figure 4.41(B), the excitation frequencies as a function of the DMI exchange and uniaxial anisotropy are shown moving along the path indicated in (A): (i)–(iv).

4.6   Magneto-Plasmonics

Magneto-plasmonics is a relatively new area of study combining the study of plasmonic properties of nanostructures with magnetic functionality. One of the principal ideas of this field is the possibility of controlling plasmonic behavior using a magnetic field. As a first approach, a hybrid system is envisaged, in which the plasmonic structures are combined with a ferromagnetic layer. The former will support localized plasmon resonances, while the latter exhibit magnetic behavior, and in many cases, it is the magneto-optic properties, which are probed.

Early work by Chiu and Quinn (1973) in this area was conducted on the study of changes in the SPP dispersion relation due to an external magnetic field. The Lorentz force, which acts on the electrons in the metal, causes their trajectories to change and alters the magnitude of the SPP wave vector. Largest effects are expected for in-plane magnetic fields perpendicular to the SPP wave vector. Despite this, effects are rather small, though in agreement with the Drude theory for Au and Ag (Haefner et al., 1994), which expresses the off-diagonal elements of the linear dielectric permittivity tensor as:

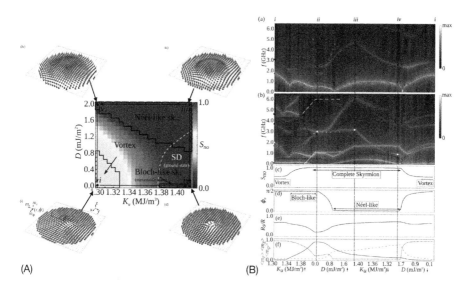

FIGURE 4.41 (A) Skyrmion number N as a function of the parameters $\mathcal{D}$ and K_u. The black line indicates the area of the vortex, Bloch-like, and Néel-like skyrmion stability. The white dashed line indicates the region where a single domain (SD) state is found with the energy lower than the magnetic soliton energies. The corners represent: (i) vortex, (ii) Bloch-like skyrmion, (iii) Néel-like skyrmion, and (iv) Néel-like skyrmion with low Q as defined in Eq. (4.121). (B) The soliton resonance frequencies are plotted along the path presented in (A) between the four (i)–(iv) inhomogeneous magnetic configurations presented at the vertices. The frequency values are obtained with the spatially uniform microwave magnetic field excitation of the sinc type in the time domain with (a) out-of-plane dynamic magnetization component and (b) in-plane magnetization component. The SW excitations in the SD ground state are indicated with white solid, dotted, and dashed lines. The static properties of the solitons are presented by the skyrmion number (c) skyrmion number N, (d) skyrmion phase, (e) skyrmion radius, and (f) the averaged magnetization components, $\langle m_r \rangle$, $\langle m_\phi \rangle$, $\langle m_z \rangle$, with red dashed, green dot-dashed, and blue continuous lines, respectively. Reprinted figure with permission from M. Mruczkiewicz, M. Krawczyk, & K. Y. Guslienko, (2017). *Phys. Rev. B, 95*, 094414, ©(2017) by the American Physical Society.

$$\varepsilon_{Drude} = i \frac{\omega_c \omega_p^2}{\omega[(1 - i\omega\tau)^2 + (\omega_c\tau)^2]} \qquad (4.124)$$

where $\omega_c = eB/m^*$ is the cyclotron resonance frequency and τ is the electron relaxation time. For noble metals in moderate magnetic fields, the cyclotron frequency is much smaller than the plasmon frequency, consequently, the size of magneto-optic (MO) effects are very small. On the other hand, in the case of ferromagnetic metals, such as Fe, Co and Ni, spin-orbit coupling, exchange, and the specific band structure determine the MO activity, and these give significantly larger responses in the visible portion of the electromagnetic spectrum than the noble metals. The problem is, that ferromagnetic metals are very lossy and are thus characterized by weak and broad plasmon resonances, while the noble metals have lower losses and well-defined plasmon resonances, but have weak magnetic effects. Therefore, a proposed combination of both systems was expected to provide a promising compromise and was pursued by many researchers in the area. Such systems are referred to as magneto-plasmonic. The intermixing of magnetic and plasmonic properties can be either manifested as a tunable plasmon resonance upon the application of a magnetic field, referred to as *active plasmonics* (Temnov et al., 2010), or as an increase or enhancement of MO effects caused by the plasmon resonance (Armelles et al., 2009).

One candidate system has been Co/Au layers deposited on Au, where an SPP geometry is used, see Figure 4.42. In this system, the strong dielectric contrast at the air–metal interface produces a large negative real part of the metal dielectric function, which determines the spatial distribution of the exponentially decaying SPP electric field, $E_z(z), E_x(z) \propto e^{-|z|/2\delta_{skin}}$, in the film and $E_z(z), E_x(z) \propto e^{-|z|/2\delta_d}$ on the air side. Here, the skin depth is given as δ_{skin} inside the metal and δ_d is the decay length above the metallic layers. To interact effectively with the SPP, the ferromagnetic layer must be in close proximity to the surface, such that the SPP electric field is sufficiently large, and since the skin depth is of the order of 10 nm in the visible range, the distance h should be $\sim \delta_{skin} \ll \lambda_{SPP} \simeq \lambda$, λ_{SPP} being the wavelength of the SPP.

Using an effective medium approximation, the SPP wave vector dependence on the magnetic field can be described in a reliable manner. In the case, where the magnetization of the ferromagnetic layer (Co) points along the y-axis, the effective MO tensor can be expressed as (Temnov et al., 2016):

$$\hat{\varepsilon}_{eff}(\pm m_y) = \begin{pmatrix} \varepsilon_{eff} & 0 & \pm \varepsilon_{eff}^{xz} m_y \\ 0 & \varepsilon_{eff} & 0 \\ \mp \varepsilon_{eff}^{xz} m_y & 0 & \varepsilon_{eff} \end{pmatrix} \tag{4.125}$$

where $m_y = M_y/M$ and the off-diagonal elements are given by:

$$\varepsilon_{eff}^{xz} = \frac{1}{\delta_{skin}} \int_0^\infty \varepsilon^{xz}(z) e^{-z/\delta_{skin}} dz \tag{4.126}$$

This leads to the SPP dispersion relation:

$$k_{SPP}(\pm m_y) = k_0 \sqrt{\frac{\varepsilon_{\text{eff}}}{1 + \varepsilon_{\text{eff}}}} \left(1 \pm \frac{i\varepsilon_{\text{eff}}^{xz} m_y}{(1 - \varepsilon_{\text{eff}}^2)\sqrt{\varepsilon_{\text{eff}}}} \right) \qquad (4.127)$$

For the case of a ferromagnetic layer of thickness $h_{FM} \ll \delta_{skin}$, sandwiched between two noble metals, the effective medium approximation leads to off-diagonal components:

$$\varepsilon_{\text{eff}}^{xz} \simeq \frac{h_{FM}}{\delta_{skin}} \varepsilon_{FM}^{xz} e^{-h/\delta_{skin}} \qquad (4.128)$$

resulting in a magneto-plasmonic modulation, $\Delta k_{mp} = [k_{SPP}(+m_y) - k_{SPP}(-m_y)]/2$, of the SPP wave vector of the form:

$$\Delta k_{mp} \simeq i\varepsilon_{FM}^{xz} m_y \frac{2h_{FM}\varepsilon_{\text{eff}} k_0^2}{(1 + \varepsilon_{\text{eff}})(1 - \varepsilon_{\text{eff}}^2)} e^{-h/\delta_{skin}} \qquad (4.129)$$

where $\delta_{skin} = \Im[\sqrt{1 + \varepsilon_{\text{eff}}}/\varepsilon_{\text{eff}}]/2k_0$ is used as the skin depth. In Figure 4.43, we illustrate the comparison between experiment and theory for the Au-Co-

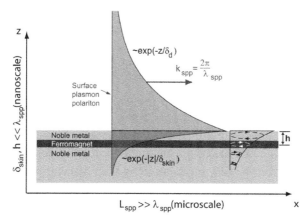

FIGURE 4.42 SPPs in metal-ferromagnet-metal multilayer structures. A ferromagnetic layer measuring just a few nanometers thick located at a position h within an SPP skin depth of δ causes little disturbance to the skin spatial distribution of SPP intensity inside a noble metal, and can therefore be used to control SPP propagation through the magneto-optical effect. A SPP with a frequency of $\hbar\omega = 1.55$ eV (corresponding to $\lambda = 800$ nm in free space) at the gold-air interface is characterized by an SPP wavelength of $\lambda_{SPP} = 794$ nm, a propagation distance of $L_{SPP} = 45$ m, a skin depth of $\delta_{skin} = 13$ nm and a decay length in air of $\delta_d = 307$ nm. The shaded area shows $|E_x(z)|^2$, and dashed elliptical contours represent the trajectories of electrons moving in the SPP electric field. Reprinted by permission from V. V. Temnov, *Nat. Photon.*, 6, 728 (2012), ©(2012). Nature.

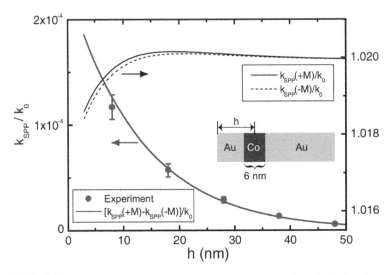

FIGURE 4.43 In-plane magnetization reversal (in y-direction) in ferromagnetic cobalt changes the SPP wave vector $k_{SPP}(M)$. Its magnetic modulation, $k_{SPP}(+M) - k_{SPP}(-M)$, is obtained within the effective medium approximation and is in agreement with experimental measurements. Figure courtesy of Vasily Temnov, IMMM CNRS 6283, Université du Maine, Le Mans, France.

Au system as a function of the depth of the Co layer, h. This shows, that the influence of the magnetization on the SPP decays with increasing depth as would be expected as the ferromagnetic layer descends to beyond the skin depth. In effect, this method can be used as a direct measurement of the skin depth of a noble metal. A modified version of a surface plasmon interferometer (Temnov et al., 2009) has been constructed using the sandwich type structure discussed here. The SPP modulation by the FM layer is measured using ultrafast pump-probe methods to determine the variation of $\Delta k_{mp}/k_{SPP}$. Results suggest changes of unto 12% in Au-Fe-Au using a dielectric substrate at a wavelength of 950 nm, indicating possible applications for magneto-plasmonic switching (Martin-Becerra et al., 2012).

Magneto-optic effects have formed a core area of interest in magneto-plasmonics research, where particular interest has been focused on the enhancement of magneto-optic properties of a ferromagnetic component found at the plasmon frequency of a metallic component in close proximity. The use of metallic grating systems for the plasmonic component, deposited on a magnetic layer, has been extensively researched.

A simple approach would be to use the Faraday effect with a transparent magnetic layer. In this case, the Faraday rotation is directly proportional to the magnetization, M, of the film multiplied by its thickness L and the Verdet constant of proportionality, $\mathscr{V}$, i.e., $\Theta_F = \mathscr{V}LM$. In the study by Chin et al. (2013), a BIG (bismuth iron garnet) was used as a (ferri)magnetic layer since it has a rather large Verdet constant. The experimental set-up and sample are schematically illustrated in Figure 4.44. In this case, localized plasmon resonance is excited in the Au grating by the incident light which is purposely polarized in the direction perpendicular to the wires. The periodicity of the wires forming the grating means, that this structure is also a photonic crystal and acts as a waveguide for photonic modes. This ensemble means that the localized plasmon resonance and waveguide interact strongly with the underlying magnetic layer. This interaction is controlled by the applied field. As a result, the Faraday rotation was shown to be enhanced by almost an order of magnitude, leading to increased transparency of the BIG film. These are shown in Figure 4.45. By altering the periodicity of the grating, the LSPR shifts to higher wavelengths. The position of the maximum of the Faraday rotation shifts with the LSPR, Figure 4.45(a), and a corresponding shift in the transmission characteristics as also observed, Figure 4.45(b). This type of structure is referred to as a magneto-plasmonic crystal (MPC). This type of MPC has also been studied in reflection geometry using the transverse magneto-optic Kerr effect (TMOKE). The sample geometry, in this case, is as shown in Figure 4.46(a). TMOKE and Faraday measurements made on the Au-BIG systems, at a fixed magnetic field of 170 mT, as a function the angle of incidence allowed the plasmon dispersion relation to be studied in the first Brillouin zone. A summary of the results is illustrated in Figure 4.46(b) and (c). The shifts in the spectra as a function of the angle incidence on the incoming light indicate, that the SPP dispersion is transferred into the MO signal, as was the case for the transmission geometry, Figure 4.45. The presence of the air/metal SPPs and metal/magnetic dielectric SPPs means that both signals can be observed simultaneously. When these two modes occur at similar photon energies, the SPP must be modeled as a coupled oscillator. In this model, the two SPPs are considered as two oscillators, with coupling between then arising from the finiteness of the grating thickness and the slits. Such a model leads to coupled modes of frequencies (M. Pohl et al., 2015):

$$\omega_{\pm}^2 = \frac{1}{2}(\omega_a^2 + \omega_d^2) \pm \sqrt{(\omega_a^2 - \omega_d^2)^2 + 4\eta_1\eta_2} \qquad (4.130)$$

where ω_a and ω_d are the eigenfrequencies of the two oscillators (i.e., at the air/metal and the metal/magnetic dielectric interfaces, respectively), $\eta_{1,2}$ denote the coupling coefficients, which depend on the grating parameters. The

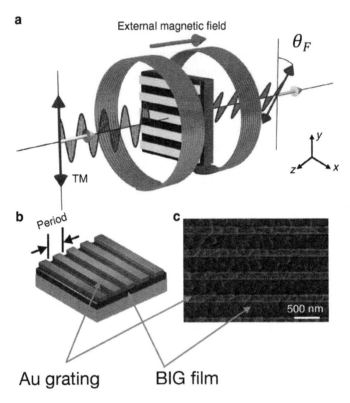

FIGURE 4.44   Magneto-plasmonic photonic crystal. (a) Faraday rotation by a magneto-plasmonic photonic crystal for TM-polarized incident light, where f is the Faraday rotation angle. At normal incidence, TM-polarized light has the electric field perpendicular to the gold wires, and TE-polarized light has the electric field parallel to the wires. (b) Schematic of the magneto-optical photonic crystal, where the BIG film (dark red) is deposited on a glass substrate (blue) and periodic gold nanowires (golden) are sitting atop. (c) A scanning electron microscopy image of a typical sample. Reprinted from J. Y. Chin, T. Steinle, T. Wehlus, D. Dregely, T. Weiss, V. I. Belotelov, B. Stritzker, & H. Giessen, (2013). *Nat. Comm., 4*, 1599, under a Creative Commons Attribution License.

resonance frequency at the metal/magnetic dielectric interface will depend on the magnetization of the magnetic layer, which can be expressed in terms of the wave vector as: $k = k_0(1 + \alpha g)$, where α is a frequency-dependent parameter and $g \propto M$, $k_0 = k(M = 0)$.

In addition to work on 1D gratings, there has also been interest in the magneto-optical response of magnetic nanodot arrays and with magnetic nanoparticles. In the latter, magnetic nanoparticles (magnetite, γ-Fe_2O_3)

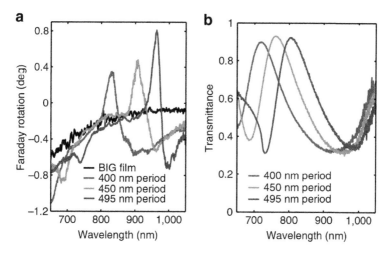

FIGURE 4.45 (a) Measured Faraday rotation of the three samples at normal incidence (TM polarization), compared with measured Faraday rotation of the bare BIG film. (b) Measured transmittance of the three samples at normal incidence (TM polarization). Reprinted from J. Y. Chin, T. Steinle, T. Wehlus, D. Dregely, T. Weiss, V. I. Belotelov, B. Stritzker, & H. Giessen, (2013). *Nat. Comm., 4,* 1599, under a Creative Commons Attribution License.

have been coated with gold. In this case, the spectral overlap of the localized surface plasmon resonance and the electronic transitions involved in the MO spectral response appear to have an important influence. An enhancement of the Faraday rotation is observed in these particles with respect to the uncoated particles, see Figure 4.47. In a similar manner, the presence of gold nanodisks on a Co film was also seen to enhance the TMOKE signal. The angle of incidence dependence is also observed, where a crossing of SPP and LSPR modes is evidenced (Torrado et al., 2010). It is shown, that only the SPP mode was affected by a magnetic field.

Changes of the refractive index around a magnetic nanoparticle can have a significant effect on the observed MO signal. Indeed, Bonanni et al. (2011) showed, that it was possible to reverse the MOKE loop just by altering the refractive index surrounding a Ni nanodisk using PMMA and air. Such findings demonstrate, that the magneto-optic and magneto-plasmonic response of a material can be manipulated in a very sensitive manner. This could allow the development of highly sensitive sensors and biosensors. Such systems have also exhibited dipolar plasmon modes. Comparison of near and far-field resonances are seen to be stronger than simple gold nano-antennae and

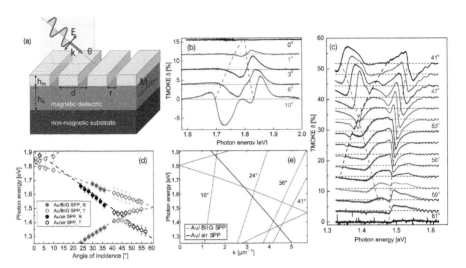

FIGURE 4.46 (a) The magneto-plasmonic heterostructures consist of a gold grating on top of a planar ferromagnetic-dielectric (bismuth iron garnet film) grown on a non-magnetic substrate (gadolinium gallium garnet). The sketch shows the TMOKE geometry, i.e., M is perpendicular to the plane of incidence and is in the sample plane. Incident light is p-polarized, i.e., its electric field $\mathbf{E}$ lies in the plane of incidence. The TMOKE is measured in transmission at incidence angles (b) close to $k = 0$ (from $0°$ to $10°$) and (c) from $41°$ to $61°$, i.e., close to the edge of the first Brillouin zone. The dashed lines are guides to the eye representing the shift of the TMOKE features. Surprisingly, close to the intersection of the SPP dispersions, the curves show features whose shift can be attributed not only to the Au-BIG magneto-plasmon (red dashed lines), but also to the Au/air SPP (black dashed line, compare figures (d) and (e)). At both $k = 0$ as well as at the first Brillouin zone boundary edge, the TMOKE amplitude is zero due to symmetry considerations. Dispersion relations of the SPPs at the Au/BIG (red) and Au/air (black) interfaces within the first Brillouin zone (d) and compared to the Drude model calculation (e) for a grating period d and the dielectric constant of the magnetic film $\varepsilon_d = 4.5$. The intersection of the SPP dispersion curves leads to splitting, i.e., anti-crossing around an incidence angle of $40°$ in (d). Reprinted from M. Pohl, L. E. Kreilkamp, V. I. Belotelov, I. A. Akimov, A. N. Kalish, N. E. Khokhlov, V. J. Yallapragada, A. V. Gopal, M. Nur-E-Alam, & M. Vasiliev, (2015). *New J. Phys., 15*, 075024, under a Creative Commons Attribution 3.0 license.

imply a possible application in the use of Ni nanostructures as nano-antennae (Chen et al., 2011).

We mentioned earlier, that one of the common applications for surface plasmon resonances (SPR) were in sensing for gas detection and in biosensing, see Section 3.4.4. In principle, changes of refractive index are one of

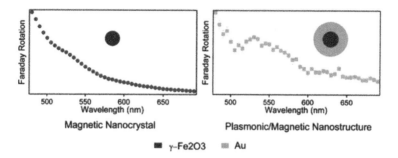

FIGURE 4.47 (a) Faraday rotation spectrum of γ-Fe$_2$O$_3$ nanoparticles and (b) same for gold-coated γ-Fe$_2$O$_3$ nanoparticles, showing the effect of the LSP resonance of the Au shell. Reprinted (adapted) with permission from P. K. Jain, Y. Xiao, R. Walsworth, & A. E. Cohen, (2009). *Nano Lett., 9*, 1644. ©(2009) American Chemical Society.

the easiest to measure, as long as we can ensure, that the element requiring sensing is in close proximity to the measurement cell such that the change of environment causes the alteration of the dielectric constant, in which the SPP or LSPR is active. Magneto-optics, as we have seen, are also very sensitive to changes in the environment and can be adapted to these applications. The basic set-up for both SPR and magneto-optic SPR (MOSPR) is shown in Figure 4.48(a). A shift in the surface plasmon resonance due to a change of environment will be manifested as a shift in the refractive index, being directly related to the dielectric constant. This will cause the reflectivity R_{pp} to also suffer a change in characteristics, as illustrated in Figure 4.48(b) - upper panel. At a specific value of the angle, there will be a large shift in the measured refractive index (see right panel of Figure 4.48(b)), this change is illustrated as a function of time. The corresponding variation of the relative change in reflectivity for the MOSPR is shown in Figure 4.48(b) - lower panel. So even for a modest shift in the SPR frequency, large changes in the reflectivity (SPR) and reflectivity change (MOSPR) illustrate, that these provide sensitive responses for detection purposes. A comparison of the two methods (SPR and MOSPR) is shown in Figure 4.48(c). The relative changes in signal for the MOSPR are more sensitive to changes in the refractive index.

4.7 Summary

Magnetism, as with other physical properties, has several characteristic length scales. These are usually expressed in terms of the energy contri-

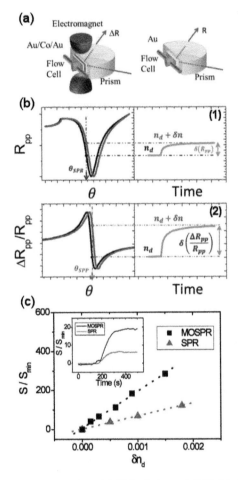

FIGURE 4.48 (a) Sketch of the MOSPR (left) and SPR (right) sensors. The experimental parts are very similar, the differences being the transducer, an Au layer for the SPR and an Au-Co-Au trilayer for the MOSPR, together with the application of an external magnetic field for the MOSPR. (b) Operating principle of the SPR (1) and MOSPR (2) biosensors: sensing response generated in a SPR sensor, δR_{pp}, or in a MOSPR one, $\delta(\Delta R_{pp})/R_{pp}$, by a refractive index change of the dielectric medium, δn_d. (c) Comparison of the experimental normalized signals of the MOSPR and the SPR sensors due to refractive-index changes, and evaluation of their experimental sensitivities. The inset shows the normalized signal of the detection of the physical adsorption of bovine serum albumin proteins. Reprinted from G. Armelles, A. Cebollada, A. García-Martín and M. U. González, (2013). *Adv. Opt. Mater., 1*, 10, under a The Creative Commons Attribution License, Wiley Publishes Open Access.

butions to magnetism and regard the exchange interaction and the magnetic anisotropies of a material. The most commonly used of the characteristic lengths is the exchange length. However, the domain wall width is also a characteristic of the material properties of magnetic materials and is also used as a standard of length in magnetism. When discussing the properties of magnetic materials at the nanoscale, the critical size for a single domain spherical particle is also a characteristic length that is of interest. All of these length scales are typically on the order of nanometers. As we have noted at various points, the reduced coordination at surfaces and boundaries will lead to an alteration of the magnetic properties of the material in question, this is particularly important when considering the magnetocrystalline anisotropy.

Magnetic nanoparticles have unique magnetic properties that can be harnessed for a number of applications. Of particular importance are magnetic particles, which have had their surface functionalized to which specific molecules can be attached. These can be exploited to perform targeted functions such as filtering and detection and are often considered as an ideal candidate for drug delivery. They can also be used as vehicles for cancer treatment via hyperthermia and can also be used as a method for enhancing magnetic resonance imaging. At a more fundamental level magnetic nanoparticles can be considered as model systems for manipulating the properties of materials with low spatial dimensions. The modeling of magnetic behavior is frequently made via a consideration of the magnetic energies of the system. We will generally use a free energy approach with contributions from the exchange energy, magnetic anisotropy (including magnetocrystalline, shape and surface effects) and the Zeeman energy. This allows us to determine the stable energy configuration of a system under specific conditions of the applied field as a function of orientation. A simplified model accounting for the anisotropy and applied field energies are found in the Stoner-Wohlfarth model. For the case of a uniaxial anisotropy, the system can be well understood and simulated. This model allows the determination of the magnetic configuration to be made as a function of the strength and orientation of the applied magnetic field and was one of the early models used to explain hysteresis phenomena in ferromagnetic systems.

Since the magnetic anisotropy of a particle is directly proportional to its size, the magnetic stability of the particle comes into question when the anisotropy energy is of the same order of magnitude as the thermal energy. In such a case, the magnetization of the particle can spontaneously switch while remain in the ferromagnetic state. This phenomenon is known as superparamagnetism. Its manifestation is further dependent on the way in which we observe the magnetic system and the measurement time will be an important factor when describing the superparamagnetic state of a system. An applied magnetic field can also be used to stabilize the system in a specific orientation. Any magnetic switching and dynamics can be understood in terms of the energy landscape, which depends on a number of factors and importantly the size and relative orientations of any magnetic anisotropies.

Other forms of the magnetic nanostructure, such as magnetic dots, rings, and wires have also received much attention from the research community. Such systems can be designed to be a single domain with well-defined domain structures. One of the consequences of the spin configurations in such systems is the appearance of magnetic vortex states, which are characterized by polarity and chirality and have four equivalent but distinct states. The control of switching and the dynamics of their movement has formed the subject of many studies, and in particular the gyrotropic motion of the vortex as a function of the frequency and applied fields. The magnetically stable state of a magnetic dot will depend on its geometric shape and size. In magnetic wires, it is possible to inject domain walls and under certain conditions, anti-vortex configurations can be obtained. As well as the nanodot structure, antidot arrays can also be produced which consist of a magnetic film with a series of periodically manufactured voids. The specific magnetic properties of these systems are rather unique and can be thought of as a magnetic analogy to the photonic crystal, also referred to as a magnonic crystal, which the propagation of spin waves can be specifically manufactured via the dispersion relation which depends on the properties of the magnetic layer and the size and periodicity of the voids. Artificial spin-ice systems are another form of periodic magnetic nanostructure. Here the interactions between magnetic dots, with specific periodicity, control the magnetic response of the system by introducing magnetic frustrations and can be used to define a chiral structure, which depends on a ratchet potential inherent in the system.

The study of the dependence of the plasmonic properties of nanosystems coupled to their magnetic behavior is referred to as magneto-plasmonics. This is a relatively new area of research and aims at using magnetic functionalities to control the plasmonic response of a system. Conversely, we can also consider how magnetic behavior depends on the optical stimulation of a system. A number of effects have been observed, such as the enhancement of the Faraday rotation of a magnetic layer at the plasmon frequency of a metallic nanostructure placed in close proximity. Magneto-plasmonic heterostructures, typically consisting of gold gratings deposited on a magnetic layer have provided model systems for such work. It has also been shown that magneto-optic surface plasmon resonances are very sensitive to their environment and can be used as precision biosensors for example.

4.8 Problems

(1) Evaluate the ground state of a Fe^{2+} ion. In crystalline Fe, the orbital angular momentum is quenched. Taking this into account, evaluate the magnetic moment of Fe, assuming it forms Fe^{2+} ions in solid. What is the corresponding magnetization of Fe? Note that Fe has a bcc structure and lattice parameter of 2.87 Å.

(2) Show that the magnetocrystalline anisotropy for tetragonal systems to fourth order in α, given by:

$$E_{crys}^{tetra} = K_0 + K_1 \alpha_3^2 + K_2 \alpha_3^4 + K_3 (\alpha_1^4 + \alpha_2^4)$$

can be expressed as:

$$E_{crys}^{tetra} = K_0' + K_1' \sin^2 \theta + K_2' \sin^4 \theta + K_3' \sin^4 \theta \cos 4\phi$$

where the α terms have been replaced by their directional cosines. Find expressions for the K_i' in terms of K_i.

(3) Consider the equilibrium condition for a surface spin, which is described by the Néel surface anisotropy:

$$e_K^{NSA} = -K_S \sum_i \sum_j (\mathbf{S}_j \cdot \mathbf{e}_{ij})^2$$

If the sample has a simple cubic structure, in which direction will the surface spins lie at equilibrium without an applied magnetic field? N.B. The summation j is over nearest neighbors and $\mathbf{e}_{ij} = \mathbf{r}_{ij}/r_{ij}$.

(4) Consider the energies involved in a Bloch domain wall for a ferromagnetic particle with uniaxial magnetic anisotropy. Find an expression for the critical diameter of a spherically shaped particle in terms of the anisotropy strength and the exchange stiffness constant. Determine a value for the size of an iron single-domain particle. (Look up any constants that you may require for this estimate.)

(5) Consider the Stoner–Wohlfarth model for uniaxial single domain particles with energy:

$$\varepsilon = K \sin^2 \theta - \mu_0 H M_z \cos(\theta - \phi)$$

Show that the parallel and perpendicular components of the applied magnetic field at the switching condition can be expressed as:

$$H_{\parallel} = -H_K \cos^3 \theta \quad \text{and} \quad H_{\perp} = H_K \sin^3 \theta$$

where

$$\frac{d\varepsilon}{d\theta} = \frac{d^2\varepsilon}{d\theta^2} = 0 \quad \text{and} \quad H_K = \frac{2K}{\mu_0 M}.$$

(6) From this, derive the Stoner–Wohlfarth astroid condition for switching:

$$H_{SW} = \frac{H_K}{(\sin^{2/3} \phi + \cos^{2/3} \phi)^{3/2}}$$

Sketch this in the $H_{\parallel}$, $H_{\perp}$ plane. Further show that for the field-aligned along the easy axis, the energy barrier for switching can be expressed as:

$$E_B = KV \left(1 - \frac{H}{H_K}\right).$$

(7) A magnetic measurement is made using a SQUID magnetometer on a monodisperse assembly of Fe nanoparticles with a diameter of 8 nm. The results indicate a blocking temperature of $T_B = 33$ K, where it was assumed that the characteristic measuring time is 100 s. Evaluate the corresponding blocking temperature for these nanoparticles using ferromagnetic resonance, where the measuring time can be approximated as around 10^{-10} s. The anisotropy constant for Fe is about 0.48×10^6 erg cm^{-3}. State any assumptions made in the calculation.

(8) Calculate the domain-wall width and the effective anisotropy constant $< K >$ in an assembly of exchange-coupled grains with $D = 20$ nm having $K_1 = 10^4$ J m^{-3}, $A = 10$ pJ m^{-1} and $M_S = 1$ MA m^{-1}. Give an upper limit to the expected coercivity.

References and Further Reading

Akbarzadeh, A., Samiei, M., & Davaran, S., (2012). *Nanoscale Res. Lett.,* 7(2), 144.

Armelles, G., Cebollada, A., A. García-Martín, J. García-Martín, M., González, M. U., González-Díaz, J. B., Ferreiro-Vila, E., & Torrado, J. F., (2009). *J. Opt. A: Pure Appl. Opt., 11,* 114023.

Armelles, G., Cebollada, A., García-Martín, A., & González, M. U. (2013). *Adv. Opt. Mater., 1,* 10.

Berry, C. C., & Curtis, A. S. G., (2003). *J. Phys. D: Appl. Phys., 36,* R198.

Berry, C. C., (2009). *J. Phys. D: Appel. Phys., 42*, 224003.

Bonanni, V., Bonetti, S., Pakizeh, T., Pirzadeh, Z., Chen, J., J. Nogués, Vavassori, P., Hillenbrand, R., Akerman, J., & Dmitriev, A., (2011). NanoLett., *11*, 5333.

Brown, W. F., Jr., (1963). *Phys. Rev., 130*, 1677.

Budrikis, Z., Morgan, J. P., Akerman, J., Stein, A., Politi, P., Langridge, S., Marrows, C. H., & Stamps, R. L., (2012). *Phys. Rev. Lett., 109*, 037203.

Castán-Guerrero, C., Herrero-Albillos, J., Bartolomé, J., Bartolomé, F., Rodríguez, L. A., Magén, C., Kronast, F., Gawronski, P., Chubykalo-Fesenko, O., Merazzo, K. J., Vavassori, P., Strichovanec, P., Sesé, J., & García, L. M., (2014). *Phys. Rev. B, 89*, 144405.

Chen, J., Albella, P., Pirzadeh, Z., Huth, F., Bonetti, S., Bonanni, V., Akerman, J., Nogués, J., Vavassori, P., Dmitriev, A., Aizpurua, J., R., (2011). *Hillenbrand and Small, 16*, 2341.

Chin, J. Y., Steinle, T., Wehlus, T., Dregely, D., Weiss, T., Belotelov, V. I., Stritzker, B., & Giessen, H. (2013). *Nat. Comm., 4*, 1599.

Chiu, K. W., & Quinn, J. J., (1973). *Il Nuovo Cimento B, 10*, 1.

Costa, B. V., & Lima, A. B., (2012). *J. Magn. Magn. Mater, 324*, 1999.

Cullity, B. D., & Graham, C. D., (2009). *Introduction to Magnetic Materials* (2e), Wiley-IEEE Press.

Farhan, A., Kleibert, A., Derlet, P. M., Anghinolfi, L., Balan, A., Chopdekar, R. V., Wyss, M., Gliga, S., Nolting, F., & Heyderman, L. J., (2014). *Phys. Rev. B, 89*, 214405.

Gliga, S., Hrkac, G., Donnelly, C., BŸchi, J., Kleibert, A., Cui, J., Farhan, A., Kirk, E., Chopdekar, R. V., Masaki, Y., Bingham, N. S., Scholl, A., Stamps, R. L., & Heyderman, L. J., (2017). *Nat. Mat., 16*, 1106.

Guo, T., Lin, M., Huang, J., C.Zhou, W.Tian, Yu, H., Jiang, X., J.Ye, Shi, Y., Y.Xiao, Bian, X., & Feng, X., (2018). *J. Nanomater.*, Article ID 7805147, https://doi.org/10.1155/2018/7805147 (accessed on 30 March 2020).

Guslienko, K. Yu. (2008). *J. Nanosci. Nanotech., 8*, 2745.

Haefner, P., Luck, E., & Mohler, E., (1994). *Phys. Stat. Sol., 185*, 289.

Hola, K., Markova, Z., Zoppellaro, G., Tucek, J., & Zboril, R., (2015). *Biotechnol. Adv., 185*, 1162. doi: 10.1016/j.biotechadv.2015.02.003.

Huber, D. L., (1982). *Phys. Rev. B, 26*, 3758.

Jain, P. K., Xiao, Y., Walsworth, R., & Cohen, A. E., (2009). *Nano Lett., 9*, 1644.

Kamionka, T., Martens, M., Chou, K. W., Curcic, M., Drews, A., SchŸtz, G., Tyliszczak, T., Stoll, H., Van Waeyenberge, B., & Meier, G., (2010). *Phys. Rev. Lett., 105*, 137204.

Kapaklis, V., Arnalds, U. B., Harman-Clarke, A., Papaioannou, E. Th. Karimipour, M., Korelis, P., Taroni, A., Holdsworth, P. C. W., Bramwell, S. T., & Hjörvarsson, B. (2012). *New J. Phys., 14*, 035009.

Krupinski, M., Mitin, D., Zarzycki, A., Szkudlarek, A., Giersig, M., Albrecht, M., & Marszalek, M., (2017). *Nanotechnol., 28*, 085302.

Linemann, T., Thomsen, L. B., du Jardin, K. G., Laursen, J. C., Jensen, J. B., Lichota, J., & Moos, T., (2013). *Pharmaceutics, 5*, 246.

Maier-Hauff, K., Rothe, R., Scholz, R., Gneveckow, U., Wust, P., Thiesen, B., Feussner, A., von Deimling, A., Waldoefner, N., Felix R. R., & Jordan, A., (2007). *J. Neuro-Oncol., 81*, 53.

Martin-Becerra, D., Temnov, V. V., Thomay, T., Leitenstorfer, A., Bratschitsch, R., Armelles, G., García-Martín, A., & González, M. U. (2012). *Phys. Rev. B, 86*, 035118.

Monnier, C. A., Burnand, D., Rothen-Rutishauser, B., Lattuada, M., & Petri-Fink, A., (2014). *Eur. Nanomedicine, J., 6*, 201.

Morgan, J. P., Stein, A., Langridge, S., & Marrows, C. H., (2011). *Nat. Phys., 7*, 75.

Morup, S., & Tronc, E., (1994). *Phys. Rev. Lett., 72*, 3278.

Mruczkiewicz, M., Krawczyk, M., & Guslienko, K. Y., (2017). *Phys. Rev. B, 95*, 094414.

Nagaosa, N., & Tokura, Y., (2013). *Nat. Nanotech., 8*, 899.

O'Handley, R. C. (2000). *Modern Magnetic Materials—Principles and Applications*, John Wiley and Sons, New Jersey.

Pankhurst, Q. A., Connolly, J., Jones, S. K., & Dobson, J., (2003). *J. Phys. D: Appl. Phys., 36*, R167.

Pankhurst, Q. A., Thanh, N. T. K., Jones, S. K., & Dobson, J., (2009). *J. Phys. D: Appl. Phys., 42*, 224001.

Pohl, M., Kreilkamp, L. E., Belotelov, V. I., Akimov, I. A., Kalish, A. N., Khokhlov, N. E., Yallapragada, V. J., Gopal, A. V., Nur-E-Alam, M., & Vasiliev, M., (2015). *New J. Phys., 15*, 075024.

Roca, A. G., Marco, J. F., del Puerto Morales, M., & Serna, C. J., (2009). *J. Phys. D: Appl. Phys., 42*, 224002.

Rocha, J. C. S., Coura, P. Z., Leonel, S. A., Dias, R. A., & Costa, B. V., (2010). *J. Appl. Phys. 107*, 053903.

Rohart, S., Repain, V., Thiaville, A., & Rousset, S., (2007). *Phys. Rev. B,* *76*, 104401.

Rosensweig, R. E., (2002). *J. Magn. Magn. Mater., 252*, 370.

Schmool, D. S., & Kachkachi, H., (2015). *Single Particle Phenomena in Magnetic Nanostructures*, in Solid State Physics, Volume 66, Stamps, R. L., & R. Camley (Eds.), pp. 301–423, Elsevier.

Schmool, D. S., & Kachkachi, H., (2016). *Collective Effects in Assemblies of Magnetic Nanoparticles*, in Solid State Physics, Volume 67, Stamps, R. L., & R. Camley (Eds.), pp. 1–101, Elsevier.

Schmool, D. S., & Markó, D. (2018). *Magnetism in Solids: Hysteresis*, in Reference Module in Materials Science and Materials Engineering, S. Hashmi (Ed.), Elsevier.

Tannous, C., & Gieraltowski, J., (2008). *Eur. J. Phys. 29*, 475–487.

Tartaj, P., del Puerto Morales, M., Veintemillas-Verdaguer, S., González-Carreño, T., & Serna, C. J., (2003). *J. Phys. D: Appl. Phys., 36*, R182.

Temnov, V. V., Nelson, K. A., Armelles, G., Cebollada, A., García-Martín, A., Thomay, T., Leitenstorfer, A., Bratschitsch, R., (2009). *Opt. Exp., 10*, 8423.

Temnov, V. V., Armelles, G., Woggon, U., Guzatov, D., Cebollada, A., García-Martín, A., García-Martín, J. M., Thomay, T., Leitenstorfer, A., Bratschitsch, R., (2010). *Nat. Photonics, 4*, 107.

Temnov, V. V., (2012). *Nat. Photon., 6*, 728.

Temnov, V. V., Razdolski, I., Pezeril, T., Makarov, D., Seletskiy, D., Melnikov, A., & Nelson, K. A., (2016). *J. Opt., 18*, 093002.

Thiele, A. A., (1973). *Phys. Rev. Lett., 30*, 230.

Torrado, J. F., J. B. González-Díaz, M. U. González, A. García-Martín, Armelles, G., (2010). Opt. Exp., *15*, 15635.

Vaz, C. A. F., Hayward, T. J., Llandro, J., Schackert, F., Morecroft, D., Bland, J. A. C., Kläui, M., Laufenberg, M., Backes, D., Rüdiger, U., Castaño, Ross, C. A., Heyderman, L. J., Nolting, F., Locatelli, A., Faini, G., Cherifi, S., & Wernsdorfer, W., (2007). *J. Phys.: Condens. Matter, 19*, 255207.

Veiseh, O., Gunn, J. W., & Zhang, M., (2010). *Adv. Drug Deliv. Rev., 62*, 284.

Wang, C. C., Adeyeye, A. O., & Singh, N., (2006). *Nanotechnol., 17*, 1629.

Wang, L., Chen, L., Wang, Q., Wang, L., Wang, H., Shen, Y., Li, X., Fu, Y., Shen, Y., & Yu, Y. (2014). *Oncol. Rep., 32*, 2007.

Woo, S., Litzius, K., Krüger, B., Im, M.-Y., Caretta, L., Richter, K., Mann, M., Krone, A., Reeve, R. M., Weigand, M., Agrawal, P., Lemesh, I., Mawass, M.-A., Fischer, P., Kläui, M., & Beach, G. S. D., (2016). *Nat. Mater. 15*, 501.

Yang, J., Lee, H., Hyung, W., Park, S.-B., & Haam, S., (2006). *Microencapsulation, J., 23*, 203.

Yu, X. Z., Mostovoy, M., Tokunaga, Y., Zhang, W., Kimoto, K., Matsui, Y., Kaneko, Y., Nagaosa, N., & Tokura, Y. (2012). *Proc. Natl. Acad. Sci., 109*, 8856-8860.

Zhang, W., Singh, R., Bray-Ali, N., & Haas, S., (2008). *Phys. Rev. B, 77*, 144428.

Zhang, W., & Haas, S., (2010). *Phys. Rev. B, 81*, 064433.

Zhang, W., Xiao, G., & Carter, M. J., (2011). *Phys. Rev. B, 83*, 144416.

Chapter 5

Spintronics and Device Applications

5.1 Introduction to Magnetic Thin Films and Multilayers

Magnetic thin films are of crucial technological importance. This is also an area of strong development, which is heavily linked to the numerous applications and the potential for creating model systems for the study of specific physical effects. The latter is obviously linked to the advanced capabilities in film preparation techniques, which allow a large degree of control over the preparation of magnetic thin films and layered structures with planar geometry.

One particular area of research related to thin magnetic films is the surface modification of the local anisotropy, which is related to the reduced coordination of the atoms at the outermost atomic layer. The fact that a particular crystallographic layer can be chosen using the orientation and type of crystalline substrate means, that we can control the surface conditions of a magnetic layer to a large degree. In doing so, we can use this as a model system to study the orientational dependence of the magnetic anisotropy. To perform in-depth studies, it is necessary to make a detailed study of the crystalline properties of the surface as well as to characterize the modifications of the lattice parameters and surface roughness. For the case of ultrathin magnetic films of a few monolayers, the magnetic anisotropy can lie in the direction normal to the surface, which will reorient the magnetization into the out-of-plane configuration and hence must overcome the magnetostatic energy, that tends to align the magnetization in-plane. There have been many experimental studies, which show that the thickness dependence of the effective anisotropy can be expressed as:

$$K_{\text{eff}} = K_V - 2\pi M_S^2 + \frac{2K_S}{t} \qquad (5.1)$$

Here, t represents the film thickness, K_V the bulk anisotropy and K_S the surface anisotropy. From Eq. (5.1), we note that the surface contribution diminishes as the thickness of the film increases. The second term arises from the magnetostatic energy for the thin film geometry. In Figure 5.1, we show some experimental data for permalloy (NiFe) films with different capping layers and fits to Eq. (5.1). This demonstrates the influence of the capping layer, which affects the strength of the surface anisotropy.

As the film becomes thicker, the magnetization will display a spin reorientation transition (SRT). Transitions of this type are typically structurally related, where changes in structure due to relaxation as a function of film thickness play a central role. Such is the case in Fe and Co thin films in the region of 10 ML. As we noted above, the nature of the substrate is also a central issue, since the magnetic moment can be enhanced or suppressed, depending on the underlying substrate (or overlayer in certain cases). Some examples of the variation of spin and orbital magnetic moments in Co and

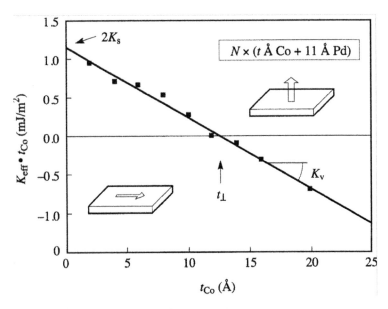

FIGURE 5.1 $K_{eff} t_{Co}$ of a Co thin film layer in a Co/Pd multilayer as a function of the Co thickness. The straight lines are the result of the linear fit of the experimental data, as expressed in Eq. (5.1). The intercept with the vertical axis gives the magnitude of the surface anisotropy, K_S. Reprinted from F. den Broeder, W. Hoving, and P. Bloemen, (1991). Magnetic anisotropy of multilayers, *J. Magn. Magn. Mat.*, *93*, 562–570, Copyright (1991), with permission from Elsevier.

Fe are illustrated in Figure 5.2, which have been measured using the XMCD (X-ray magnetic circular dichroism) method.

It is important to remember the length scales that we introduced for magnetism at the beginning of this chapter. Indeed, when the film thickness is below, say the exchange length, we can expect strongly thickness-dependent properties, such as a paramagnetic–ferromagnetic phase transitions, which will also be temperature-dependent. Other effects due to modified band structures at a surface or interface will also play a role, such as we saw for the modified magnetic moments in ultrathin films. The thickness at which the Curie temperature begins to differ from that of the bulk values is also a measure of

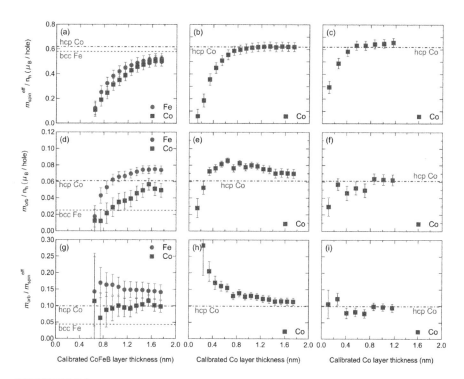

FIGURE 5.2   Magnetic layer thickness dependence of spin and orbital magnetic moments. Spin magnetic moment per hole, (m_{spin}^{eff}/n_h), (a-c), orbital magnetic moment per hole (m_{orb}/n_h), (d-f) and their ratio $m_{orb}/m_{spin}^{eff})$, (g–i) are plotted as a function of magnetic layer thickness, t, for Fe and Co in Ta/ t CoFeB/MgO (a,d,g), Co in Pt/ t Co/AlO$_x$ (b,e,h) and Co in Pt/ t Co/Pt (c,f,i). The horizontal red and blue dashed lines represent corresponding values of bulk bcc Fe and hcp Co, respectively. Reprinted from T. Ueno, J. Sinha, N. Inami, Y. Takeichi, S. Mitani, K. Ono, & M. Hayashi, (2015). *Sci. Rep.*, 5, 14858; doi: 10.1038/ srep14858, under a Creative Commons Attribution 4.0 International License.

the spin–spin correlation length, ξ, describing the critical spin fluctuations in the vicinity of T_C. This effect has been observed in thin films of a few tens of nm, with the behavior being described by a so-called shift exponent λ, which governs the thickness-dependent Curie temperature, $T_C(N)$, where N is the thickness in ML (monolayers), with:

$$\frac{T_C - T_C(N)}{T_C} \sim N^{-\lambda} \tag{5.2}$$

Assuming a direct proportionality between coverage and film thickness, the magnetization of a thin film develops according to the power law:

$$M \sim \left(1 - \frac{t}{t_c}\right)^\beta \tag{5.3}$$

where $\beta \sim 0.26$ for Fe. In many cases, the film does not grow in a layer-by-layer fashion. In Eq. (5.3), t_c represents a critical thickness, which is related to percolation. As may be expected, the growth of islands and percolation will also have a strong influence on the Curie temperature of a thin film. The value of this parameter will be dependent on the nature of the thin film, substrate and the details of the growth itself.

The anisotropy of a thin film was discussed above and can be expressed in the form of an effective anisotropy. In a more general manner, we can account for this due to magnetocrystalline effects as a thickness-dependent effective anisotropy with non-symmetric interfaces as:

$$K_{\text{eff}}(t) = \left(K_V + \frac{K_{S1} + K_{S2}}{t}\right)\sin^2\theta \tag{5.4}$$

where we consider only the case of uniaxial bulk anisotropy, which is aligned with that of the two surface anisotropy constants.

In this introduction to thin films, we have considered some of the main aspects of the effect of reduced dimensions on the magnetic properties. A more in-depth discussion is beyond the scope of this chapter and a number of good reviews are available. A good starting point would be the series of books by J. A. C. Bland and B. Heinrich (1994–2005).

The study of magnetic layered systems, which we view as a one-dimensionally size reduced system, can be viewed as an extension to the study of thin magnetic films. While at once allowing the magnetic signal to be amplified by increasing the quantity of magnetic material, magnetic multilayered structures rapidly came to the forefront of research in magnetic materials in the late 1980s as a result of the discovery of the giant magnetoresistance (GMR) observed in metallic layered systems. We will carry this topic over to the following section.

5.2 Magnetoresistance and Giant Magnetoresistance

The discovery of the giant magnetoresistance effect in magnetic multilayers can now be seen as a pivotal point in magnetic research. In this and some of the following sections, we will discuss in some detail the physics of the GMR effect and see how the theory has been developed and ultimately lead to the spintronic revolution, which has paved the way for many technological developments. The importance of GMR was not lost on the Nobel committee, which awarded Albert Fert and Peter Grünberg the 2007 Nobel prize for their joint, but independent discoveries of this effect, (A. Fert, 2008) and (P. Grünberg, 2008).

As we know, electrons have the properties of charge and spin, but until recently, these have been treated separately. The transport of the charge of the electron is at the root of conventional electronics and is manipulated via the application of electric fields, while the spin of the electron is ignored. The spin of the electron was considered as the property of the electron being central to magnetic phenomena, with technologies such as magnetic recording being dependent on this property via the ferromagnetic and ferromagnetic description of materials. The discovery of GMR in 1988 lead to a fundamental rethinking of this description with the advent of magnetic multilayer research. It is important to note that the context of this work lies in advances in the fundamental technologies for the preparation techniques for the deposition of thin-film structures, and most importantly molecular beam epitaxy, which proved so successful in the development of semiconductor research.

With regard to the GMR phenomenon, there are two related points that we need to consider. Firstly, the exchange coupling between two ferromagnetic layers via an intermediate metallic and nonmagnetic layer displays an oscillatory dependence as a function of the thickness of the intervening film. When we describe this exchange coupling as oscillatory, we mean that the coupling oscillates between ferromagnetic (FM) coupling, with the magnetizations of the two films being parallel, and antiferromagnetic (AFM) coupling, where they are oppositely aligned. This variation of the exchange coupling between the layers leads to the oscillatory variation of the magnetoresistance as a function of the interlayer thickness. We note that when the layers are FM coupled, the electrical resistance is low, while when they are AFM coupled, this resistance is high in the absence of an applied magnetic field. The GMR effect lies in the application of a magnetic field, which will align the magnetizations of the two ferromagnetic layers from the AFM state to the FM state. This leads to a large drop in electrical resistance. In Figure 5.3, we show the first publication of the GMR effect in 1988 in magnetic multilayers, or superlattices, of Fe/Cr alternating layers. In this figure, we note that the magnetoresistance (or MR) increases as a function of the Cr interlayer thickness, but also as a function of the number of repeats of the Fe/Cr unit.

In a more general manner, magnetoresistance is the phenomenon associated with the change of the electrical resistance under the action of an applied magnetic field. This effect is observed in a majority of metals, but tends to be quite a small effect in non-magnetic metals and positive. Negative magnetoresistance, where the electrical resistance reduces with increasing applied magnetic field, is observed in ferromagnetic metals. In layered and multicomponent magnetic systems, there are a number of different effects that can be distinguished. These can be summarized as: giant magnetoresistance (GMR), tunnel magnetoresistance (TMR), colossal magnetoresistance

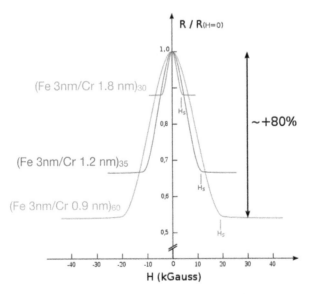

FIGURE 5.3   The founding results of Albert Fert and Peter Grünberg (1988): change in the electrical resistance of Fe/Cr superlattices at 4.2 K in the external magnetic field *H*. The current and magnetic field were parallel to the [110] axis. The arrow to the right shows maximum resistance change. H_s is the saturation field. Reprinted figure with permission from M. N. Baibich, J. M. Broto, A. Fert, F. Nguyen Van Dau, F. Petroff, P. Etienne, G. Creuzet, A. Friederich, & J. Chazelas, (1988). *Phys. Rev. Lett., 61*, 2472, ©(1988) by the American Physical Society.

(CMR), and extraordinary magnetoresistance (EMR). We will outline most of these in this and the following sections.

The presence of a magnetic field can cause the variation of the electrical resistance due to the Hall effect, where the charge carriers are deflected from their normal trajectories by the Lorentz force in the solid and give rise to the appearance of a voltage in the direction perpendicular to the electrical current and the applied field. Once a charge carrier begins to orbit around the magnetic field, it no longer contributes to the current density ($\langle v_x \rangle = 0$ over a complete cyclotron orbit) until it is scattered. After scattering, it will commence a new cyclotron orbit with an initial velocity biased towards the applied magnetic field. Thus, the longer the relaxation time (lower resistivity), the larger can be the effect of the field on the electrical resistance. In a ferromagnetic material, the Hall resistivity can be expressed as:

$$\rho_H = \frac{E_H}{J} = \rho_{OH} + \rho_{SH} = \mu_0 (R_0 H + R_S M) \qquad (5.5)$$

where the first term is referred to as the ordinary Hall effect, which is proportional to the external magnetic field. The second term is the spontaneous effect, which is proportional to the sample magnetization. Just as the Hall effect has ordinary and spontaneous terms for a ferromagnetic material, so too will its magnetoresistance. This latter is usually referred to as the anisotropic magnetoresistance (AMR) due to the angular dependence on θ, the angle between the magnetization, **M** and the current density, **J**, with a variation that can be expressed in the form:

$$\rho(\theta) = \rho_\perp + (\rho_\parallel - \rho_\perp) \cos^2 \theta \qquad (5.6)$$

where $\rho_\perp$ and $\rho_\parallel$ denote the perpendicular and parallel components of the resistivity with respect to the current density. Such effects in ferromagnetic materials can account for changes in the magnetoresistance of up to a few percentages. As we see from Figure 5.3, the giant magnetoresistance effect in magnetic multilayers is over an order of magnitude larger than this, with a GMR ratio of up to 80% reported at low temperatures. These values correspond to the maximum values of the GMR at specific values of the non-magnetic spacer layer. The variation of GMR as a function of the interlayer thickness is seen to oscillate, as illustrated in Figure 5.4. Also shown in this figure is the variation of the saturation field, which exhibits an oscillating behavior.

The particular contribution of the groups of Albert Fert and Peter Grünberg was to understand the correlation of the antiferromagnetic coupling

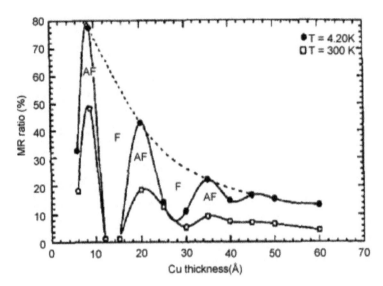

FIGURE 5.4   GMR ratio $\Delta \rho / \rho$ as a function of the interlayer thickness for Cu in a [Co/Cu] superlattice. Reprinted from G. W. Fernando, (2008). *Handbook of Metal Physics, Vol 4*, T GMR in Metallic Multilayers A Simple Picture, 1–32, ©(2008), with permission from Elsevier.

between the ferromagnetic layers with the maxima in the giant magnetoresistance. This lead to the early models of the GMR theory and the association with the RKKY exchange coupling mechanism. These subjects will be treated in more detail in the following sections. Furthermore, other magnetic measurements such as scanning electron microscopy with polarization analysis (SEMPA), Brillouin light scattering (BLS) and ferromagnetic resonance (FMR) were all applied to the study of exchange coupling and provided strong support for the oscillatory coupling observed in the magneto-transport measurements discussed above. Indeed, the group of Peter Grünberg commenced their studies using BLS measurements. We will return to the GMR effect in a later section, where we will outline some of the theory and also show how this led the way to the development of spintronics (or spin electronics) and evolved into work on spin valves and magnetic tunnel junctions.

A rather nice visual demonstration of the oscillatory coupling between adjacent ferromagnetic layers through a non-magnetic spacer layer was provided by SEMPA measurements on a system with a wedge-shaped spacer layer. This means that the interlayer separation between the magnetic layers will vary as a function of the position along with the wedge. Using the SEMPA method allowed researchers to visualize the magnetic domain pattern on the upper ferromagnetic film, while the underlying ferromagnetic layer is saturated along the axis defined by the wedge. This is illustrated in Figure 5.5. In this case, the underlayer is a Fe whisker, which has two oppositely oriented magnetic domains, as illustrated in the upper panel of Figure 5.5.

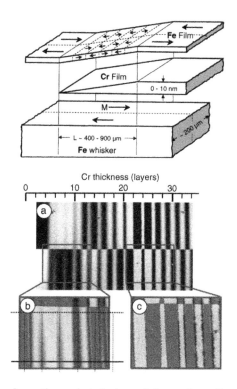

FIGURE 5.5 A schematic exploded view of the wedge trilayer sample structure showing the Fe(001) single-crystal whisker substrate, the evaporated Cr wedge, and the Fe overlayer. The arrows in the Fe show the direction of magnetization in each domain. The vertical scale is expanded many times. (a) SEMPA image showing the oscillatory magnetic coupling in a Fe/Cr/Fe(001) trilayer. (b) and (c) Enlarged angle maps from the regions outlined in (a). The colors give the direction of the magnetization. Canted non-collinear coupling is evident in (b). 90° biquadratic coupling regions, shown as red and blue, of varying width, are seen in (c). Reprinted from D. T. Pierce, J. Unguris, R. J. Celotta, & M. D. Stiles, (1999). *J. Magn. Magn. Mater., 200,* 290, ©(1999), with permission from Elsevier.

5.3   The RKKY Interaction

The local magnetic moments in metal are coupled via an indirect exchange interaction which is mediated by the conduction electrons in the metal. The theoretical framework for this coupling was initially proposed by Ruderman and Kittel in 1954. This was added to by Kasuya in 1956 and Yosida a year later. This interaction was subsequently dubbed RKKY, in honor of its proponents. In the original formulation, two local moments embedded in a gas of electrons are considered. The calculation of the exchange interaction between the local moments is treated by perturbation theory. In simple terms, we understand the RKKY as arising from the spin density oscillations, which ripple out from the magnetic moments. The polarization of the conduction electrons in the vicinity of the magnetic impurity gives rise to the perturbation of the potential in which the conduction electrons are immersed. The spin density oscillations allow the presence of the magnetic moments to permeate the electron gas and if another magnetic impurity is in the region of influence of this perturbation, its moment can become coupled (mutually) to the first magnetic moment. We can envisage the emanation of the spin density wave as something analogous to the ripples caused by a stone dropped into water. In terms of the coupling, the separation between the magnetic impurities will determine, whether the magnetic moments align in a parallel or an antiparallel configuration. The RKKY interaction is characterized by a periodic component and a decay length. We will discuss these

We saw in the previous section, that the relative alignment of the magnetization vectors in a magnetic multilayer is a sensitive function of the separation of the magnetic layers via the intervening non-magnetic spacer, where the variation of parallel or ferromagnetic and antiparallel or antiferromagnetic alignment leads to the oscillatory exchange coupling observed. This similarity to the oscillatory behavior as described by the RKKY interaction immediately led to attempts by researchers to describe the interlayer coupling using this indirect exchange mechanism. In the following, we will describe some of the main elements of RKKY theory. At the heart of this form of exchange interaction is the contact potential:

$$V(\mathbf{r}, \mathbf{s}) = A \sum_i \mathbf{s} \cdot \mathbf{S}_i \delta(\mathbf{r} - \mathbf{R}_i) \tag{5.7}$$

in which $\mathbf{S}_i$ is the spin of atom i at position $\mathbf{R}_i$ and $\mathbf{s}$ the spin of a conduction electron at position $\mathbf{r}$. The constant A is an adjustable parameter characterizing the exchange interaction strength. Since the exchange interaction of interest is between two atomic spins, one from each of the magnetic layers, we can express the interaction in the form:

$$\hat{\mathcal{H}}_{ij} = J(\mathbf{R}_{ij})\mathbf{S}_i \cdot \mathbf{S}_j \tag{5.8}$$

The form of this equation means that $J(\mathbf{R}_{ij})$ is an effective exchange integral. Indeed, we note that Eq. (5.8) takes the general form of the Heisenberg–Dirac Hamiltonian. Since the coupling between the spins $\mathbf{S}_i$ and $\mathbf{S}_j$ is mediated by conduction electrons, the RKKY interaction is an example of an indirect coupling mechanism. The role of the conduction electrons is to mediate the interaction between the spins in each ferromagnetic layer. The evaluation of the effective exchange integral $J(\mathbf{R}_{ij})$ is thus the principal goal of the calculation and this can be performed by means of perturbation theory using the Bloch spinors $\psi_{\mathbf{k}}(\mathbf{r},s) = \psi_{\mathbf{k}}(\mathbf{r})\chi(s)$ of the conduction electrons, where the Bloch functions take the form:

$$\psi_{\mathbf{k}}(\mathbf{r}) \equiv \psi_{n\mathbf{k}}(\mathbf{r}) = u_{n\mathbf{k}}(\mathbf{r})e^{i\mathbf{k}\cdot\mathbf{r}} \tag{5.9}$$

The contribution from first order perturbation drops out, leaving an expression for the total energy to second order written as:

$$E_{TOT} = E_{TOT}^{(0)} + \sum_{ks} f_{\mathbf{k}} \sum_{\mathbf{k}'} \left[\frac{\sum_{ij}|V_{\mathbf{k}',\mathbf{k}}|^2 \mathbf{S}_i \cdot \mathbf{S}_j}{\varepsilon_{\mathbf{k}} - \varepsilon_{\mathbf{k}'}} \right] \tag{5.10}$$

where the factor $\mathbf{S}_i \cdot \mathbf{S}_j$ is obtained by manipulating the spin part of $V(\mathbf{r},s)$, and

$$|V_{\mathbf{k}',\mathbf{k}}|^2 = A^2 \psi_{\mathbf{k}}^*(\mathbf{R}_i)\psi_{\mathbf{k}'}(\mathbf{R}_i)\psi_{\mathbf{k}'}^*(\mathbf{R}_j)\psi_{\mathbf{k}}(\mathbf{R}_j) \tag{5.11}$$

We note that $f_{\mathbf{k}}$ denotes the Fermi–Dirac distribution function. From this, the exchange integral can be obtained as:

$$J(\mathbf{R}_{ij}) =$$

$$2A^2 \Re \left\{ \sum_{n\mathbf{k}} \sum_{n'\mathbf{q}} \frac{f_{n\mathbf{k}}(1 - f_{n'\mathbf{k}+\mathbf{q}})}{\varepsilon_{n'\mathbf{k}+\mathbf{q}} - \varepsilon_{n\mathbf{k}}} e^{i\mathbf{q}\cdot\mathbf{R}_{ij}} u_{n\mathbf{k}}^*(\mathbf{R}_i)u_{n'\mathbf{k}+\mathbf{q}}(\mathbf{R}_i)u_{n'\mathbf{k}+\mathbf{q}}^*(\mathbf{R}_j)u_{n\mathbf{k}}(\mathbf{R}_j) \right\} \tag{5.12}$$

This is not a simple calculation and will be generally solved by making some simplifying assumptions. Using plane waves allows us to replace the product of four wave functions by the factor $1/\Omega^2$, where Ω is the volume in which the plane waves are normalized. This allows us to simplify the above expression to:

$$J(\mathbf{R}_{ij}) = \left(\frac{A}{\Omega}\right)^2 \Re \sum_{\mathbf{q}} \chi_0(\mathbf{q}) e^{i\mathbf{q}\cdot\mathbf{R}_{ij}} \tag{5.13}$$

where

$$\chi_0(\mathbf{q}) = \sum_{nn'\mathbf{k}} \frac{f_{n\mathbf{k}} - f_{n'\mathbf{k}+\mathbf{q}}}{\varepsilon_{n'\mathbf{k}+\mathbf{q}} - \varepsilon_{n\mathbf{k}}} \tag{5.14}$$

is the magnetic susceptibility (in units of μ_B^2). Assuming that the magnetic moments are immersed in a gas of free electrons allows us to rewrite the magnetic susceptibility as:

$$\chi_0(\mathbf{q}) = \frac{\Omega}{(2\pi)^3} \int \frac{f_{\mathbf{k}} - f_{\mathbf{k}+\mathbf{q}}}{\varepsilon_{\mathbf{k}+\mathbf{q}} - \varepsilon_{\mathbf{k}}} d\mathbf{q} \tag{5.15}$$

Using the free electron eigenvalues, $\varepsilon_{\mathbf{k}} = \hbar^2 k^2/2m^*$, it is possible to show:

$$\chi_0(\mathbf{q}) = \frac{3N}{4\varepsilon_F}\left[1 + \frac{4k_F^2 - q^2}{4qk_F}\ln\left|\frac{2k_F + q}{2k_F - q}\right|\right] \tag{5.16}$$

where N is the number of electrons in the system and ε_F is the Fermi energy, expressed as: $\varepsilon_F = \hbar^2 k_F^2/2m^*$, where k_F is the Fermi wave vector, which we can further express as: $k_F = (3\pi^2 n)^{1/3}$, with $n = N/V$ being the electron density. It is interesting to note that for $q \to 0, \chi_0 \to 3N\mu_B^2/2\varepsilon_F$, which is exactly the result obtained for the Pauli susceptibility of a free electron gas in a uniform magnetic field. Taking the Fourier transformation of $\chi_0(\mathbf{q})$ allows us to express the RKKY exchange coupling between a magnetic moment taken at the origin and a magnetic moment on a spherical shell of radius R as:

$$J(R) = \frac{16A^2 m^* k_F^4}{(2\pi)^3 \hbar^2} \mathscr{F}(2k_F R) \tag{5.17}$$

The function $\mathscr{F}(x)$ is given by:

$$\mathscr{F}(x) = \frac{x\cos x - \sin x}{x^4} \simeq \frac{\cos x}{x^3}; \quad \text{for} \quad x \to +\infty \tag{5.18}$$

The period of $\mathscr{F}(x)$, given as λ, is thus related to the Fermi wavelength, $\lambda_F = 2\pi/k_F$, with $\lambda = \lambda_F/2$. In the case of a magnetic multilayer system, we consider the moments of the two adjacent ferromagnetic films separated by the non-magnetic spacer, where we sum over pairs ij running over the two layers (i in one layer and j in the other). We can take into account the angle ϑ between the magnetization vectors of the two layers by writing the exchange coupling energy as:

$$\varepsilon_C = I_C \cos\vartheta \tag{5.19}$$

where

$$I_C = \frac{d}{\Omega_0} S^2 \sum_j J(\mathbf{R}_{0j}) \tag{5.20}$$

The subscript 0 indicates that the origin is taken as the surface of one of the layers, d is the spacer thickness between the ferromagnetic layers and Ω_0 is the volume of the atomic unit cell. The sign convention is chosen such that positive (negative) J gives ferromagnetic (antiferromagnetic) coupling. For the simple model for free electrons in the spacer layer, the summation can be replaced by an integral:

$$\sum_j \to \int_{layer2} d^2 \mathbf{R}_{\parallel} \tag{5.21}$$

where $\mathbf{R}_{\parallel}$ is the in-plane projection of $\mathbf{R}_{0j} = \mathbf{R}_0 - \mathbf{R}_j$. From the form of $J(\mathbf{R})$, we can write:

$$\int_{layer2} \mathscr{F}\left(2k_F\sqrt{z^2 + R_{\parallel}^2}\right) d^2 \mathbf{R}_{\parallel} = \frac{2\pi}{(2k_F)^2} \int_{2k_F}^{\infty} \mathscr{F}(x) dx \tag{5.22}$$

where z is the distance between the planes of the two layers. It is then possible to express the coupling as:

$$I_C(z) = 2I_0 d^2 k_F^2 \left[\frac{\pi}{2} - \text{Si}(2k_F z) + \frac{\sin(2k_F z)}{(2k_F z)^2} - \frac{\cos(2k_F z)}{2k_F z}\right] \tag{5.23}$$

with

$$\text{Si}(2k_F z) = \int_0^t \frac{\sin x}{x} dx \tag{5.24}$$

and

$$I_0 = \left(\frac{A}{V_0}\right)^2 S^2 \frac{m^*}{4\pi^2\hbar^2} \tag{5.25}$$

With an asymptotic value of $(\sin x/x + \cos x/x)$ for $(\pi/2 - \text{Si}(x))$, Eq. (5.25) yields:

$$I_C = I_0 \left(\frac{d}{z}\right)^2 \sin(2k_F z); \qquad \text{as} \quad z \to \infty \tag{5.26}$$

This coupling expression has a single oscillation period of $\lambda = \lambda_F/2$ and decays as $1/z^2$. One of the principal factors determining the oscillation period is the Fermi (spanning) vector, k_F. It is material and structure dependent. In Figure 5.6, we show the form of the Cu Fermi surface for the reduced zone scheme in the first Brillouin zone (a) and a cross-section of this structure is illustrated in (b), showing the Fermi spanning vectors in the [001] direction. We note that in this direction, there are two extremal vectors, which would give rise to a double period. Other directions will have different periods depending on the size of the k_F vector. The interlayer exchange for multilayers of $Ni_{80}Co_{20}$ with a Ru spacer is illustrated in Figure 5.7.

Early work on fitting experimental data with theory suffered difficulties, however, the realization that aliasing effects can distort the observed oscillation period allowed a better agreement. The relation between the effective period, Λ, the interlayer spacing thickness d and the original period, λ, due to aliasing can be expressed as:

$$\Lambda = \frac{1}{|1/\lambda - n/d|} \tag{5.27}$$

with $n \geq d/\lambda - 1/2$, such that $\Lambda \geq 2d$.

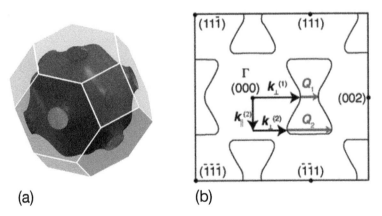

(a)   (b)

FIGURE 5.6 Fermi surface of copper with the famous belly-neck or dog-bone shape. (a) Reduced zone scheme in the first Brillouin zone. (b) Cross-sectional view of the same structure cut across the (111) plane. The arrows show the extremal or nesting vectors, which determine the period of the RKKY oscillations.

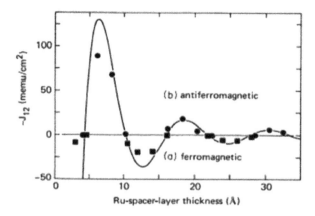

FIGURE 5.7   Interlayer exchange coupling strength J_{12} for coupling of $Ni_{80}Co_{20}$ layers through a Ru spacer layer. The solid line corresponds to a fit to the data of RKKY form. Reprinted figure with permission from S. S. P. Parkin, D. Mauri, (1991). *Phys. Rev. B, 44,* 7131, ©(1991) by the American Physical Society.

5.4   Other Mechanisms of Exchange Coupling

The apparent agreement between the experimentally observed oscillations in magnetic coupling in magnetic multilayers (MML) systems and the theory for the RKKY interaction gave strong support that this was indeed the physical mechanism to describe such behavior. Many research papers have been dedicated to using the RKKY model to describe the interlayer coupling in MML (Bruno, 2002). Despite this apparent success, the RKKY theory was not able to provide a fully quantitative description for the amplitude and phase of the oscillatory coupling. In this section, we will describe some of the other coupling schemes that have been applied to this problem.

One approach to the problem was the application of the quantum well model, as we have outlined in Chapter 2. In this framework, the coupling arises between adjacent magnetic layers via the reflection of spin-polarized electrons (or holes) at the interfaces between the magnetic and non-magnetic metallic layers. In the case of free electrons, the interfaces are envisaged as a simple spin-dependent step potential and the reflected amplitudes calculated in the usual manner. Remarkably, the quantum well approach yields the same oscillatory behavior and decay as the RKKY interaction. We shall outline some of the elements of this approach is the following.

The crux of the problem can be stated as follows: if we consider a simple trilayer (FM/NM/FM), the electrons reflected from the FM/NM interfaces are multiply reflected and will interfere. The amplitude for one round-trip, with the NM spacer layer of thickness d, will be: $e^{ikd}R_Re^{ikd}R_L$, where e^{ikd} is the phase accumulated in one passage of the spacer, while $R_{R,L}$ denote the reflected amplitudes of the left and right interfaces. Therefore, for all possible round-trips, the total amplitude will be:

$$\sum_{n=1}^{\infty}\left[e^{i2kd}R_RR_L\right]^n = \frac{e^{i2kd}R_RR_L}{(1-e^{i2kd}R_RR_L)} \tag{5.28}$$

For the condition $2kd + \phi_R + \phi_L = 2n\pi$, where n is an integer and $\phi_{R,L} = \Im\{\ln R_{R,L}\}$ is the phase change induce upon reflection at the right or left interface, constructive interference will occur and the denominator becomes small. Such a condition leads to resonances in the structure, otherwise called resonant or quantum well states. When the reflection probability is unity, these states correspond to true bound states. With increasing thickness, d, the resonances, and bound states will shift in energy. If there is a resonant state at the Fermi energy for a thickness d, then the resonance will also cross the Fermi energy for thicknesses satisfying the condition: $d + 2n\pi/2k_F$. The oscillations in the interlayer coupling are related to the crossing of the Fermi energy, see Figure 5.8. In this one-dimensional model, the Fermi surface consists of just two points, situated at $k = \pm k_F$. The period of oscillation will be determined by the spanning vector $2k_F$.

The quantum well states affect the density of states in a trilayer giving rise to a cohesive energy. The change in the density of states for each spin can be expressed as:

$$\Delta N(E,d) = -\frac{1}{\pi}\Im\left\{\frac{d}{dE}\ln(1-e^{i2kd}R_RR_L)\right\} \tag{5.29}$$

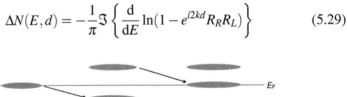
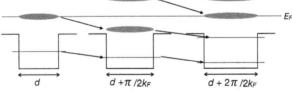

FIGURE 5.8   Evolution of quantum-well resonant states as a function of the well (spacer) layer thickness. The three different thicknesses illustrate the bound states (lines) and resonances (indicated by ellipses) for the quantum wells. Arrows indicate the shift in the resonant states with thickness.

The cohesive energy of the quantum well is given by the sum of the single particle energies, or the integral over the change in DOS:

$$\Delta E_{QW} = - \int_{-\infty}^{\infty} dE\,(E - E_F)\Delta N(E,d) = \frac{1}{\pi}\Im\left\{ \int_{-\infty}^{E_F} dE \ln(1 - e^{i2kd}R_R R_L)\right\}$$

(5.30)

This result is valid for the case where there is only one state in either direction of the spacer layer.

For a given thickness, the integrand will oscillate via the energy dependence of k. These oscillations cancel out in the integration, except those close to the Fermi energy, where there will be a sharp cut-off. The only contribution remaining will be from a range of states near E_F, with a width proportional to $\hbar v_F/d$, v_F being the Fermi velocity. For this energy range, the energy dependence of $R_{R,L}$ can be ignored and the wave vector is assumed to vary linearly with energy; $k \simeq k_F + E/\hbar v_F$, so that in the limit of thick interlayer spacings:

$$\lim_{d\to\infty} \Delta E_{QW} = \frac{\hbar v_f}{2\pi d}\sum_n \frac{1}{n}\Re\{(R_R R_L)^n e^{i2k_F nd}\}$$

(5.31)

For the case of small reflection amplitudes, only the first term in the summation over n will contribute and we can write:

$$\lim_{d\to\infty} \Delta E_{QW} \simeq \frac{\hbar v_f}{2\pi d}|R_R R_L|\cos(2k_F d + \phi_R + \phi_L)$$

(5.32)

For higher powers of reflection amplitudes, we obtain higher harmonics in the oscillatory behavior. The coupling strength for this one-dimensional model depends on the two reflection amplitudes and the oscillation period is dependent on the spanning vector $2k_F$. The interlayer coupling is then the sum and difference of the cohesive energies of the four quantum wells for the two spin states in the parallel and antiparallel configurations, see Figure 5.9. All four quantum well configurations have cohesive energies with the same period, which is determined by the spanning vector of the spacer later material. The oscillations do not cancel out because they have different amplitudes and possibly phases. For the one-dimensional case considered here, for large d and small R, the result can be expressed as:

$$\lim_{d\to\infty} J \simeq \frac{\hbar v_F}{4\pi d}\Re\{(R_\uparrow R_\downarrow + R_\downarrow R_\uparrow - R_\uparrow^2 - R_\downarrow^2)e^{i2k_F d}\} \simeq -\frac{\hbar v_F}{4\pi d}\Re\{(R_\uparrow - R_\downarrow)^2 e^{i2k_F d}\}$$

(5.33)

In the first case, the contribution from each of the quantum wells, with two asymmetric quantum wells for antiparallel alignment minus the majority and minority quantum wells for parallel alignment of the magnetization. The

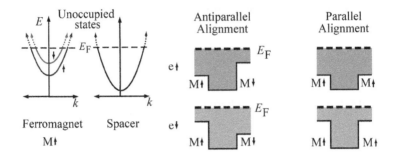

FIGURE 5.9   Quantum wells used to compute interlayer exchange coupling. On the left, the two panels give typical band structures for free-electron models of interlayer exchange coupling. The majority and minority bands in the ferromagnetic layer are illustrated in blue and red, respectively. On the right, the four panels give the quantum wells for a spin up and spin down electrons for parallel and antiparallel alignment of the magnetization. The shaded regions designate the occupied states.

second form illustrates that the interlayer exchange coupling is determined by the spin dependence of the reflection amplitudes.

For real systems, it is necessary to take into account the three-dimensional nature of the system. In this case, we need to consider any defects, disorder variations in the spacer layer thickness, including the interface roughness. Such effects can be incorporated into the model of the exchange coupling by summing the constant over a range of interlayer thicknesses:

$$J(d) = \sum_n P(n,d)J(n) \tag{5.34}$$

Here, $J(n)$ is the coupling constant for an ideal flat layer of n atomic layers, with thickness d_0 and $P(n,d)$ represents the probability of having a thickness nd_0 for a nominal deposited thickness of d.

In the case of an ideal magnetic multilayer with perfectly flat interfaces, the magnetic coupling is generally considered as having a bilinear form, which we express as the energy per unit area of the film as:

$$\frac{E}{A} = -J\frac{\mathbf{M_1}}{M_1} \cdot \frac{\mathbf{M_2}}{M_2} = -J(\hat{\mathbf{m}}_1 \cdot \hat{\mathbf{m}}_2) \tag{5.35}$$

The non-ideal case in which there is a thickness variation of the interlayer spacing can introduce more complex forms of coupling, such as the case where the magnetization vectors in adjacent magnetic layers are aligned perpendicularly with respect to one another. This can be expressed algebraically as:

$$\frac{E}{A} = -J_2(\hat{\mathbf{m}}_1 \cdot \hat{\mathbf{m}}_2)^2 \qquad (5.36)$$

This form of coupling is referred to as biquadratic coupling. All measured values of J_2 are negative, which favors the perpendicular alignment of the magnetization vectors.

Of the other forms of coupling, the magnetostatic energy is also an important mechanism for interlayer coupling. The principal effect arising from the magnetostatic interaction is the demagnetizing field, which generally produces an in-plane anisotropy for thin-film geometries. However, on a microscopic scale, interlayer roughness can lead to other important complications. The first is related to a ferromagnetic coupling, which was first described by Néel in the early 1960s and is referred to as "orange-peel" coupling. Roughness can also lead to a form of biquadratic coupling which, like the orange-peel effect, is derived from the fringing fields that exist outside the surface.

A rough magnetic surface produces magnetic poles at the interface. This is because the intralayer exchange is sufficiently strong to prevent the magnetization from rotating to follow the profile of the film. We consider the case where the roughness is smaller in amplitude than the interlayer thickness, with a slow variation, such that the local normal does not wander too far from the average surface normal. In this case, we can consider the surface roughness from the distribution of magnetic charges on a flat surface:

$$\sigma(\mathbf{R}) = \mathbf{M} \cdot \hat{n}(\mathbf{R}) \qquad (5.37)$$

where the magnetization, $\mathbf{M}$, is considered to be uniform and $\hat{n}(\mathbf{R})$ is the local normal direction to the interface, which varies as a function of position on the surface. For the case where the interlayer is of uniform thickness, the interface normals will be locally opposite; $\hat{n}_1 = -\hat{n}_2$. This gives a bilinear coupling. The strength of the coupling due to the orange-peel effect will depend on the orientation of the magnetization (in the plane of the film) with respect to the corrugation direction. For $\mathbf{M}$ perpendicular to the direction of the corrugation, the coupling will be a maximum and for the case where $\mathbf{M}$ is parallel to the corrugation, the coupling will be zero, i.e., for the magnetostatic energy. Figure 5.10 shows both parallel and antiparallel alignments of the magnetization, illustrating the fringing fields for each case. As the interface roughness increases, the coupling energy will increase quadratically due to the magnetostatic interaction. This model will break down when the thickness of the interlayer tends to zero. For uncorrelated roughness, the orange

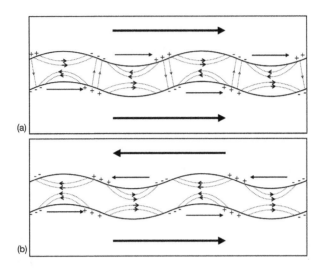

FIGURE 5.10   Orange peel coupling. (a) The fringing field (dashed lines) outside the rough surface of a layer with uniform magnetization couples with that of an adjacent layer, giving rise to a parallel configuration of the magnetization vectors for the two ferromagnetic layers. (b) Antiparallel magnetizations can also be produced in which none of the field lines cross the interlayer.

peel coupling will also tend to zero, though a biquadratic effect can occur on the microscopic level. For gaps in the interlayer, typically for the case of very thin spacers, the contact between the magnetic layers will tend to favor a ferromagnetic coupling. This is referred to as pin-hole coupling.

Irrespective of the coupling mechanism, the interlayer coupling has been shown to have a profound effect on the magnetic properties of magnetic multilayer systems and can be used as a method for tailoring the magnetic properties for a range of applications. In the following section, we will discuss some of these effects and discuss the evolution of magnetic coupling in layered systems to the development of spintronics.

5.5   Aspects of GMR Theory

As we discussed above, the exchange coupling between ferromagnetic layers is sensitively dependent on the non-magnetic metallic spacer layer thickness between adjacent layers. This is seen to oscillate from ferromagnetic to antiferromagnetic alignment with a periodicity that depends on the spacer layer orientation and Fermi vector, and decays rapidly with thickness. For a specific thickness of the metallic spacer, the normally antiferromagnetic alignment of the ferromagnetic layers can be brought to align in parallel under the action of a sufficiently strong applied magnetic field. Indeed, the high-resistance state in zero field is significantly reduced under the application of this field and gives rise to GMR. This effect has been observed in both current in-plane (CIP) and current perpendicular-to-plane (CPP) geometries. In both cases, the resulting change in electrical resistance produces a GMR. The GMR is typically expressed as the ratio of the change in electrical resistance in zero field (for the antiparallel configuration) to the low resistance state (for the parallel alignment), see Figure 5.11. We can write this as follows:

$$GMR_1 = \frac{R^{AP} - R^P}{R^P} = \frac{R(0) - R(H)}{R(H)} = \frac{\Delta R}{R} = \frac{\Delta \rho}{\rho_{\uparrow\uparrow}} \tag{5.38}$$

or alternatively in the form:

$$GMR_2 = \frac{R^{AP} - R^P}{R^{AP}} = \frac{R(0) - R(H)}{R(0)} = \frac{\Delta R}{R'} = \frac{\Delta \rho}{\rho_{\uparrow\downarrow}} \tag{5.39}$$

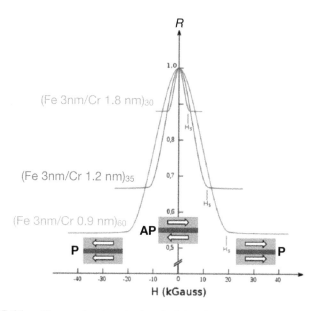

FIGURE 5.11 Change in electrical resistivity from antiparallel (AP) alignment in zero applied field to low resistance state for parallel (P) alignment of the magnetization vectors. The number of bilayers in the MML stack increases the measured GMR ratio. (Reprinted with permission from M. N. Baibich, J. M. Broto, A. Fert, F. Nguyen Van Dau, F. Petroff, P. Etienne, G. Creuzet, A. Friederich, & J. Chazelas, (1988). *Phys. Rev. Lett., 61,* 2472, ©(1988) by the American Physical Society.)

The use of the GMR formulae is a matter of convention, with different authors choosing different references for normalizing the GMR ratio. The more common choice appears to be that for GMR_1. The measured GMR ratio can be amplified by increasing the number of bilayers in the structure, as we saw previously and again illustrated in Figure 5.11.

The electrical resistance in ferromagnetic metals is dependent on the spin state of the electrons, which for electrons with a spin parallel to the magnetization is different from that for the electrons in the antiparallel state. This problem was initially considered by Neville Mott in 1936 in which the current for the spin-up and the spin-down states are taken separately in what is called the *two-current model*. It is necessary to consider the spin-mixing and spin-flip scattering.

In general, the conduction for a metal can be treated using the Boltzmann equation, which expresses the non-equilibrium energy (or wave vector) distribution, $f(E)$ or $f(\mathbf{k})$, much in the same way as the equilibrium energy distribution is expressed by the Fermi-Dirac function $f_0(E)$. For the case of a band i, the non-equilibrium distribution can be expressed as an extension of the Dirac distribution:

$$f_i(\mathbf{k}) \simeq f_0[E_i(\mathbf{k})] - \Phi_{i\mathbf{k}} \frac{\partial f_0[E_i(\mathbf{k})]}{\partial E} \tag{5.40}$$

Here, the second term represents the scattering, where for spherical bands the scattering probability depends on $|\mathbf{k} - \mathbf{k}'|$, and the scattering angle between the initial, $\mathbf{k}$, and final, $\mathbf{k}'$, states is θ. This shows that the deviation of the distribution $f(\mathbf{k})$ from its equilibrium value depends on $\Phi_{i\mathbf{k}}$, which can be expressed in terms of the applied electric field, $\mathscr{E}$, the group velocity, $\mathbf{v}_g$ and the relaxation time, τ:

$$\Phi_{i\mathbf{k}} = e\tau\mathscr{E} \cdot \mathbf{v}_g \tag{5.41}$$

For the case of spin dependent scattering, we need to establish the transport equation for each of the majority and minority spin subbands. This leads to the following:

$$e\mathscr{E} \cdot \mathbf{v}_g \frac{\partial f_0[E_i(\mathbf{k})]}{\partial E} = -\frac{f_\uparrow - f_0}{\tau_\uparrow} - \frac{f_\uparrow - f_\downarrow}{\tau_{\uparrow\downarrow}} \tag{5.42}$$

and

$$e\mathscr{E} \cdot \mathbf{v}_g \frac{\partial f_0[E_i(\mathbf{k})]}{\partial E} = -\frac{f_\downarrow - f_0}{\tau_\downarrow} - \frac{f_\downarrow - f_\uparrow}{\tau_{\uparrow\downarrow}} \tag{5.43}$$

Here, $\tau_\uparrow$ and $\tau_\downarrow$ are the spin-dependent relaxation times, while $\tau_{\uparrow\downarrow}$ character-izes the spin-flip and electron–electron collision processes. The latter allows us to couple the Boltzmann equations (5.42) and (5.43), which has a solution of the form:

$$\rho = \frac{\rho_\uparrow \rho_\downarrow + \rho_{\uparrow\downarrow}(\rho_\uparrow + \rho_\downarrow)}{\rho_\uparrow + \rho_\downarrow + 4\rho_{\uparrow\downarrow}} \qquad (5.44)$$

where $\rho_\uparrow = m^*/ne^2\tau_\uparrow$, $\rho_\downarrow = m^*/ne^2\tau_\downarrow$ and $\rho_{\uparrow\downarrow} = m^*/ne^2\tau_{\uparrow\downarrow}$. Spin-flip scat-tering, which is represented by the relaxation time $\tau_{\uparrow\downarrow}$, causes spin mixing via electron-magnon scattering, which becomes important at higher temperatures and tends to equalize the populations of two spin states. At low temperatures, the spin mixing term becomes negligible and we have:

$$\rho = \frac{\rho_\uparrow \rho_\downarrow}{\rho_\uparrow + \rho_\downarrow} \qquad (5.45)$$

It is worth noting that the spin dependence of the conduction in ferro-magnetic metals can be understood in terms of the band structure. The spin splitting of the electron bands into *majority spin* and *minority spin*, as illus-trated in Figure 5.12, means that the electrons at the Fermi level, which carry the electrical current, will be in different states for opposite spin directions and exhibit different conductivities. The degree of spin polarization at the

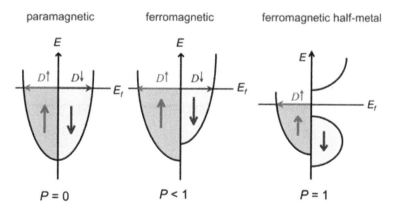

FIGURE 5.12 Spin bands in metals. For a paramagnetic metal (left), there is no splitting of the bands and the spin polarization, as defined in Eq. (5.46), is zero. In a ferromagnetic metal, the spin splitting of the bands leads to a difference in the density of states at the Fermi energy and gives a positive spin polarization, $0 < P < 1$. In the case of a half-metal, the spin splitting is such that the minority band sits below the Fermi edge and the spin polarization is complete, $P = 1$.

Fermi energy can be characterized using the following definition:

$$P = \frac{D_\uparrow(E_F) - D_\downarrow(E_F)}{D_\uparrow(E_F) + D_\downarrow(E_F)} \tag{5.46}$$

The above illustrates the treatment for a ferromagnetic metal, which can now be applied to the case of the magnetic multilayer structure. The spin dependent scattering of the two spin currents will affect the total resistance of the magnetic multilayer and in particular for the cases where the interlayer coupling is parallel or antiparallel. To demonstrate this, we will consider the case of a magnetic bilayer with a non-magnetic interlayer. We schematically illustrate these two situations in Figure 5.13. In the parallel configuration, one of the spin channels will be parallel to the magnetization vectors of the two ferromagnetic layers (Figure 5.13, left). In this case, the spin channel encounters a low resistance in both magnetic layers (denoted as r). For the opposing spin channel, both magnetic layers will present a high resistance state (denoted by R), since in addition to the intrinsic resistivity of the metal, the electrons will suffer spin flip/scattering processes. In the lower panel of Figure 5.13, we represent the effective resistance of the structure as a two branch parallel composite resistance, with one branch, spin-up, having a resistance of $2r$, while the spin-down channel has a resistance of $2R$. This allows us to represent the resistance as a to parallel resistances:

$$\frac{1}{R^P} = \frac{1}{2r} + \frac{1}{2R} = \frac{1}{2}\left(\frac{R+r}{Rr}\right) \tag{5.47}$$

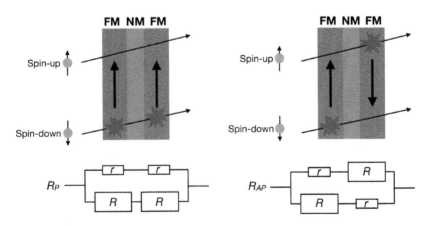

FIGURE 5.13   Resistance states for the parallel, R_P, and antiparallel, R_{AP}, configurations for the two current model in magnetic multilayer systems.

which yields the parallel resistance of:

$$R^P = \frac{2Rr}{R+r} \tag{5.48}$$

In a similar manner, we can see for the antiparallel configuration that both spin channels encounter a high and a low resistance in traversing the multi-layer, hence each channel can be represented as a resistance of $R + r$. From this, we obtain the resistance for the antiferromagnetic coupled multilayer (antiparallel configuration) as:

$$R^{AP} = \frac{R+r}{2} \tag{5.49}$$

We now substitute these results into Eq. (5.38):

$$GMR = \frac{R^{AP} - R^P}{R^P} = \frac{R+r}{2Rr} \left(\frac{R+r}{2} - \frac{2Rr}{R+r} \right) \tag{5.50}$$

With a little manipulation, we obtain:

$$GMR = \frac{(R-r)^2}{4Rr} \tag{5.51}$$

The above considers the situation for just two ferromagnetic layers. For a larger stack of layers, we can extend this argument, where the resistance for the AP configuration will be higher than that of the P configuration and will increase the total GMR ratio, as illustrated earlier in Figure 5.11.

A more rigorous treatment is given in the Valet–Fert (VF) model of GMR, who applied the Boltzmann equation to the case of CPP MR. One of the principal findings of the VF model is that for the CPP geometry, an additional potential drop occurs at the interface between the ferromagnetic and non-magnetic layers due to spin accumulation. This affects the electrochemical potential across the interface and depends therefore on the spin orientation in the magnetic layer. The interface potential, ΔV_I, is proportional to the current density, J, with $\Delta V_I = Jr_{SI}$, where r_{SI} is the spin-coupled interface resistance. This contribution is absent for the CIP geometry, since there will be no net charge or spin transport through the interface. The spin accumulation in the CPP geometry leads to spin injection from the ferromagnetic to the non-magnetic metal and hence a transfer of spin angular momentum. The degree of spin accumulation is dependent on the rate at which spins are injected and their decay via the spin-flip scattering process as they move away from the interface. This leads to the spin diffusion length, l_{sf}, which is used to define the profile of the spin accumulation, that decays exponentially; $n(x) = n_0 e^{-x/l_{sf}}$, where $n(x)$ is the spin electron density at a distance x from the interface, where the spin electron density is n_0. The VF model demonstrated that the spin diffusion length is much shorter than the electron means free path. In the case where the layers in a multilayer stack are thin with respect to the spin-diffusion length, it is necessary to take into account the overlap of the spin accumulation at adjacent interfaces to account for the CPP MR. In Figure 5.14, we illustrate the electrochemical potential and current densities traversing the interface between a ferromagnetic and a normal metal, where a constant current passes through the system. In the ferromagnetic metal, the difference in populations of electrons means that the electrical current will be split into a majority and minority spin current, with say $J_\uparrow > J_\downarrow$. This produces a spin current, which is given by $J_S = J_\uparrow - J_\downarrow$, while the total electrical current, given by $J = J_\uparrow + J_\downarrow$, remains constant throughout the structure. The accumulation of spins at the interface is illustrated in the variation of the chemical potential in the region of the boundary between the two layers. This illustrates the spin injection from the FM metal to the non-magnetic conductor. For electrons traversing in the opposite direction (spin extraction), the situation is analogous, with the exception that the spin accumulation will be in the opposite direction, leading to the spin polarization in the non-magnetic layer. In the case of spin injection into (or extraction from) a non-magnetic semiconductor, the larger density of states in the metal results in a larger spin accumulation density and a greater number of spin-slips on the metallic side of the interface. This will lead to a faster depolarization of the spin in the ferromagnetic metal and an almost complete depolarization upon entering the semiconductor. This undesirable effect can be remedied by introducing a spin-dependent interface resistance and thus allow the proportion of spin polarization to increase on the semiconductor side.

5.6   Exchange Bias

The *exchange bias* mechanism refers to a direct coupling interaction between an antiferromagnetic material and a ferromagnetic layer in direct contact and is not directly a spintronic effect. However, we introduce this because of its importance in spintronic device development and its exploitation in *spin valves*, which we will describe in the following section. The exchange bias effect was first discovered in 1956 by Meiklejohn and Bean when they studied the magnetic properties of Co particles embedded in the native antiferromagnetic oxide, CoO. Exchange bias is characterized by a shift of the mag-

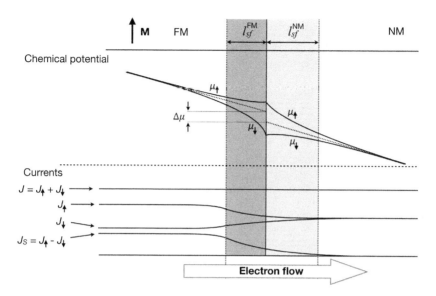

FIGURE 5.14 Chemical potential and current densities for majority (↑) and minority (↓) spins across the interface between a ferromagnetic (FM) and nonmagnetic (NM) metallic interface. The central region illustrates the spin diffusion lengths in the ferromagnetic and non-magnetic metals either side of the interface.

netic hysteresis loop along the field axis once the sample has been prepared under the correct conditions. This requires the sample, which consists of a ferromagnetic and antiferromagnetic interface, to be cooled through the Néel temperature, T_N, of the antiferromagnetic layer in the presence of an applied magnetic field. We note that the Curie temperature, T_C, of the ferromagnetic layer, is greater than the initial temperature, such that; $T_N < T < T_C$. This will give rise to the unidirectional anisotropy that characterizes the exchange bias effect, and is illustrated in Figure 5.15. As is shown schematically in this figure, the entire M-H loop is shifted along the field axis by the unidirectional anisotropy. In this case, the magnetic state in zero field is adjusted such that it will always point in the same direction. The shift can be expressed from the loop center shift from zero as given by the expression:

$$H_{EB} = \frac{H_{C1} + H_{C2}}{2} \qquad (5.52)$$

while the coercive field will be measured as:

$$H_C = \frac{H_{C1} - H_{C2}}{2} \qquad (5.53)$$

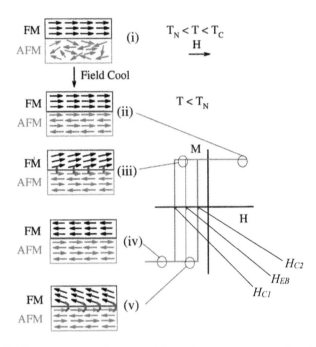

FIGURE 5.15   Schematic diagram of the spin configuration of an FM/AFM bilayer at different stages (i)–(v) of an exchange biased hysteresis loop. Note that the spin configurations are just a simple cartoon to illustrate the effect of the coupling and they are not necessarily accurate portraits of the actual rotation of the FM or AFM magnetizations. Reprinted from J. Nogués & I. K. Schuller, (1999). *J. Magn. Magn. Mater., 192*, 203–232, ©(1999), with permission from Elsevier.

In the simple case, where the ferromagnetic anisotropy is assumed to be negligible, under the condition $K_{FM}t_{FM} \ll K_{AFM}t_{AFM}$, the free energy of the ferromagnetic (FM)/antiferromagnetic (AFM) bilayer can be expressed as:

$$E = -HM_{FM}t_{FM}\cos(\theta_H - \vartheta) + K_{AFM}t_{AFM}\sin^2\alpha - J_{INT}\cos(\vartheta - \alpha) \quad (5.54)$$

Here, the angles refer to the direction of the magnetic field, θ_H, the easy axis of the ferromagnetic film ϑ, and that for the anti-ferromagnetic layer, α. After minimizing the free energy with respect to ϑ and α, the exchange bias field can be expressed as:

$$H_{EB} = \frac{J_{INT}}{M_{FM}t_{FM}} \quad (5.55)$$

An important result from the energy minimization results in the condition:

$$K_{AFM}t_{AFM} \geq J_{INT} \quad (5.56)$$

which is required for the observation of exchange bias, also referred to as exchange anisotropy. For the case of $K_{AFM}t_{AFM} \gg J_{INT}$, the energy will be minimized by keeping α small independently of ϑ. On the other hand, if $K_{AFM}t_{AFM} \ll J_{INT}$, it will be energetically favorable to keep the quantity $(\vartheta - \alpha)$ small, in which case, the antiferromagnetic and ferromagnetic spins rotate together. Therefore, if the condition (5.56) is not satisfied, the antiferromagnetic spins will follow the rotation of the ferromagnetic layer and therefore, no exchange bias will be observed and only an increase of the coercivity will be expected.

5.7 Spin Valves and Magnetic Tunnel Junctions

One of the principal applications of the exchange bias effect is its use as a pinning device for a magnetic layer in a multilayer structure in which the separation between ferromagnetic layers is set to a distance which ensures an antiparallel alignment with the pinned ferromagnetic layer. This is schematically illustrated in Figure 5.16(a), where the pinned layer is exchange coupled to the antiferromagnetic layer, while a free layer can rotate on the direction of an external field, such as that from a magnetic bit in close proximity. A slightly more sophisticated structure is shown in Figure 5.16(b), where the reference (fixed) layer is now held via a synthetic antiferromagnetic (SAF) structure due to the antiferromagnetic coupling with an additional magnetic layer which is exchange coupled to the antiferromagnetic film. One of the advantages of this SAF structure is that it allows us to control the strength of the pinning in the reference layer. The magnetic hysteresis loop of a spin

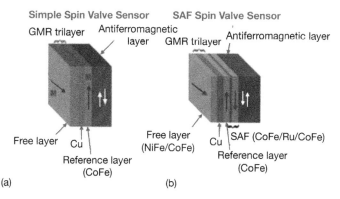

FIGURE 5.16 Schematic diagram of the spin valve structure, (a) simple spin valve with a pinned layer and a free magnetic layer, and (b) a SAF spin valve.

valve structure is illustrated in Figure 5.17 along with its magnetoresistance behavior. It is noted, that the shift in the hysteresis means that the soft or free layer part of the loop rapidly switches near zero field, which results in large changes in the MR, making the device very sensitive to small magnetic fields and hence very suitable for sensing applications. Early applications

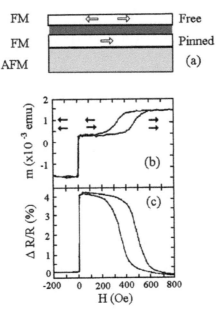

FIGURE 5.17   (a) Schematic diagram of a spin valve device. (b) Hysteresis loop, $m(H)$, and (c) magnetoresistance, $\Delta R/R(H)$, of FeNi/Cu/FeNi/FeMn GMR spin-valve at room temperature. Reprinted figure with permission from B. Dieny, V. S. Speriosu, S. S. P. Parkin, B. A. Gurney, D. R. Wilhoit, & D. Mauri, (1991). *Phys. Rev. B, 43*, 1297. ©(1991) by the American Physical Society.

of such devices were in fact in the read head of magnetic bits in hard disk drives (HDD). This contributed enormously to the miniaturization of HDDs and consequently to the boom in the areal density of HDDs, as we will illustrate in the following. The sensitivity of the device near the zero field point can be measured from its slope:

$$S = \frac{\partial(\Delta R/R)}{\partial H} \tag{5.57}$$

with values of up to 18% (kA/m) being reported in the literature at room temperature.

An alternative approach to the GMR devices outlined above is the use of an insulating barrier between the ferromagnetic layers. This will effectively eliminate the possibility of classical electron transport through the device. If the insulating layer is sufficiently thin, of the order of a few atomic layers, then electrons can tunnel through the barrier region into available states on the other side. Such a structure is referred to as a *magnetic tunnel junction* or MTJ. The tunneling process being dependent on the available states will be sensitive to the spin configurations for the two ferromagnetic electrodes, as shown in Figure 5.18. The tunneling process can be seen to be more probable, and hence have a lower electrical resistance, when the two magnetic layers have a parallel alignment of their magnetizations, Figure 5.18(a). This is precisely what is observed experimentally, as illustrated in Figure 5.18(b). At lower temperatures, the tunnel resistance increases, but the tunnel magnetoresistance (TMR) ratio also increases. The TMR ratio is defined in a similar manner to the GMR ratio, and can also be expressed in terms of the spin polarization of the two ferromagnetic layers, P_1 and P_2:

$$TMR = \frac{R^{AP} - R^P}{R^P} = \frac{2P_1 P_2}{1 - P_1 P_2} \tag{5.58}$$

where the spin polarization, P, is expressed using Eq. (5.46). The TMR effect was observed well before the GMR in all metal systems, with M. Jullière (1975) having studied the Fe/GeO/Fe tunnel junction, for which 14% MR ratio was found for low temperatures. Jullière's model for the TMR was based on the work of Meservey and Tedrow (1971), where the tunnel current

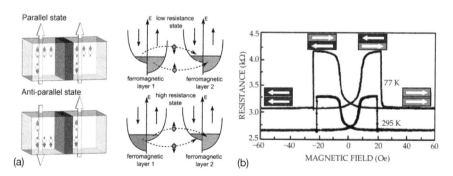

(a)   (b)

FIGURE 5.18 (a) Schematic illustration of the tunnel magnetoresistance (TMR) due to the availability of empty states of the same spin orientation when two ferromagnetic layers are aligned parallel and antiparallel. (b) Electrical resistance for a magnetic tunnel junction at liquid nitrogen and room temperatures as a function of the applied magnetic field.

is proportional to the product of the density of spin states on either side of the tunnel barrier. The Jullière model assumes that spin is conserved in the tunnel process and is expressed in the form:

$$\frac{\Delta R}{R} = \frac{R^{AP} - R^P}{R^{AP} + R^P} = \frac{2P_1 P_2}{P_1 + P_2} \tag{5.59}$$

The Jullière model of spin-dependent tunneling is rather simple and while providing a reasonable first approximation is incapable of explaining the finer details of experimental observations. It is found that, for example, the combination of the metal/insulator is very important and that the tunneling electrons come from the first few atomic layers of the metal, which are then hybridized with the insulator and hence do not have the bulk band structure. Since spin electronic effects are almost always an interface related effect, the interface band structure needs to be considered for a correct understanding of the tunneling properties.

One approach to the question of the tunneling between ferromagnetic electrodes through an insulator was made by Slonczewski. Under the realization that the barrier is relatively permeable, it is necessary to consider the overlap of the wave functions within the barrier region and therefore, the wave function matching is an important consideration across the entire device. It turns out, that one of the principal findings is that the spin polarization, as used in Eq. (5.59), depends on the barrier height, V_b, through the imaginary wave vector in the barrier:

$$\kappa = \frac{\sqrt{2m(V_b - E_F)}}{\hbar} \tag{5.60}$$

and can be expressed in the form:

$$P = \left(\frac{k_\uparrow - k_\downarrow}{k_\uparrow + k_\downarrow}\right) \frac{\kappa^2 - k_\uparrow k_\downarrow}{\kappa^2 + k_\uparrow k_\downarrow} \tag{5.61}$$

The term in brackets is similar to that expressed earlier and used by Jullière. However, the additional multiplicative term will modify the Jullière result. Since κ ranges from zero for low barriers to infinity in the case of large barriers, the spin polarization will thus be dependent on the barrier height.

A detailed consideration of the tunneling between the ferromagnetic electrodes can be performed on the basis of the Kubo/Landauder formalism, which was discussed in Chapter 2. In this case, it is necessary to take into account each spin channel. The general approach can be visualized from Figure 5.19, where a bias voltage, V, is applied across the tunnel junction. The rate of tunneling, $\Gamma_\sigma(V)$, from the left to the right electrode can be expressed using Fermi's golden rule:

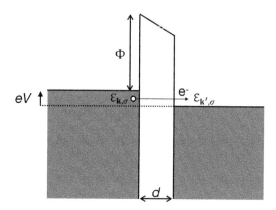

FIGURE 5.19 Band diagram for the tunneling of an electron with wave vector, k, and spin state σ through an insulating barrier of width, d.

$$\overrightarrow{\Gamma}_\sigma(V) = \frac{2\pi}{\hbar} \sum_{k,k',\sigma} |T_{kk'}^\sigma|^2 f(\varepsilon_{k,\sigma})[1 - f(\varepsilon_{k',\sigma})]\delta(\varepsilon_{k,\sigma} - \varepsilon_{k',\sigma} + eV) \quad (5.62)$$

where ε_k and $\varepsilon_{k'}$ are one electron energies measured from the Fermi levels on either side of the insulating barrier, $f(\varepsilon_{k,\sigma}) = [e^{\varepsilon_k/k_B T} + 1]^{-1}$ is the Fermi distribution function and $T_{kk'}^\sigma$ are the tunneling matrix elements, which take into account that the tunneling probability decreases exponentially as the barrier thickness, d, of the insulating layer increases. The steady-state current through the junction is determined by the difference between the forward and reverse tunneling rates:

$$I_\sigma = e[\overrightarrow{\Gamma}_\sigma(V) - \overleftarrow{\Gamma}_\sigma(V)] \quad (5.63)$$

Here, $\overleftarrow{\Gamma}_\sigma(V)$ is the reverse tunneling rate for electrons with spin, σ, going from right to left. The relation between the two tunneling rates is given by: $\overleftarrow{\Gamma}_\sigma(V) = \overrightarrow{\Gamma}_\sigma(-V)$. Using the above expressions, we can obtain the tunneling current, I_σ, for the spin channel σ as:

$$I_\sigma = \frac{2\pi e}{\hbar} \langle |T_{kk'}^\sigma|^2 \rangle \int_{-\infty}^{\infty} D_{1\sigma}(\varepsilon - eV)D_{2\sigma}(\varepsilon)[f(\varepsilon - eV) - f(\varepsilon)]d\varepsilon \quad (5.64)$$

where $D_{1\sigma}(\varepsilon)$ and $D_{2\sigma}(\varepsilon)$ are the tunneling density of states for spin σ, in the left and right electrodes, respectively. $\langle |T_{kk'}^\sigma|^2 \rangle$ is the averaged tunneling probability and can be taken to be a constant proportional to $e^{2\kappa d}$, where $\kappa = \sqrt{2m\Phi}/\hbar$ is the decay constant of the wave function penetrated into the insulating barrier with height Φ. The total electrical current will be given by the sum of the spin current for up and down states: $I = I_\uparrow + I_\downarrow$.

In the low bias regime, where the density of states is nearly constant, the spin dependent conductance, expressed as $G_\sigma = dI_\sigma/dV$, for each spin channel can be written in the form:

$$G_\sigma(V) \simeq \frac{2\pi e^2}{\hbar} \langle |T_{kk'}^\sigma|^2 \rangle D_{1\sigma} D_{2\sigma} \tag{5.65}$$

If we assume that the magnetic moments of the electrodes FM_1 and FM_2 are aligned antiparallel (AP) in zero magnetic field and parallel (P) at a certain value of the applied field, the total conductance, $G = G_\uparrow + G_\downarrow$, for parallel alignment will be:

$$G_P = G_P^\uparrow + G_P^\downarrow \propto D_{M1}D_{M2} + D_{m1}D_{m2} \tag{5.66}$$

while for antiparallel alignment we have:

$$G_P = G_{AP}^\uparrow + G_{AP}^\downarrow \propto D_{M1}D_{m2} + D_{m1}D_{M2} \tag{5.67}$$

where D_{Mi} and D_{mi} are the density of states for the majority (M) and minority (m) spin bands in the ith electrode. From this, we can write the TMR ratio in the form:

$$TMR = \frac{\Delta R}{R_P} = \frac{R_{AP} - R_P}{R_P} = \frac{G_P - G_{AP}}{G_{AP}} = \frac{2P_1 P_2}{1 - P_1 P_2} \tag{5.68}$$

Now, we express the spin polarization, much as given above, in the form:

$$P_i = \frac{D_{Mi} - D_{mi}}{D_{Mi} + D_{mi}} \tag{5.69}$$

For identical magnetic layers, we find:

$$TMR = \frac{\Delta R}{R_P} = \frac{2P^2}{1 - P^2} \tag{5.70}$$

With improved growth methods, by the mid-1990s several research teams had observed TMR ratios of around 18% and later $\sim$ 50%, that were measured at room temperature using insulating junctions of Al_2O_3 and Fe electrodes. By 2004, epitaxial Fe/MgO MTJs were yielding TMR values in the region of 200%. These enhanced TMR ratios were a result of purely ballistic transport, which can have values of over 1000% at room temperature. This allows for further reduction in the read head size for HDDs.

Granular TMR has also been reported by many researchers. In this case, ferromagnetic nanoparticles are dispersed in an insulating matrix, though TMR ratios at room temperature are rather small and typically well below 10%. An enhancement is possible, giving a TMR of around 20% at low temperature, which is induced by co-tunneling across nanoparticles in the Coulomb blockade regime. The theoretical enhancement is given as $2/(1 - P^2)$. Recent work shows the increasing complexity of device structures at the gain of improved TMR ratios; an example is illustrated in Figure 5.20 for a perpendicular anisotropy system or p-MTJ.

As we mentioned earlier, one of the main applications of the spin valve and the magnetic tunnel junction is the read-write head in hard disk drive systems. This is a multi-billion dollar industry, and provides a huge stimulus for research and development in data storage. This has provided a major impetus for the rapid miniaturization and increase of areal density in data storage systems. In Figure 5.21(a), we show this progress as an example of Moore's law. Current (at the time of writing 2018) products are already over the 1 Tbit in^{-2} range. This means that this industry is true nanotechnology. It will be noted that at the time of the GMR discovery, both read and write components used inductive methods. This changed in the 1990s,

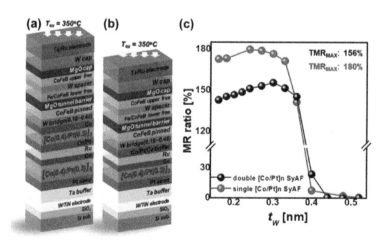

FIGURE 5.20 Dependency of the TMR ratio on p-MTJ spin-valve structure. Schemes of double MgO based p-MTJ spin-valve with a top $Co_2Fe_6B_2$ free layer using (a) a double SAF $[Co/Pt]_n$ layers, (b) a single SAF $[Co/Pt]_n$ layer, (c) TMR ratio depending on W bridge-layer thickness (t_w) and p-MTJ spin-valve structure. Reprinted from J.-Y. Choi, D.-G. Lee, J.-U. Baek, & J.-G. Park, (2018). *Sci. Rep. 8*, 2139, under a Creative Commons Attribution 4.0 International License.

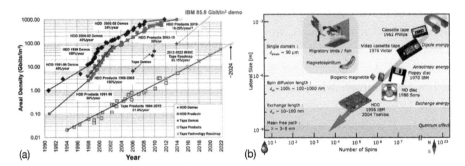

FIGURE 5.21   (a) Moore's law for the areal density of HDD (red), showing the comparison with tape products (green) and HDD demonstration devices (blue). (b) Magnetic length scales for magnetic recording devices. Reprinted from A. Hirohata & K. Takahashi, (2014). *J. Phys. D: Appl. Phys., 47*, 193001. Reproduced with permission of IOP Publishing.

when MR heads and GMR heads were introduced. The effect is seen as a knee in the areal density Moore's law, where an increased gradient is clear. TMR heads were introduced about a decade later, in 2005, and have allowed growth to continue. Further developments include the use of perpendicular recording, where the magnetic bits have a magnetic anisotropy which lies in the direction perpendicular to the surface. Such technologies require very strong magnetocrystalline anisotropies to overcome the strong effect of the magnetostatic energy which generally favors in-plane configurations of the magnetization. This method has shown areal densities of over 300–400 Gbit in^{-2}. Using an exchange-coupled composite soft layer coupled to the hard perpendicular anisotropy material allows the reversal field to be reduced and coherent reversal can be achieved. The proposed improvement for increasing the HDD areal density will need to consider the so-called magnetic recording trilemma, the competing requirements of the readability, writability, and stability. Heat-assisted magnetic recording or HAMR, will incorporate a small laser, which will heat the magnetic bit, reducing its anisotropy which will allow its magnetization to be reversed at lower magnetic fields applied to the write head. This has lead to areal densities in the region of 1.5 Tbit in^{-2}. The patterning on the nanoscale will further improve this to around 10 Tbit in^{-2}. This is not the end of the story since much research is also being dedicated to other techniques, such as microwave-assisted magnetic recording (MAMR) and 3D magnetic recording. In MAMR, a microwave field excites the magnetization at a frequency corresponding to its resonance and assists in the magnetic switching of the bit. This will both reduce the field necessary for magnetic reversal and also reduce the switching time and thus make the writing process faster. In the 3D recording, the magnetic bits are not a single magnetic layer, but are distributed over several layers and will significantly increase the areal density for HDDs.

The development of MRAM (magnetic random access memory) provides an interesting method, which has attracted much attention in recent years. MRAM can achieve non-volatility, similar to the HDD, without the need for mechanical read and write heads, which can be performed electronically on the order of nanoseconds, with high-density and low power consumption. This makes MRAM an attractive alternative to the HDD. In the basic MRAM device, the memory cell is based on a magnetic tunnel junction, as illustrated in Figure 5.22. The binary information "0" and "1" are stored as parallel and antiparallel orientations of the magnetization of the free layer in the MTJ. These are connected at the cross points of the "bit" and "digit" (or "word") lines. Information can be written by sending current pulses through one line of each array and only at the crossing points of these lines will the resulting magnetic field, created via the Ampère field, be sufficient to orient the free magnetic layer. The reading process is performed by measuring the resistance between the two lines connecting the addressed cell. The cross-point architecture allows for very high areal density. Clearly, thermal stability is an issue that needs to be addressed and typically, a thermal stability of $KV/k_BT > 100$ is required.

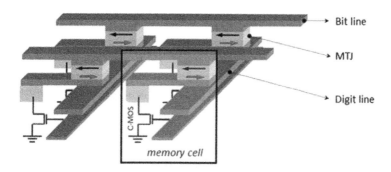

FIGURE 5.22 Schematic illustration of the magnetic random access memory (MRAM) device.

In the late 1990s, it was realized that a spin-polarized current is able to induce a spin-transfer torque (STT). This means that the passage of electrons also imparts angular momentum transfer since there are more of one type of spin than the other. This concept was introduced by Slonczewski and

Berger independently in 1996 and was first observed experimentally in 2000 for multilayered nanopillar structures. The STT was seen to traverse from one magnetic layer to another through an intervening non-magnetic spacer. The spin-transfer refers to the conservation of the angular moment in this passage from one ferromagnetic layer to the other. Experimentally, the switching of the magnetic element in the nanopillar structure can be observed by the resistance change due to the GMR changes, hence making it also electrically detectable.

The spin-transfer torque effect can be exploited as a method of magnetic switching in an MRAM device and could, therefore, negate the necessity of an Ampère field and hence could reduce the risk of cross-talk and reduce power consumption. Interestingly, STT can be used as a method of microwave generation, where the spin torque induces a magnetic precession rather than irreversible switching. Indeed, the spin-transfer can also be used as a method of control for the damping in magnetization excitations.

5.8   Spintronic Devices

In the previous section, we considered the development of the spin valve and tunnel junction. These are essentially two-terminal devices that were an extension of the GMR structures. One of the key concepts of spintronics is the controlled injection of electrons with a given spin into a device. We will outline two methods for spin injection in the following. The first technique was in fact devised before the discovery of the GMR effect. It was based on a three-terminal device using non-equilibrium spin accumulation generated by a spin polarized current from a ferromagnetic metal (FM_1) into a non-magnetic metal (NM). The spin injection signal is then detected in another ferromagnetic metal (FM_2). The spin diffusion length, λ_S, of the NM metal can be obtained by varying its thickness. The spin injection and accumulation can then be examined, as illustrated in Figure 5.23, using a double tunnel junction. The electrodes of FM_1 and FM_2 are made from different ferromagnetic metals and the central NM metal has a thickness d. With the orientation of the first FM layer chosen in a specific direction, the second layer, FM_2, can be either parallel or antiparallel, Figure 5.23(a). The bias injection current, I_{inj}, flows through the junction (in the CPP geometry). A measurement of the potential differences, V_1 and V_2, can be used to evaluate the tunnel resistance for the two junctions, R_1 and R_2. A second arrangement, shown in Figure 5.23(b), can be used to induce a spin polarized current into the NM metal layer. When there is a gradient in the electrochemical potential, $\mu_\sigma(x)$, for electrons with spin σ in the NM, the electrical current, j_σ, will flow according to the relation:

$$j_\sigma(x) = -\frac{\sigma_{NM}}{2e}\frac{\partial}{\partial x}[\mu_\sigma(x)] \tag{5.71}$$

where σ_{NM} is the conductivity in the non-magnetic metal. In the case of spin-flip scattering in the NM, the divergence of the current of the spin σ channel will be balanced with the difference between the spin relaxation rate from σ to $-\sigma$, and in the steady-state we have:

$$\frac{\partial}{\partial x}[j_\sigma(x)] = -\frac{e}{2\tau_s}[n_\sigma(x) - n_{-\sigma}(x)] \tag{5.72}$$

where $n_\sigma(x)$ is the local density of spin σ in the NM and τ_s is the spin relaxation time. Taking the difference between the current densities for the two spin channels, we obtain:

$$\frac{\partial}{\partial x}[j_\uparrow(x) - j_\downarrow(x)] = -\frac{eD_{NM}}{\tau_s}[\mu_\uparrow(x) - \mu_\downarrow(x)] \tag{5.73}$$

Here, D_{NM} is the density of states per spin in the NM. Summing the two spin currents, we maintain the current conservation condition:

$$\frac{\partial}{\partial x}[j_\uparrow(x) + j_\downarrow(x)] = 0 \tag{5.74}$$

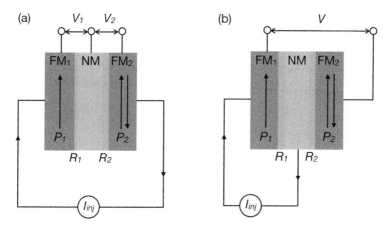

FIGURE 5.23 (a) Double tunnel junction for the FM$_1$/NM/FM$_2$ structure in the CPP geometry. (b) Spin injection into the NM layer.

We can introduce the spin splitting and average of the chemical potentials for the two spin channels as follows:

$$\delta\mu(x) = \frac{1}{2}[\mu_\uparrow(x) - \mu_\downarrow(x)] \tag{5.75}$$

and

$$\bar{\mu}(x) = \frac{1}{2}[\mu_\uparrow(x) + \mu_\downarrow(x)] \tag{5.76}$$

Now, the spin density, $S(x)$, induced by the spin splitting of the chemical potential, $\delta\mu(x)$, and spin current $j_{spin} = j_\uparrow(x) - j_\downarrow(x)$ flowing in the NM can be expressed as:

$$S(x) = D_{NM}\delta\mu(x) \tag{5.77}$$

and

$$j_{spin}(x) = -\frac{\sigma_{NM}}{eD_{NM}}\frac{\partial}{\partial x}[S(x)] \tag{5.78}$$

Using the above expressions, we can establish the following equations for $\delta\mu(x)$ and $\bar{\mu}(x)$:

$$\frac{\partial^2}{\partial x^2}[\delta\mu(x)] = \frac{\delta\mu(x)}{\lambda_s^2} \tag{5.79}$$

and

$$\frac{\partial^2}{\partial x^2}[\bar{\mu}(x)] = 0 \tag{5.80}$$

where $\lambda_s = \sqrt{D\tau_s}$ is the spin diffusion length with the spin diffusion coefficient, which satisfies the Einstein relation, $D = \sigma_N/(2e^2 D_{NM})$, in the NM layer. These equations have the following solutions:

$$\bar{\mu}(x) = \left(\frac{ej}{\sigma_{NM}}\right)x + \bar{\mu}(0) \tag{5.81}$$

where j denotes the current density in the NM.

$$\delta\mu(x) = \alpha_1 e^{x/\lambda_s} + \alpha_2 e^{-x/\lambda_s} \qquad ; -d/2 < x < d/2 \tag{5.82}$$

where α_1 and α_2 are coefficients determined by the spin-dependent tunnel currents, $I_{i\sigma}$:

$$I_{1\sigma} = \frac{G_{1\sigma}}{e}[eV_1 - \sigma\delta\mu(-d/2)] \tag{5.83}$$

and

$$I_{2\sigma} = \frac{G_{2\sigma}}{e}[eV_2 + \sigma\delta\mu(d/2)] \tag{5.84}$$

Here, the $G_{i\sigma}$ denote the tunnel conductances for electrons of spin σ at the junctions $i = 1,2$, where $G_{i\sigma} \propto D_N D_{FMi}^{\sigma}$ and $V_{1,2}$ being the voltage drops across the junctions 1, 2. The spin polarizations are expressed in the form:

$$\tilde{P}_i = \frac{G_{i\uparrow} - G_{i\downarrow}}{G_{i\uparrow} + G_{i\downarrow}} = \frac{D_{FMi}^{\uparrow} - D_{FMi}^{\downarrow}}{D_{FMi}^{\uparrow} + D_{FMi}^{\downarrow}} \tag{5.85}$$

We note that $\tilde{P}_i$ changes sign depending on whether the magnetizations are parallel or antiparallel. The equation for $\delta\mu(x)$ will therefore give the spin polarization profile in the NM layer. This, in effect, constitutes the spin injection.

It is worth noting that the form of spin polarization can be distinguished depending on the nature of the transport of the charge carriers and their specific spin states. These are outlined in Figure 5.24. In the first case, we have an equal number of spin-up and spin-down charge carriers moving in a certain direction. This is referred to as an unpolarized current, representing the majority of cases for charge transport. In the second case, we have an unequal population of spin-up and spin-down charge carriers moving in a specific direction. This is a general case of a spin-polarized current. If only one type of charge carrier is present, say for spin-up electrons, then we say the current is fully spin-polarized. In the final, case we have spin-up carriers moving in one direction, while the spin-down electrons move in the opposite sense. In this case, there is no charge current whatsoever and we have an example of pure spin current.

The generation of non-equilibrium spin polarization can also be realized using optical methods, known as optical orientation or optical pumping. This is frequently used in semiconductors, where optical irradiation changes the relative populations within the Zeeman and hyperfine levels of the ground states of atoms. In the case of optically induced spin injection, angular momentum is transferred to the medium via the absorption of circularly polarized light. The electron orbital momenta are directly oriented by the photons and through the spin-orbit interaction, the electron spins become polarized. In the case of injection of polarized carriers in a direct bandgap semiconductor, the degree of spin polarization of the carriers can be detected via the (circular) polarization of the light created in the recombination of electrons

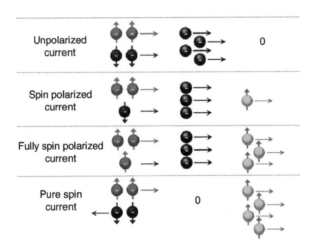

FIGURE 5.24　Classification of spin currents in spintronic devices.

and holes. The degree of circular polarization is directly proportional to the spin polarization of the carriers.

In a semiconductor, the photo-excited spin-polarized electrons and holes will exist for a specific time, τ_r, before recombining. If a fraction of the carriers' initial orientation survives for longer than the recombination time, i.e., if $\tau_r < \tau_s$, the electroluminescence (recombination radiation) will be partially polarized. Thus, by measuring the circular polarization of the luminescence, it is possible to study the spin dynamics of the non-equilibrium carriers. This will also permit the extraction of useful physical quantities such as the spin orientation, the recombination time or the spin relaxation time of the charge carriers.

In Figure 5.25, we show the principal transformations for the direct semiconductor GaAs. Here, we note that the maximum of the valence band and the minimum of the conduction band are both found at the Γ-point. The valence band (p-symmetry) splits into four degenerate $P_{3/2}$ states at Γ_8 and a two-fold degenerate $P_{1/2}$ state at Γ_7, which lies 0.43 eV below the $P_{3/2}$ state (Γ_8). The conduction band (s-symmetry) is two-fold $S_{1/2}$ degenerate at Γ_6. The $P_{3/2}$ band consists of two-fold degenerate bands for heavy and light holes.

When circularly polarized light with $\hbar\omega = E_g$ illuminates the GaAs, electrons are excited from $P_{3/2}$ to $S_{1/2}$. According to the selection rule; $\Delta m_j = \pm 1$, two transitions for each photon helicity (σ^+ and σ^-) are possible. The relative transition probabilities for light and heavy holes are different, resulting in net spin polarization of excited electrons. This arises because the transitions from the heavy hole band ($m_j = \pm 3/2$) are three times as likely than those from the light hole band ($m_j = \pm 1/2$) for each polarization of light. Theoretically, a maximum of 50% spin polarization can be expected, though in reality measurements show a value of around 40%, due to the limitations encountered such as spin depolarization at the surface or within the GaAs layer.

For illumination with photons of energy of $\hbar\omega \geq E_g + \Delta$, the polarization decreases with increasing energy due to simultaneous excitation of electrons from light and heavy hole states and the split of valence band states ($P_{1/2}$). By tuning the energy of the circularly polarized light to the band gap of GaAs, a spin polarization of up to 50% can be theoretically achieved. Denoting the density of electrons in the $m_j = \pm 1/2$ states as $n_\pm$, we can define the spin polarization as:

$$P_n = \frac{n_+ - n_-}{n_+ + n_-} \tag{5.86}$$

We note that a reversal of polarization will reverse the sign of P_n. In GaAs, we have $P_n = (1-3)/(1+3) = -1/2$. The circular polarization of the luminescence is given by:

$$P_{circ} = \frac{I_+ - I_-}{I_+ + I_-} \tag{5.87}$$

where $I_\pm$ is the radiation intensity for the helicity $\sigma^\pm$. The polarization of the σ^+ photoluminescence will be given by:

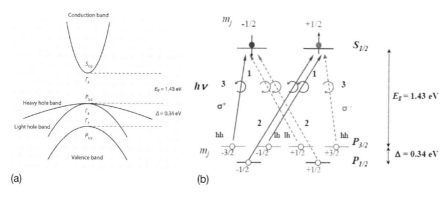

(a)   (b)

FIGURE 5.25 (a) Schematic band diagram for GaAs around the Γ point. (b) Radiative combination/absorption transitions for spin-up and spin-down electrons according to the optical transition selection rules.

$$P_{circ} = \frac{(n_+ + 3n_-) - (3n_+ + n_-)}{(n_+ + 3n_-) + (3n_+ + n_-)} = -\frac{P_n}{2} = \frac{1}{4} \qquad (5.88)$$

There have been numerous studies on the effects of spin polarization. One of the earlier applications was the construction of a spin LED. The principal difference between an ordinary LED and a spin LED is that, as a consequence of the radiative recombination of spin-polarized carriers, the emitted light will be circularly polarized and this can be used to trace the degree of polarization of carriers upon spin injection into a semiconductor.

In Figure 5.26, we show an example for a device, which consists of a magnetic layer, in this case, the ferrimagnetic vanadium tetracyanoethylene (V[TCNE]$_2$), which acts as a spin filter for the quantum well-formed by n-AlGaAs/i-GaAs/p-AlGaAs. Spin-polarized electrons are injected across the tunnel barrier into the conduction band of an n-doped AlGaAs layer, which then recombines with holes in the GaAs quantum well, Figure 5.26(a). Since the electrons are spin-polarized, the light emitted is circularly polarized due to the conservation of angular momentum from the electron to the photon. Emission from the quantum well region is illustrated in Figure 5.25(b), where the strong polarization is shown from the asymmetry between the heavy hole (right-circularly polarized light) and the light hole (left-circularly polarized light) recombinations, shown respectively in red and blue shading. In Figure 5.26(c), we see the correlation between the magnetic state of the V[TCNE]$_2$ layer and the heavy hole transitions as a function of the applied magnetic field.

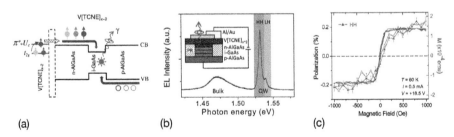

(a)   (b)   (c)

FIGURE 5.26 (a) Schematic band diagram of the spin LED. (b) Electroluminescence (EL) of spin-LED. (c) Circular polarization of spin-LED (red triangles) and magnetization of V[TCNE]$_2$ (vanadium tetracyanoethylene)(green line). HH, LH: heavy-hole and light-hole, respectively. i-GaAs: Undoped gallium arsenide. Reprinted figure with permission from L. Fang, K. Deniz Bozdag, C.-Y. Chen, P. A. Truitt, A. J. Epstein, & E. Johnston-Halperin, (2011). *Phys. Rev. Lett., 106*, 156602, ©(2011) by the American Physical Society.

Spin transport dependent transistors have also been the subject of many studies and mark a shift in spintronics towards the use of semiconductors. This also led to a boom in the study of magnetic semiconductors, which has not yet reached its full potential. One of the first devices proposed was the Datta–Das spin-polarized field effect transistor, or spin-FET in 1990. The principle of operation relies on the Rashba effect, in which relativistic effects cause an electric field to have magnetic components for an electron. Under certain conditions, this field is capable of spin splitting the conduction band of a semiconductor and of causing spin precession. Datta and Das exploited this idea in their conception of the spin FET. The device has a similar structure to a conventional FET, being a three-terminal device with a source, a drain, and a gate, see Figure 5.27.

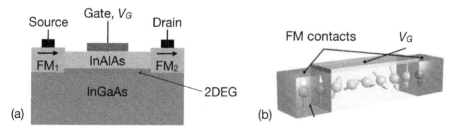

FIGURE 5.27 (a) Schematic diagram of the Datta–Das spin FET. A spin-polarized current is injected from the source and analyzed by the drain. Spin precession of the carriers will occur via the spin-orbit coupling due to the Rashba effect and is controlled by the gate voltage, V_G. (b) Schematic illustration of the rotation of the spin of charge carriers below the gate region under the action of the gate voltage.

The main difference in the construction of the device is that the source and drain are made from ferromagnetic materials, which means that they provide the source for the spin injection and detection, respectively. The spin-polarized electrons are injected from the source (FM_1) into a high-mobility two-dimensional electron gas (2DEG) channel. The electric field applied to the channel under the gate region adjusts the effective magnetic field in the channel via the Rashba effect, causing the spins to rotate as they move along the 2DEG channel. The amount of rotation will be controlled by the amplitude of the gate voltage. When the carriers reach the drain (FM_2), they will either be transmitted or reflected with a certain probability, depending on the relative orientation of the carrier spin with respect to the drain magnetization. Therefore, the conductance of the spin FET is partially controlled by the orientation of the carrier spin in the 2DEG channel under the control of the gate voltage.

In the discussion above, we considered the junction between a ferromagnetic and a non-magnetic metal, where we considered the spin accumulation and injection at the interface. For the spin FET device, we are now confronted with a ferromagnetic metal and a semiconductor interface. This can frequently develop a space-charge region with a Schottky barrier and depletion region. This situation is well-known for the usual non-magnetic metal/semiconductor interface, where it is found that the interface forms a rectifying junction or diode. We are concerned as to whether the spin polarization at the interface between the FM metal and the semiconductor will result in spin injection at the interface. Naively, we might expect this to be the case since the spin is carried by the electrons and a spin split interface should result in some spin injection. Reality can be a little more complex and typically we require a forward bias to promote the injection. On the metal (or n-) side of the interface, we have say more spin-up electrons than spin-down; $n_\uparrow > n_\downarrow$. Denoting the spin-split energy as $2e\zeta$, we can write:

$$n_\uparrow(\zeta) = n_\uparrow(0)e^{e\zeta/k_B T} \qquad (5.89)$$

Under forward bias electrons, will flow from the metal into the semiconductor (p-type, say). However, the flow will be limited by the thermal activation over the barrier region, which is for spin-up electron greater by a factor $e\zeta$. Using Boltzmann statistics, the rate of transmission of spin-up electrons over the barrier is $\propto e^{-e\zeta/k_B T}$. Since the current is proportional to both the carrier density and the transmission rate, the two experimental factors will cancel out. The same argument holds for the spin-down carriers. Therefore, the spin current will be unaffected by $2e\zeta$ and no spin current or spin accumulation will occur for low bias voltages. The spin injection is expected for larger bias voltages, when it is driven by electric drift leading to non-equilibrium spin populations. In addition to the spin injection, spin extraction is also predicted for magnetic p-n junctions, where for large biases, spins are extracted (depleted) from the non-magnetic region. It is worth noting that spin alignment can be achieved using a magnetic layer or a magnetic semiconductor to promote spin injection. This process is the same as described for the operation of the spin LED structure.

The experimental studies of Schottky barrier structures of FM metal/semiconductor such as GaAs/Fe have shown rather a poor spin injection efficiency, where spin LED structures have exhibited values of P_{circ} of around

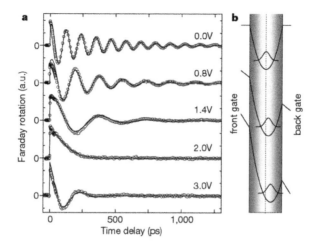

FIGURE 5.28 Voltage controlled spin precession. (a) Time-resolved Kerr rotation measurements of the electron spin resonance (ESR) in a quantum well structure at different applied gate voltages, V_G at 5 K in an applied field of 6 T. Note that the symbols represent the experimental data and the solid line is a fit to Eq. (5.90). (b) Displacement of the electron wave function towards one side of the quantum well due to the gate voltage applied to the back and front gate. Reprinted by permission from G. Salis, Y. Kato, K. Ensslin, D. C. Driscoll, A. C. Gossard, & D. D. Awschalom, (2001). *Nature, 414,* 619, ©(2001). Nature.

only 2%. The existence of spin splitting in semiconductors can strongly influence the spin dynamics of carriers and often resonant effects occur. In Figure 5.28, we show the precessional dynamics, observed using time-resolved magneto-optic Kerr effect (or pump-probe) measurements. The application of a gate voltage, V_G, causes a shift in the effective field. This causes a shift in the electron wave function and a subsequent change in the g-factor.

The change of precession frequency, Ω, can be described using a generic relation of the form:

$$\Theta_K(t) = \Theta_0 e^{-\Delta t/T_2} \cos(\Omega \Delta t) \tag{5.90}$$

where Δt is the delay time between the arrival times of the pump and probe pulses at the sample and T_2 is the transverse electron spin lifetime. The angular frequency of precession is expressed by:

$$\Omega = \frac{g}{\hbar} \mu_B B \tag{5.91}$$

which can be used to determine the value of g. The anisotropy of the factor g, which is a tensor quantity, permits the control of the size and direction of the spin precession vector, Ω, using the gate voltage.

With regards to the spin FET, the device operation can be understood based on the spin injection and detection (between source and drain), taking into account spin precession, all of which is controlled via the gate voltage. The effective field consists of an applied component and an internal component. This latter is dependent on V_G. The precessional angular frequency can be written in the form:

$$\Omega(\mathbf{k}) = \frac{e}{2m^*} \mathbf{B}_i(\mathbf{k}) \tag{5.92}$$

The quantity $\mathbf{B}_i(\mathbf{k})$ is an intrinsic $\mathbf{k}$-dependent magnetic field about which the spins precess and is which derived from the spin-orbit coupling (Rashba effect) in the band structure, giving rise to the dependence on the wave vector. The precession can be described using the Hamiltonian:

$$\mathcal{H}(\mathbf{k}) = \frac{1}{2}\hbar\hat{\sigma} \cdot \Omega(\mathbf{k}) \tag{5.93}$$

where $\hat{\sigma}$ denotes the Pauli matrices. The expression for $\Omega(\mathbf{k})$ comes from a consideration of the band structure of the semiconductor. Many researchers have modeled different systems, such as the III-VI semiconductors, where the lack of inversion symmetry leads to an intrinsic Larmor frequency, as expressed by the Dyakonov-Perel model:

$$\Omega(\mathbf{k}) = \frac{\alpha_c \hbar^2}{\sqrt{2m_c^{*3} E_g}} \kappa \tag{5.94}$$

The quantity κ is expressed as $\kappa = [k_x(k_y^2 - k_z^2), k_y(k_z^2 - k_x^2), k_z(k_x^2 - k_y^2)]$, α_c is the Dresselhaus spin-orbit coupling constant ($\alpha_c = 0.07$ for GaAs), E_g is the band gap energy and m_c^* is the conduction band effective mass of an electron. From Eq. (5.94), we can obtain the dephasing time of the electron spin. Inversion symmetry can be an important consideration in the evaluation of $\Omega(\mathbf{k})$, and can be analyzed by treating the wave vectors as operators of the form: $\hat{\mathbf{k}} = -i\nabla$ and obtaining the expectation values of $\Omega(\mathbf{k})$ for the confined states. This leads to a momentum quantization along the confinement unit vector $\hat{n}$ and $\hat{\mathbf{k}}$ for a Bloch state in the plane, where $k_n^2 = \langle(\hat{\mathbf{k}} \cdot \hat{n})^2\rangle$ is the expectation value for the wave vector squared; we note that for a rectangular quantum well of width a this has a value of $k_n^2 = (\pi/a)^2$. The form of $\Omega(\mathbf{k})$ will depend on the bulk directions of the planes in the QW. A further structural inversion asymmetry gives the Bychkov–Rashba spin-splitting with the following expression :

$$\Omega(\mathbf{k}) = 2\alpha_{BR}(\mathbf{k} \times \hat{\mathbf{n}}) \tag{5.95}$$

where α_{BR} is a parameter which depends on the spin-orbit coupling and the asymmetry of the confining electrostatic potentials. The BR field always lies in the plane and has a constant magnitude. It is interesting to note about the structural asymmetry that α_{BR} can be tuned electrostatically, potentially providing an effective spin precession control without the need for an external applied field. The variation of the spin precession and its relaxation can be obtained and evaluated. In the Datta–Das spin FET, the 2DEG is confined along the plane defined by the unit vector $\hat{\mathbf{n}}$. The precession axis of Ω always lies along the axis of the channel plane, see Figure 5.29, and is insensitive to the relative orientation of $\hat{\mathbf{n}}$ and the principal crystalline axes. The quantity $\Omega(\mathbf{k})$, given in Eq. (5.96), will govern the expectation value. For spins perpendicular to the plane which vary in time due to precession; $s_n = \mathbf{s} \cdot \hat{\mathbf{n}}$, and for a spin parallel to the in-plane $\mathbf{k}$; $s_\| = \mathbf{s} \cdot \mathbf{k}/k$. From this, we can establish the following relations:

$$\frac{ds_n}{dt} = 2\alpha_{BR}ks_\|; \quad \text{and} \quad \frac{ds_\|}{dt} = -2\alpha_{BR}ks_n; \tag{5.96}$$

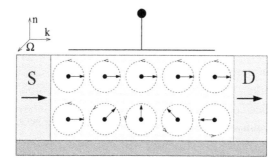

FIGURE 5.29 Schematic illustration of the spin FET. Source S and drain D are made of ferromagnetic materials. The spins of the injected spin polarized electrons precess with angular frequency, Ω, in the transfer medium. This frequency can be manipulated through the spin-orbit coupling with the upper gate. If the spin arriving at D has parallel orientation with D (upper row), then a large current will flow, however, if the spin is antiparallel (bottom row) a small current will flow in the transistor. Therefore, the current through the device is unable via the gate voltage. The spin orientation is highly influenced by the transfer medium as well, because the injected spins lose their orientation due to spin relaxation caused by the spin-orbit coupling.

The average spin component along Ω, $s_\perp = \mathbf{s} \cdot (\mathbf{k} \times \hat{\mathbf{n}})/k$, is constant. As a result, $s_\parallel = s_{0\parallel} \cos(\omega t)$, where $\omega = 2\alpha_{BR}k$ and the injected spin at the source is labeled with 0. If we denote the angle between $\mathbf{k}$ and the source–drain axis as ϕ, the electron will reach the drain in a time $t' = Lm_c^*/(\hbar k \cos\phi)$, where L is the length of the channel between source and drain, with the spin $s_\parallel$ precessing at an angle, $\psi = 2\alpha_{BR}Lm_c^*/\hbar$. The average spin at the drain in the direction of the magnetization is given by: $s_\parallel(t')\cos\phi + s_{0\perp}\mathbf{m} \cdot (\mathbf{k} \times \hat{\mathbf{n}})$, so the current is modulated by the factor $1 - \cos^2\phi \sin^2(\psi/2)$, the probability of finding the spin in the direction of the magnetization, $\mathbf{m}$. The relative orientation of the spin at the drain will therefore be important and will govern the channel resistance (as is the case for GMR), with the reflection and transmission coefficients being determined by the relative spin phase.

More recent work on the spin FET envisages the use of 2D materials to define the channel region, as illustrated in Figure 5.30. Here, a silicon layer is introduced to form a layer between the gate and substrate in the perpendicular direction and between the source and drain along the channel direction. Once again, the Rashba effect is exploited to provide the spin-dependent transmission along the channel region. Also shown in Figure 5.30 are the current–voltage characteristics of this device, demonstrating the good transistor-like properties, which depend on the magnetic (spin) properties and the Rashba effect.

Another application of spintronics is the addition of the spin functionality to the resonant tunnel diode (RTD) that was discussed in Section 2.9. This enables a spin-gated mode in which the device conductance can be determined by the relative alignment of the emitter and QW states in both energy and spin. If both emitter and QW, are ferromagnetic or exhibit spin split states when the resonance condition is satisfied, changing the relative orientation of the emitter/QW spin system will gate the current through the device. If the emitter is non-magnetic, the structure will then serve as a tunable spin filter: at a given bias, resonant tunneling occurs through only one of the spin state channels and the output current will exhibit the corresponding spin polarization. In Figure 5.31, we show an example of such a device in which only the QW region exhibits a spin-splitting.

An extension of this concept is seen with the development of the magnetic tunnel transistor (MTT), which we show schematically in Figure 5.32. By ensuring different coercive fields for the two magnetic layers, indicated as regions 1 and 3, we can obtain independent switching for the two magnetizations in an applied field. The magnetic dependent current $I_{MC} =$

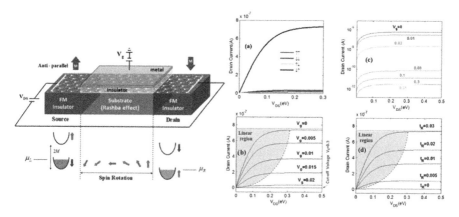

FIGURE 5.30 Left: Schematic view of the proposed spin FET in which the large navy blue arrows show the direction of leads magnetization with parallel or antiparallel configuration (here in antiparallel configuration), and the small colored arrows show the electron spin directions. The white and orange circles show the silicon atoms in sublattices A and B of silicene nanoribbon. If $E > (E < 0)$, electrons enter into the channel from the left (right) lead. The diagram of spin bands for electrons with energy $0 < E < M$ is also presented. When the leads are in antiparallel configuration, electrons with spin up are blocked in the left lead (source). Thus, electrons with spin down enter the channel and after passing the spin rotation region, only the electrons with rotated spin to up are allowed to enter the right lead (drain). Right: I–V characteristic for (a) a fixed Rashba strength and top gate voltage ($V_g = 0$), (b) and (c) a fixed Rashba strength and different values of top gate voltage (linear and logarithmic scale, respectively) and (d) a fixed top gate voltage ($V_g = 0$) and different values of Rashba strengths. Here, $M = 0.03$ eV and the dashed-red parabolic line separates the linear region from the saturation one, and the horizontal red-dotted line displays the saturation value. Reprinted from N. Pournaghavi, M. Esmaeilzadeh, A. Abrishamifar, & S. Ahmadi, (2017). *J. Phys.: Condens. Matter, 29*, 145501, Reproduced with permission of IOP Publishing.

$(I_{c\uparrow\uparrow} - I_{c\downarrow\uparrow})/I_{c\downarrow\uparrow}$ will show a non-magnetic behavior with the emitter-base voltage, V_{EB}, influenced by the conduction band structure of the collector. In GaAs, in addition to the direct transition at the Γ point, there are indirect minima at the L points at higher energies. After an initial decrease of I_{MC} with electron energy, at $V_{EB} \simeq 0.3\,V$ above the BC Schottky barrier, there will be an onset of hot electron transport into the L valleys accompanied by an increase in I_{MC}. It will be noted that regions 1, 2 and 3 form a MTJ, with the first layer FM (region 1) being fixed by an AFM layer. The current at the collector will be governed by the relative orientation of the magnetization of the base region with respect to the emitter.

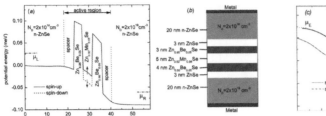
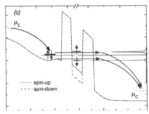

FIGURE 5.31   (a) Potential energy profile in the paramagnetic RTD for the conduction electrons with spin-up (solid line, red) and spin-down (dotted line, blue). Coordinate z is measured along the layer growth direction, $\mu_{L(R)}$ is the electrochemical potential of the left (right) contact. (b) The undoped (active) region consists of the paramagnetic quantum well made of $Zn_{1-x}Mn_xSe$ sandwiched between two $Zn_{0.95}Be_{0.05}Se$ potential barrier layers. The active region of the nanodevice is separated from the n-doped ZnSe contacts by the two spacer layers on the left and right sides. (c) Schema of spin-dependent resonant tunneling through the paramagnetic RTD. The right scale shows the spin-dependent density of resonant states as a function of energy. The positions of peaks determine the energies of spin-dependent quasi-bound states in the nanostructure. Reprinted from P. Wójcik, B. J. Spisak, M. Woloszyn, & J Adamowski, (2012). *Semicond. Sci. Technol.*, 27, 115004, Reproduced with permission of IOP Publishing.

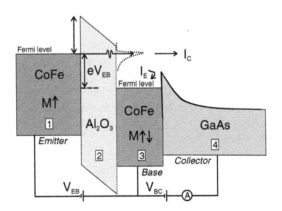

FIGURE 5.32   Schematic energy diagram of the magnetic tunnel transistor. Region 1 forms the ferromagnetic emitter, region 2 the tunnel barrier, region 3 is the ferromagnetic base, and region 4 is the semiconductor collector. Reprinted figure with permission from S. van Dijken, X. Jiang, & S. S. P. Parkin, (2003). *Phys. Rev. Lett.*, 90, 197203, ©(2003) by the American Physical Society.

Spin-dependent transistor operation can also be obtained using magnetic semiconductors in the base region of saying an n-p-n bipolar device, or a magnetic bipolar transistor. In the base region, there will be a spin splitting of the charge carriers, which means that the transport into the collector will produce a spin-polarized injection in this region of the device. The basic band diagram for this operation is shown in Figure 5.33. The excitation of specific spin populations can be achieved, as mentioned previously, using left- or right-circularly polarized light in the emitter. Since the spin splitting in the base region has a lower barrier height for the preferred orientation of spin, as set by an applied magnetic field, these charge carriers will be preferentially transmitted across the base region, creating a spin-polarized current in the collector. All other considerations for the transport areas for a conventional bipolar transistor.

In general, a semiconductor-based spintronics technology requires four essential components, which we can state as follows:

(1) Efficient spin transport and sufficiently long spin lifetimes within the semiconductor host medium.
(2) Efficient electrical injection of spin-polarized carriers from an appropriate source into the semiconductor heterostructure.
(3) Effective control or manipulation of the spin carriers to provide the desired functionality.
(4) Effective detection of the spin-polarized carriers.

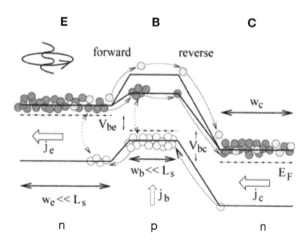

FIGURE 5.33 Schematic illustration of a magnetic n-p-n bipolar (junction) transistor. There is a spin splitting in the base for the conduction band, and the non-equilibrium spin source in the emitter.

While our discussion of spintronic devices has not been exhaustive, we have outlined some of the principal developments for the realization of working devices which depend on spin functionality as introduced using magnetic components. For further details of such issues, the reader is referred to the review articles by Gregg (2002), Thompson (2008), and Hirohata and Takahashi (2014). Details of the development of single-electron effects in spintronics can be found in the article by Seneor et al. (2007). In this case spin splitting, say on a quantum dot, will allow the spin functionalization to tune the single electron transfer to only these electrons of a specific spin.

5.9   Summary

The magnetic properties of magnetic thin films and multilayers have been the object of intense research for a number of decades. One of the reasons for the longevity of this field is the wealth of magnetic phenomena that can be observed as well as some important applications of these systems, most notably in the magnetic data storage industry. As with other low dimensional systems, thin films and multilayers have a significant proportion of the magnetic moments located at the surface or interfaces of films. Since the reduced dimensions are in one orientation, the theoretical treatment can be simplified and often can be represented as a single crystalline plane with a know associated anisotropy. As we mentioned previously, the reduced magnetic coordination means that the local anisotropy can be significantly different from that in the bulk material.

Magnetic multilayers refer to systems comprised of multiple magnetic films typically separated by a non-magnetic spacer, the thickness of which can be varied to adjust the magnetic coupling between adjacent magnetic layers. It was in this way that the variation of the electrical resistance was observed to be very sensitive to the applied magnetic field for specific thicknesses of the non-magnetic spacer layer. This phenomenon was labeled giant magneto-resistance. The origin of the effect is associated with the spin-dependent scattering of electrons as they traverse the different magnetic layers. It is found that when the magnetizations of the adjacent layers are parallel, the electrical resistance is low, while for antiferromagnetically aligned layers have a significantly higher resistance. It was found that the magnetic coupling between ferromagnetic layers is strongly dependent on the nature of the non-magnetic spacer and importantly its thickness. For metallic spacers, it was found that the coupling has an oscillatory nature and was related to the RKKY interaction in metallic systems. The giant magnetoresistance (GMR) effect can be understood as the variation of the electrical resistance as a function of the applied magnetic field, where for a system with antiferromagnetically coupled layers, the magnetic field will act to align the magnetizations of the different layers leading to the low resistance state. This discovery rapidly leads to the development of a new area of study called spintronics. Further work later demonstrated important changes to the electrical resistance state in systems with insulating non-magnetic spacers. These rely on spin-dependent tunneling between the magnetic layers, and are called magnetic tunnel junctions. It is possible to pin one of the ferromagnetic layers using the exchange bias effect. This leads to a system, which is no-longer symmetric in the applied field behavior and produces what is referred to as a spin-valve. Such devices have been developed with such high precision that they can be used as reading head devices for systems with ever-decreasing bit sizes. This has allowed profound progress in the increase of the areal density for data storage systems. This is a major industry since hard-disk drives are one of the principal methods for storing information on PCs and other devices. The importance of the discovery of GMR lead to the awarding of the 2007 Nobel prize to the principal discoverers, Albert Fert and Peter Grünberg. Other spintronic devices have been developed, such as the spin LED and spin FET, though their range of applicability has not yet become widespread.

5.10   Problems

(1) Consider the electrical resistance of a ferromagnetic metal using the Mott two-current model. Using the approximations $R_{\uparrow} = R/a$ and $R_{\downarrow} = Ra$, where a characterizes magnetic order, show that the electrical resistance of a ferromagnetic metal can be written as:

$$R_{TOT} = R\frac{a}{1+a^2}$$

What value of a would you expect for a normal metal? Based on the above equation, how would you expect magnetic order to affect the electrical resistance of a metal? Explain.

(2) Show that the magnitude of the GMR effect can be expressed as:

$$GMR = \frac{(\rho_{\uparrow} - \rho_{\downarrow})^2}{4\rho_{\uparrow}\rho_{\downarrow}}$$

Again, this will require the use of the Mott two-current model.

(3) The resistance ratio R_{ap}/R_p of a tunnel junction with identical electrodes is 400%. Use the Jullière formula to deduce the spin polarization P of the ferromagnet. What is the magnetoresistance? What will be the magnetoresistance if one of the electrodes is replaced by cobalt? ($P(Co) =$ 45%).

(4) Describe the tunnel magnetoresistance (TMR) effect in terms of the splitting of the densities of states of the two ferromagnetic layers. Also, write the form of the current between two ferromagnetic electrodes separated by a thin insulating layer using Fermiâs Golden rule and explain the various terms. How is this related to the Landauer formula?

(5) Estimate the maximum thickness of a ferromagnetic free layer, which can be switched in a nanosecond. Assume it is possible to pass a current of 10^{11} A m^{-2} without encountering severe problems of electromigration. Assume $P_e = 0.5$.

(6) Show that the polarization of the φ^+ photoluminescence excited by bandgap radiation in GaAs is 1/4.

(7) Show that the definition of spin polarization:

$$P_n = \frac{v_{F\uparrow}^n D_\uparrow - v_{F\downarrow}^n D_\downarrow}{v_{F\uparrow}^n D_\uparrow + v_{F\downarrow}^n D_\downarrow}$$

with $n = 2$ is equivalent to the polarization of the electric current, provided the spin relaxation times $\tau_\uparrow$ and $\tau_\downarrow$ are the same. $D_{\uparrow,\downarrow}$ represents the density of states for $(\uparrow,\downarrow)$ electrons at the Fermi level and v_F is the Fermi velocity.

(8) Show that only the condition of a difference in spin mobility (for spin-up and spin-down electrons) is necessary for there to exist a spin polarization.

(9) Spin injection is a crucial process in the electronics of spin. Describe how this can be achieved optically in a semiconducting device.

(10) Explain why the existence of a spin current does not necessarily have to be accompanied by an electrical current.

(11) Explain the working principles of the Datta–Das spin FET device.

References and Further Reading

Baibich, M. N., Broto, J. M., Fert, A., Nguyen Van Dau, F., Petroff, F., Etienne, P., Creuzet, G., Friederich, A., & Chazelas, J., (1988). *Phys. Rev. Lett., 61*, 2472.

Bland, J. A. C., & B. Heinrich (Eds.), *Ultrathin Magnetic Structures*, volumes I–IV, Springer, Berlin Heidelberg (1994–2005).

den Broeder, F., Hoving, W., & Bloemen, P., (1991). *J. Magn. Magn. Mat. 93*, 562.

Bruno, P., (2002). *Magnetism: Molecules to Materials III: Nanosized Magnetic Materials*, Miller, J. S., & M. Dillon (Eds.), Wiley–VCH, Weinheim.

Choi, J.-Y., Lee, D.-G., Baek, J.-U., & Park, J.-G., (2018). *Sci. Rep. 8*, 2139.

Dieny, B., Speriosu, V. S., Parkin, S. S. P., Gurney, B. A., Wilhoit, D. R., & Mauri, D., (1991). *Phys. Rev. B, 43*, 1297.

Dutta, P., Manivannan, A., Seehra, M. S., Shah, N., Huffman, G. P., (2004). *Phys. Rev. B, 70*(17), 174428.

Fang, L., Deniz Bozdag, K., Chen, C.-Y., Truitt, P. A., Epstein, A. J., & Johnston-Halperin, E., (2011). *Phys. Rev. Lett., 106*, 156602.

Fernando, G. W., (2008). Handbook of Metal Physics, *4*, 1.

Fert, A., (2008). Nobel Lecture, Rev. Mod. Phys., *80*, 1517.

Gabor, M. S., Tiusan, C., Petrisor, T., Jr., & Petrisor, T., (2014). *IEEE Trans. Magn., 50*, 2007404.

Gregg, J. F., Petej, I., Jouguelet, E., & Dennis, C., J. Phys. D: Appl. Phys., *35*, R121–R155 (2002).

Grünberg, P. (2008). *Nobel Lecture, Rev. Mod. Phys., 80*, 1531.

Hirohata, A., & Takahashi, K., (2014). *J. Phys. D: Appl. Phys., 47*, 193001.

Julliére, M. (1975). *Phys. Lett., 54*, 225.

Meservey, R., & Tedrow, P. M. (1974). *Low Temperature Physics*, K. D. Timmerhaus et al. (eds.), Plenum Press.

Nogués, J., & Schuller, I. K., (1999). *J. Magn. Magn. Mater., 192*, 203–232.

Parkin, S. S. P., & Mauri, D., (1991). *Phys. Rev. B, 44*, 7131.

Pierce, D. T., Unguris, J., Celotta, R. J., & Stiles, M. D., (1999). *J. Magn. Magn. Mater., 200*, 290.

Pournaghavi, N., Esmaeilzadeh, M., Abrishamifar, A., & Ahmadi, S., (2017). *J. Phys.: Condens. Matter, 29*, 145501.

Salis, G., Kato, Y., Ensslin, K., Driscoll, D. C., Gossard, A. C., & Awschalom, D. D., (2001). *Nature, 414*, 619.

Seneor, P., Bernand-Mantel, A., & Petroff, F., (2007). *J. Phys.: Condens. Matter, 19*, 165222.

Stiles, M. D., (2005). *Interlayer Exchange Coupling*, in *Ultrathin Magnetic Structures III: Fundamentals of Nanomagnetism*, Bland, J. A. C., & B. Heinrich (Eds.), Springer, Berlin Heidelberg.

Tedrow, P. M., & Meservey, R., (1971). *Phys. Rev. Lett., 26,* 192.

Thompson, S. M., (2008). *J. Phys. D: Appl. Phys., 41,* 093001.

Ueno, T., Sinha, J., Inami, N., Takeichi, Y., Mitani, S., Ono, K., & Hayashi, M., (2015). *Sci. Rep., 5,* 14858; doi: 10.1038/ srep14858.

van Dijken, S., Jiang, X., & Parkin, S. S. P., (2003). *Phys. Rev. Lett., 90,* 197203.

Wójcik, P., Spisak, B. J., Woloszyn, M., & Adamowski, J. (2012). *Semicond. Sci. Technol., 27,* 115004.

Chapter 6

Spin Dynamics in Magnetic Nanostructures

6.1 Introduction

Spin or magnetization dynamics and magnetic resonance phenomena are intimately related phenomena. They relate to the temporal evolution of the magnetization vector, or the assembly of spins in a magnetic system. In the previous sections, we have been concerned with the magnetic properties of low dimensional systems in which the ground state energy is defined by free energy. This described the equilibrium state of the system and can be deduced from the minimization of the free energy, taking into account the various contributions due to the magnetic anisotropies, the exchange, and Zeeman terms. Dynamics, on the other hand, consider the non-equilibrium state when a magnetic system is excited, typically via a time-varying field. This can be an alternating field in the GHz regime to provoke a resonant response or a rapidly changing pulse, which can cause the magnetization to be abruptly shifted from the equilibrium orientation and to naturally relax with some characteristic lifetime. In either case, the magnetization dynamics will be governed by the local effective field in the spin system and will usually have a precessional nature, with the magnetization precessing around the direction of the effective magnetic field. The direction and strength of the exciting field are of central importance to the trajectory of the precessional dynamics. The relaxation back to equilibrium will also depend on the nature of the sample, as we will discuss in due course. The relaxation process ends with the magnetization in a local energy minimum, which is not necessarily the same point of departure. In this case, the process would correspond to a form of magnetic switching.

Magnetization dynamics has grown and evolved in many ways over recent years. However, the origin of the phenomenon of ferromagnetic resonance absorption can be traced back to the period 1911–1913, when

V. K. Arkad'yev discovered the selective absorption of radiowaves in iron and nickel wires. The first attempts at a quantum theory description was given by Ya. G. Dorfman in 1923. It was, however, the seminal work of L. D. Landau and E. M. Lifshitz in 1935, that paved the way to a more global general theory of ferromagnetic resonance and is still the basis for the interpretation of the phenomena related to magnetodynamics today. Further work by E. K. Zavoiskii and J. H. E. Griffiths in 1946 led to the discovery of ferromagnetic resonance absorption in pure metals. The Landau–Lifshitz theory was applied to experimental data and generalized by C. Kittel and D. Polder in the late 1940s.

Many developments on experimental methods have allowed the study of dynamics to evolve into more convenient forms and have also led to the discovery of ultrafast dynamics on the femtosecond time scale using ultrafast laser technologies. This method is, while being rather complex, very versatile and uses the laser excitation of the sample and the subsequent detection of the magnetic state via the MOKE signal to plot the temporal evolution of the magnetization. The detection is performed via the controlled delay of an optical pulse. This methodology, known as pump–probe spectroscopy, has been applied to the study of many temporal effects in solid-state physics and chemistry. The original method for the study of FMR was developed around microwave spectroscopy, in which the absorption of the microwave field by the sample can be observed as a function of an applied magnetic field. This method has been largely superseded by high-frequency broadband techniques using a vector network analyzer (VNA), which permits frequency-dependent studies. Another technique that emerged uses Brillouin light scattering (BLS) to detect the dispersion of light in the creation and annihilation of spin waves. Spin transfer torque (STT) also allows the dynamics of magnetic nanostructures to be performed using electrical detection via the change in the tunnel resistance for example of layered structures. This is of particular interest for the development of MRAM devices. We will outline some of the main methods in the following sections.

6.2   General Theory of Magnetic Resonance

We will start by looking at the general description of the resonance of a generalized magnetic moment $\mathbf{m} = -\gamma\mathbf{L}$, where $\gamma = -e/2m_e$ is referred to as the gyromagnetic ratio and $\mathbf{L}$ is the angular momentum. The variation of the magnetic moment can be expressed in the form of the differential:

$$dm = -\gamma dL - Ld\gamma \tag{6.1}$$

Since we take γ to be a constant, we have

$$dm = -\gamma dL \tag{6.2}$$

Applying a static magnetic field, **H**, to the moment, we can establish the precessional motion of **m** around the direction of the field. This we can express in the form:

$$\frac{dm}{dt} = -\gamma \frac{dL}{dt} = -\gamma T \tag{6.3}$$

Here, **T** represents a torque of the form, $\mathbf{T} = \mu_0(\mathbf{m} \times \mathbf{H})$. Taking the magnetic field to be along the z direction of our coordinate system, we can represent the geometry of the precession as illustrated in Figure 6.1.

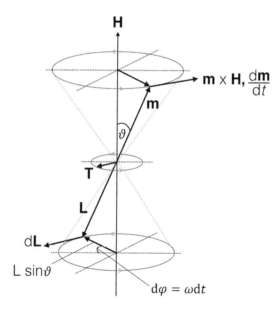

FIGURE 6.1 Schematic representation of the coordinate system for spin precession, showing the vectors for the magnetic field, **H**, the magnetic moment, **m**, the orbital angular momentum, **L** and the torque, **T**.

The equation of motion for the magnetic moment is now written as:

$$\frac{dm}{dt} = -\gamma \mu_0 (\mathbf{m} \times \mathbf{H}) \tag{6.4}$$

Thus, we see that the variation of the magnetic moment vector depends on the vector product of the magnetic moment and the applied magnetic field. With reference to Figure 6.1, we see that this is the precessional motion of the magnetic moment around the external field. The magnetic moment is a classical vector of constant length. This is the well-known *Larmor precession* and we will now derive the characteristic Larmor frequency.

The variation of the orbital angular momentum can be expressed in the form:

$$dL = |\mathbf{dL}| = L \sin\theta \, dt = |\mathbf{T}| dt \tag{6.5}$$

We can express the torque as:

$$|\mathbf{T}| = \mu_0 mH \sin\theta = \mu_0 \frac{e}{2m_e} LH \sin\theta \tag{6.6}$$

and thus:

$$dL = \mu_0 \frac{e}{2m_e} LH \sin\theta \, dt \tag{6.7}$$

The precessional angular frequency can be expressed in terms of the variation the azimuthal angle with time:

$$\omega = \frac{d\phi}{dt} = \frac{1}{L \sin\theta} \frac{dL}{dt} = \frac{e}{2m_e} \mu_0 H \tag{6.8}$$

This then establishes the equation for the angular frequency for the Larmor precession as:

$$\omega = \gamma\mu_0 H = \gamma B \tag{6.9}$$

This equation can also be derived from the equation of motion, Eq. (6.4), by taking the vector product of the magnetic moment and magnetic field, which can then be expressed in component form. It is then a simple matter to obtain the x and y components, which are then differentiated with respect to time. We note that the z component will be zero since there is no variation in this direction, which would alter the length of the magnetic moment. The results are equated with the second derivatives of the x and y components of the magnetic moment with respect to time, where it is noted that $\cos\phi = \cos\omega t$. The result given in Eq. (6.9) is then obtained; $\omega = 2\pi\nu = \gamma\mu_0 H = \gamma B$.

It is worth noting that the multiplication of the Larmor frequency by Planck's constant, $\hbar\omega = h\nu$, provides the energy unit of the precession. The work done by the applied field in varying the orientation of the magnetic moment, resulting in the torque, changes the orientation by an amount equal to $d\theta$. We can thus write:

$$dW = |\mathbf{T}|d\theta - -\mu_0(mH\sin\theta)d\theta \tag{6.10}$$

where the negative sign indicates that the torque opposes the direction of motion of the precession, as illustrated in Figure 6.1. We can now write:

$$dW = d(\mu_0 mH\cos\theta) = d(\mu_0 \mathbf{m}\cdot\mathbf{H}) = d(\mathbf{m}\cdot\mathbf{B}) \tag{6.11}$$

The energy stored as potential energy (or magnetic potential) leads to the expression:

$$dE = -dW = -d(\mu_0\mathbf{m}\cdot\mathbf{H}) \tag{6.12}$$

or simply:

$$E = -\mu_0\mathbf{m}\cdot\mathbf{H} \tag{6.13}$$

We can now substitute for the orbital angular momentum, $\mathbf{m} = -\gamma\mathbf{L}$, from which we can write:

$$E = \mu_0\gamma\mathbf{L}\cdot\mathbf{H} = \mu_0\frac{g_L\mu_B}{\hbar}\mathbf{L}\cdot\mathbf{H} = \mu_0\frac{g_L\mu_B}{\hbar}LH\cos\theta \tag{6.14}$$

Taking into account that the Landé factor for orbital angular momentum is unity and the z component of the angular momentum is $L_z = L\cos\theta$, we obtain:

$$E = \mu_0 H\frac{\mu_B}{\hbar}L_z = \omega L_z = \hbar\omega m_L \tag{6.15}$$

where $m_L = L_z/\hbar$ is the magnetic quantum number. Thus, we have shown that the magnetic or Zeeman energy of the atomic electron depends on the magnetic quantum number and is hence quantized.

6.3 Phenomenological Theory of Ferromagnetic Resonance

The above treatment of the condition for the magnetic resonance of an atomic electron provides a basis for the general description of the case, where there is a cooperative interaction, which aligns the individual electronic moments in a ferromagnetic material. This approach leads us to consider the magnetization vector of the ferromagnetic material in an effective field, which incorporates both external and internal contributions to the magnetic field experienced by the electronic moments. Our phenomenological description can be taken from the above, where the equation of motion is expressed in terms of the magnetization:

$$\frac{\partial \mathbf{M}}{\partial t} = -\gamma(\mathbf{M} \times \mathbf{H}_{\text{eff}}) \tag{6.16}$$

We have neglected the damping term which considers the relaxation of the magnetization, which we will consider later. In the above expression, we use the effective magnetic field, $\mathbf{H}_{\text{eff}}$, which accounts for the various contributions to the internal and external fields:

$$\mathbf{H}_{\text{eff}} = \mathbf{H}_0 + \mathbf{h}(t) + \mathbf{H}_K + \mathbf{H}_{dm} + \mathbf{H}_{exch} \tag{6.17}$$

The first two terms refer to the external contributions of the applied static and the time-varying microwave fields, respectively. The following terms are the internal contributions, which are respectively the magnetic anisotropy field, the demagnetizing field, and the exchange field. It is interesting to note, that each of these terms has a corresponding term in the free energy of the system. These are linked via the general relation between the magnetic field and the energy, E, which can be expressed in the form:

$$\mathbf{H}_{\text{eff}} = -\frac{1}{M}\nabla_u E = -\frac{\partial E}{\partial \mathbf{M}} \tag{6.18}$$

Before looking at the different terms in the effective field, we will define the spherical coordinate system as illustrated in Figure 6.2. For our analysis, it is instructive to extract the exchange field term from the effective field. This will permit a more generalized solution to the resonance condition, which has a direct bearing on the situation for confined magnetic systems such as thin films and nanostructures, where we will also need to consider the specific boundary conditions. The exchange field can be derived using a Taylor expansion for the spin system around the neighboring spins. For a bulk spin away from the boundary, the exchange field can be expressed in the form:

$$\mathbf{H}_{exch} = -\frac{2A}{M^2}\nabla^2\mathbf{M} \tag{6.19}$$

where A is the exchange stiffness constant, which for the general case takes the form:

$$A = \frac{\eta z J S^2}{a_{nn}} \tag{6.20}$$

where J is the exchange integral, S the spin quantum number, z is the coordination number, a_{nn} is the nearest neighbor distance between spins and η is a

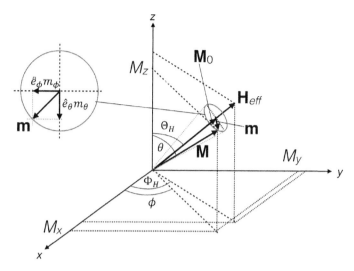

FIGURE 6.2 Schematic representation of the coordinate system for the magnetization and effective field vectors, showing polar and azimuthal angles as well as the various components of the magnetization. The inset shows the components of the transversal magnetization in the polar and azimuthal directions of the spherical coordinate system.

factor which depends on the crystalline order; for example, $\eta = 1$ for simple cubic, $\sqrt{3}$ for bcc and $2\sqrt{2}$ for bcc structures.

We can now rewrite the equation of motion, Eq. (6.16), in the form:

$$\frac{1}{\gamma}\frac{\partial \mathbf{M}}{\partial t} = -\mathbf{M} \times \left[\mathbf{H}' - \frac{2A}{M^2}\nabla^2 \mathbf{M}\right] = \frac{2A}{M^2}(\mathbf{M} \times \nabla^2 \mathbf{M}) + \mathbf{T}' \qquad (6.21)$$

In this expression, we have adjusted the terms for the field, $\mathbf{H}' = \mathbf{H}_{\text{eff}} - \mathbf{H}_{\text{exch}}$, and the torque, where $\mathbf{T}' = \mathbf{M} \times \mathbf{H}'$. We will now analyze in detail the various terms of this equation, starting with the exchange field term. The magnetization can be expressed as the vector sum of static and dynamics components: $\mathbf{M}(t) = \mathbf{M}_0 + \mathbf{m}(t)$, where $\mathbf{M}_0$ is the uniform static magnetization in the direction of the effective field, while $\mathbf{m}$ is the rotating or transversal component, which is driven by the rf or microwave field, $\mathbf{h}(t)$, which drives resonance. For linear effects with low precessional amplitudes, we write $|\mathbf{M}_0| \gg |\mathbf{m}|$. We can thus express the vector product of the first term as:

$$\mathbf{M} \times \nabla^2 \mathbf{M} = M_0(\hat{\mathbf{e}}_\phi \nabla^2 m_\theta - \hat{\mathbf{e}}_\theta \nabla^2 m_\phi) \qquad (6.22)$$

Given the periodic nature of the dynamic component of the magnetization $(\mathbf{m} \sim e^{i(\mathbf{k}\cdot\mathbf{r}-\omega t)})$, we have $\nabla^2 m_v = -k^2 m_v$, where $v = \theta, \phi$ and k represents the wave vector. We now obtain:

$$\mathbf{M} \times \nabla^2 \mathbf{M} = M_0 k^2 (\hat{\mathbf{e}}_\theta m_\phi - \hat{\mathbf{e}}_\phi m_\theta) \tag{6.23}$$

from which we can write:

$$\mathbf{M} \times \mathbf{H}_{exch} = \frac{2A}{M_0} k^2 (\hat{\mathbf{e}}_\theta m_\phi - \hat{\mathbf{e}}_\phi m_\theta) \tag{6.24}$$

We now turn our attention to the torque expression. In a general manner, we can express a generalized torque as as classical torque of the form:

$$\mathbf{T} = \mathbf{r} \times \mathbf{F} \tag{6.25}$$

where $\mathbf{r}$ represents a displacement or position vector and $\mathbf{F}$ a force vector. Applying the relation between the force and free energy of the magnetic system, we can write:

$$\mathbf{T} = -\mathbf{r} \times \nabla E \tag{6.26}$$

where we write the derivative of the free energy in spherical coordinates as:

$$\nabla E = \hat{\mathbf{e}}_r \frac{\partial E}{\partial r} + \frac{\hat{\mathbf{e}}_\theta}{r} \frac{\partial E}{\partial \theta} + \frac{\hat{\mathbf{e}}_\phi}{r \sin\theta} \frac{\partial E}{\partial \phi} \tag{6.27}$$

Writing $\mathbf{r} = \hat{\mathbf{e}}_r r$, Eq. (6.26) takes the form:

$$\mathbf{T} = -\hat{\mathbf{e}}_\phi \frac{\partial E}{\partial \theta} + \frac{\hat{\mathbf{e}}_\theta}{\sin\theta} \frac{\partial E}{\partial \phi} \tag{6.28}$$

We note, that in any consideration of the ferromagnetic resonance, it is important to take into account the equilibrium state of the system, since it is from this state that the sample will be excited, with the precession of the magnetization around the orientation of the magnetization in the equilibrium state. To obtain the equilibrium state or orientation of the magnetization of a magnetic system, it is necessary to minimize the free energy with respect to the polar and azimuthal angles. This can be achieved from the following conditions:

$$\left(\frac{\partial E}{\partial \theta} \right)_{\theta=\theta_0, \phi=\phi_0} = 0; \quad \text{and} \quad \left(\frac{\partial E}{\partial \phi} \right)_{\theta=\theta_0, \phi=\phi_0} = 0 \tag{6.29}$$

For these conditions, at the equilibrium orientation, the torque as expressed by Eq. (6.28) will clearly be zero. In the dynamic situation at resonance, we are in a non-equilibrium state. This is when the magnetization is in precessional motion, and we can define the deviation of the orientation of the magnetization as:

$$\delta\theta(t) = \theta(t) - \theta_0 \tag{6.30}$$

and

$$\delta\phi(t) = \phi(t) - \phi_0 \tag{6.31}$$

We can now express the angular deviations of these angles in terms of the magnetization geometry as follows:

$$\sin(\delta\theta) \simeq \delta\theta = \frac{m_\theta}{M_0} \tag{6.32}$$

and

$$\sin(\delta\phi) \simeq \delta\phi = \frac{m_\phi}{M_0 \sin\theta} \tag{6.33}$$

We can consider the dynamic expression for the torque, which we write as:

$$\mathbf{T} = \frac{\partial\mathbf{T}}{\partial\theta}\delta\theta + \frac{\partial\mathbf{T}}{\partial\phi}\delta\phi \tag{6.34}$$

Using Eq. (6.28), we obtain:

$$\mathbf{T} = \left(-\hat{\mathbf{e}}_\phi\frac{\partial^2 E}{\partial\theta^2} + \frac{\hat{\mathbf{e}}_\phi}{\sin\theta}\frac{\partial^2 E}{\partial\theta\partial\phi}\right)\frac{m_\theta}{M_0} + \left(-\hat{\mathbf{e}}_\phi\frac{\partial^2 E}{\partial\phi\partial\theta} + \frac{\hat{\mathbf{e}}_\phi}{\sin\theta}\frac{\partial^2 E}{\partial\phi^2}\right)\frac{m_\phi}{M_0\sin\theta} \tag{6.35}$$

This now allows us to collect the various terms and substitute back into Eq. (6.21), which yields:

$$\frac{1}{\gamma}\frac{\partial\mathbf{M}}{\partial t} = \left(-\hat{\mathbf{e}}_\phi\frac{\partial^2 E}{\partial\theta^2} + \frac{\hat{\mathbf{e}}_\phi}{\sin\theta}\frac{\partial^2 E}{\partial\theta\partial\phi}\right)\frac{m_\theta}{M_0} + \left(-\hat{\mathbf{e}}_\phi\frac{\partial^2 E}{\partial\phi\partial\theta} + \frac{\hat{\mathbf{e}}_\phi}{\sin\theta}\frac{\partial^2 E}{\partial\phi^2}\right)\frac{m_\phi}{M_0\sin\theta}$$
$$+ \frac{2A}{M_0}k^2\left(\hat{\mathbf{e}}_\theta m_\phi - \hat{\mathbf{e}}_\phi m_\theta\right) \tag{6.36}$$

Using the following relations:

$$\mathbf{M}(t) = \mathbf{M}_0 + \mathbf{m}(t) \qquad \text{and} \qquad \frac{\partial\mathbf{M}}{\partial t} = \frac{\partial\mathbf{m}}{\partial t} = i\omega\mathbf{m}$$

we can express Eq. (6.36) in its component form:

$$\frac{i\omega}{\gamma}m_\theta - \frac{1}{M_0\sin\theta}\frac{\partial^2 E}{\partial\theta\partial\phi}m_\theta - \frac{1}{M_0^2\sin\theta}\frac{\partial^2 E}{\partial\phi^2}m_\phi - \frac{2A}{M_0}k^2 m_\phi = 0 \qquad (6.37)$$

$$\frac{i\omega}{\gamma}m_\phi - \frac{1}{M_0\sin\theta}\frac{\partial^2 E}{\partial\phi\partial\theta}m_\phi + \frac{1}{M_0^2}\frac{\partial^2 E}{\partial\theta^2}m_\phi + \frac{2A}{M_0}k^2 m_\theta = 0 \qquad (6.38)$$

These equations can now be expressed in matrix form as a solution the simultaneous equation pair, yielding the general solution:

$$\left(\frac{\omega}{\gamma}\right)^2 = (Dk^2)^2 + \left[\frac{1}{\sin\theta}\frac{\partial^2 E}{\partial\phi^2} + \frac{\partial^2 E}{\partial\theta^2}\right](Dk^2) + \frac{1}{M_0^2\sin^2\theta}\left[\frac{\partial^2 E}{\partial\theta^2}\frac{\partial^2 E}{\partial\phi^2} - \left(\frac{\partial^2 E}{\partial\phi\partial\theta}\right)^2\right]$$
$$(6.39)$$

This expression is referred to as the resonance equation and represents the conditions for resonant excitation in its most general form. We can firstly note, that this equation is quadratic in (Dk^2), where $D = 2A/M_0$ is the spin wave constant. It is worth noting that we can now see why it was important to extract the exchange field term from the effective magnetic field in our above treatment. It was this manipulation that has allowed us to derive the spin wave terms, which are so dependent on the exchange interaction, as shown by the factor D. This form of the resonance equation is applicable to the case of low dimensional and confined systems, such as thin films and nanostructures, where boundary conditions may allow spin wave modes to be supported. In the case of bulk systems, the spin wave manifold will collapse and only the uniform mode, with infinite wavelength and $k = 0$, will be excited. This is strictly the ferromagnetic resonance mode, where all spins in the system precess with the same phase and amplitude. In this case, the resonance equation is simply expressed in the form:

$$\left(\frac{\omega}{\gamma}\right)^2 = \frac{1}{M_0^2\sin^2\theta}\left[\frac{\partial^2 E}{\partial\theta^2}\frac{\partial^2 E}{\partial\phi^2} - \left(\frac{\partial^2 E}{\partial\phi\partial\theta}\right)^2\right] \qquad (6.40)$$

This equation is often referred to as the Smit-Beljers equation (other variants refer to this as the Smit-Suhl equation). In terms of ferromagnetic resonance, this relation is extremely useful since it can be applied in most practical situations. We note in particular, that this equation expresses the resonance condition in terms of the second derivatives of the free energy of the system with respect to the polar and azimuthal angles. Therefore, in principle all that we require is the functional form of the free energy as a function of these angles. In most cases, this is feasible since we can take into account the different forms of magnetic anisotropy, including both magnetocrystalline and shape contributions, and also the Zeeman energy, all of which have specific forms which can be expressed in terms of the geometry and therefore the angles θ and ϕ.

6.4 Spin Wave Resonance

Spin wave resonance, as we mentioned in the previous section, refers to the excitation of higher-order modes where the wave vector $k \neq 0$. This can occur for magnetic bodies of confined dimensions and specific boundary conditions, which define the allowed wave vectors. Indeed, it is the surface anisotropy, which defines the spin freedom at the interfaces of the magnetic body and thus permits the evaluation of the various orders of the wave vectors for the system. Such considerations will allow us to define the full spectrum of the spin-wave excitations, which will depend on the dimensions of the magnetic entity and the exchange forces between spins. The simplest and most studied case is that of thin magnetic films with the external magnetic field applied in the direction perpendicular to the film plane, which is the simplest geometry. We will illustrate this case in the following discussion. Before we do so, it is instructive to consider the form of the resonance equation as expressed in Eq. (6.39), which we write in condensed form as:

$$(Dk^2)^2 + (P+Q)Dk^2 + (PQ - R^2 - \Omega^2) = 0 \qquad (6.41)$$

where we have used the following substitutions:

$$P = \frac{1}{M_0 \sin^2 \theta} \frac{\partial^2 E}{\partial \phi^2}; \quad Q = \frac{1}{M_0} \frac{\partial^2 E}{\partial \theta^2}; \quad R = \frac{1}{M_0 \sin \theta} \frac{\partial^2 E}{\partial \theta \partial \phi} = \frac{1}{M_0 \sin \theta} \frac{\partial^2 E}{\partial \phi \partial \theta};$$

$$\Omega = \frac{\omega}{\gamma}$$

In a first analysis, we can consider the general solutions to the quadratic equation presented in Eq. (6.41). This can be expressed in the following form with no loss of generality:

$$Dk^2 = -\frac{(P+Q)}{2} \pm \left[\left(\frac{P-Q}{2} \right)^2 + R^2 + \Omega^2 \right]^{1/2} \qquad (6.42)$$

We explicitly express these solutions as:

$$Dk_\alpha^2 = \left[\left(\frac{P-Q}{2}\right)^2 + R^2 + \Omega^2\right]^{1/2} - \frac{(P+Q)}{2} \tag{6.43}$$

$$Dk_\beta^2 = -\left[\left(\frac{P-Q}{2}\right)^2 + R^2 + \Omega^2\right]^{1/2} - \frac{(P+Q)}{2} \tag{6.44}$$

From Eq. (6.43), the wave vectors k_α can be (i) $k_\alpha > 0$, which indicates that the first term is greater than the second, and subsequently, the values of the wave vector will be real and hence correspond to the case of bulk volume modes, or (ii) $k_\alpha < 0$, which indicates, that the second term is greater than the first. This will give rise to imaginary wave vectors; $k_\alpha = i\kappa$. Similarly, we note from Eq. (6.44), that the wave vectors always comply to the condition $k_\beta < 0$, in which case we can write $k_\alpha = \pm i\tau$. In either case, an imaginary solution for the wave vector can only correspond to the case of a decaying mode, which is localized at an interface or boundary. In all cases, the spin-wave modes that are excited are standing spin-wave modes.

As we have already noted, the magnetic discontinuity at a surface or interface acts as a boundary. As with all boundary problems, the conditions of discontinuity can have a profound effect on the physical response of the system. In the magnetic case, this can be considered as the degree of freedom of the spins at the surface, a term we refer to as *pinning*. In the following, we will consider the relation of the pinning conditions to the surface anisotropy of the systems and then consider how this affects the allowed wave vectors for standing spin wave modes. Earlier, we discussed the exchange field with regards to the internal field of a spin system. In the case of spins at the surface/boundary, the change in the nearest neighbor environment will affect this exchange, since a full complement of neighbors is lacking. In this case, the exchange field takes the following form:

$$\mathbf{H}_{exch}^{surf} = \frac{2A}{M_0^2}\partial_n\mathbf{M} \tag{6.45}$$

where ∂_n denotes the surface normal derivative. Using this expression, we can evaluate the surface torque or the boundary condition, also referred to in the literature as the Rado–Weertman equation, as:

$$\mathbf{T}_{surf} + \mathbf{M} \times \left(\frac{2A}{M_0^2}\right)\partial_n\mathbf{M} = 0 \tag{6.46}$$

The term $\mathbf{T}_{surf}$ refers to the torque acting on the surface spins. This expression can be decomposed into static and dynamic components, since the magnetization vector contains both contributions. These are given as follows:

$$(\partial_n M)_\theta + \frac{M_0}{2A} \frac{\partial E_S}{\partial \theta} = 0; \qquad (\partial_n M)_\phi + \frac{M_0}{2A \sin \theta} \frac{\partial E_S}{\partial \phi} = 0 \qquad (6.47)$$

for the static components, and

$$\partial_n m_\theta + p m_\theta + r m_\phi = 0; \qquad \partial_n m_\phi + q m_\phi + r m_\theta = 0 \qquad (6.48)$$

for the dynamic components. Here, E_S denotes the magnetic surface energy, and the substitutions are given by:

$$p = \frac{1}{2A} \frac{\partial^2 E_S}{\partial \theta^2} - \frac{\partial_n M_0}{M_0}$$

$$q = \frac{1}{2A} \left(\frac{\cos \theta}{\sin \theta} \frac{\partial E_S}{\partial \theta} + \frac{1}{\sin^2 \theta} \frac{\partial^2 E_S}{\partial \phi^2} \right) - \frac{\partial_n M_0}{M_0}$$

$$r = \frac{1}{2A} \left(-\frac{\cos \theta}{\sin^2 \theta} \frac{\partial E_S}{\partial \phi} + \frac{1}{\sin \theta} \frac{\partial^2 E_S}{\partial \theta \partial \phi} \right)$$

If we take the simple case of a thin film (with the plane of the film in the $x - z$ plane) with uniaxial anisotropy, where:

$$E_S = K_S \sin^2 \theta \sin^2 \phi \qquad (6.49)$$

we obtain:

$$p = \frac{-K_S}{A} \sin^2 \theta - \frac{\partial_n M_0}{M_0}; \qquad q = \frac{K_S}{A} \cos 2\theta - \frac{\partial_n M_0}{M_0}; \qquad r = 0 \qquad (6.50)$$

Considering the film to be of thickness L, we can express the boundary conditions at $y = 0$ and at $y = L$ as:

$$\partial_n m_\theta + p_1 m_\theta = 0; \qquad -\partial_n m_\phi + q_1 m_\phi = 0 \qquad \text{at} \qquad y = 0 \qquad (6.51)$$

$$-\partial_n m_\theta + p_2 m_\theta = 0; \qquad \partial_n m_\phi + q_2 m_\phi = 0 \qquad \text{at} \qquad y = L \qquad (6.52)$$

For the case, where the field is applied in the direction perpendicular to the film plane, we can expect bulk volume modes to be excited in the film. In this case, the two sets of Eqs. (6.51) and (6.52) become identical of the form:

$$\partial_n m_\theta + p_1 m_\theta = 0 \qquad \text{at} \qquad y = 0 \qquad (6.53)$$

$$\partial_n m_\theta + p_2 m_\theta = 0 \qquad \text{at} \qquad y = L \qquad (6.54)$$

Using the relation for the oscillating component of the magnetization as $m \propto e^{i(\omega t - ky)} = e^{i\omega t} \sum_n \alpha_n [\cos(ky) - i\sin(ky)]$, we obtain:

$$\alpha_1 p_1 + \alpha_2 k = 0 \tag{6.55}$$

$$\alpha_1 [k\sin(kL) + p_2 \cos(kL)] + \alpha_2 [p_2 \sin(kL) + k\cos(kL)] = 0 \tag{6.56}$$

Simultaneous resolution of these equations yields the condition for the allowed wave vectors of standing spin wave modes. For bulk volume modes, we obtain:

$$\frac{k(p_1 + p_2)}{(k^2 - p_1 p_2)} = \tan(kL) \tag{6.57}$$

while for surface localized modes we have:

$$\frac{\kappa(p_1 + p_2)}{(\kappa^2 + p_1 p_2)} = \tanh(kL) \tag{6.58}$$

These expressions show that the pinning parameters p_1 and p_2 are central to the definition of the allowed wave vectors. Once these have been determined, it is possible, in conjunction with the corresponding resonance condition, Eq. (6.41), to predict the spin wave spectrum for any sample/applied field configuration. By way of example, we will consider the simple case of the thin film with an applied magnetic field along the film normal. In this case, the resonance equation yields the spin wave spectrum for the resonance fields as given by:

$$H_n = \frac{\omega}{\gamma} + H_K + \mu_0 M - D k_n^2 \tag{6.59}$$

where n indicates the spin wave mode number, which is taken to be an integer, H_K denotes the anisotropy field, where for the uniaxial case we have $H_K = 2K/\mu_0 M_0$. It should be noted, that Eq. (6.59) refers to the traditional FMR experiment with fixed microwave frequency and a field variation to measure the resonance fields. In a frequency-swept measurement (such as that using a VNA, see Section 4.7.6), we would express Eq. (6.59) as:

$$v_n = \frac{\gamma}{2\pi} (H - H_K - \mu_0 M + D k_n^2) \tag{6.60}$$

where the applied field is of a fixed value.

We will now consider the simplest cases of pinning to demonstrate the effect of the boundary conditions on the resultant spin-wave spectrum. The principal extremes will be the cases of perfect pinning and perfect freedom.

In the former, the surface or interface spins are completely anchored at the boundaries, while in the latter, the spins have the same freedom of movement as those in the bulk or interior of the sample. The more general case of partial pinning corresponds to the case between these two limits. We can also consider the situation for underpinning, in which the surface spins have greater freedom of movement than those of bulk spins. In each case, the resulting spin-wave modes will exhibit a unique spectrum of resonances, which are characterized by their field (or frequency) position and their intensity. We will discuss the intensity considerations a little later. It is common practice to use the uniform ferromagnetic resonance case as a reference for the displacement of the resonance for the spin-wave modes.

For the case of perfect pinning, the boundary spins are considered as rigidly fixed. In this case, the lowest energy mode will have the form as illustrated in Figure 6.3. Here, we note that the amplitude of the spin preces-

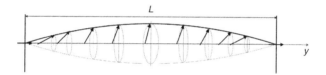

FIGURE 6.3 Lowest lying standing spin wave mode, $n = 1$, for the case of perfect boundary pinning. The scale is exaggerated for clarity. The distance L corresponds to the film thickness.

sion has a maximum value at the center of the film and gradually varies to zero amplitude at the surfaces of the magnetic layer. If we consider higher order spin wave modes with the perfect pinning condition, we can generate the modal patterns as shown in Figure 6.4. For each mode, we can determine the wavelength and wave vector, $k_n = 2\pi/\lambda_n$. From Figure 6.4 it is a simple matter to deduce these relations as a function of n:

$$\lambda_n = \frac{2L}{n}; \quad \text{and} \quad k_n = \frac{2\pi}{\lambda_n} = \frac{n\pi}{L} \tag{6.61}$$

It is worth pointing out, that the case of perfect pinning corresponds to the case of $p = p_1 = p_2 = 0$. Using Eq. (6.57) we see that this yields the condition for allowed wave vectors as $\tan(k_n L) = 0$. There is no difficulty in noting that such an expression leads directly to $k_n = n\pi/L$, as expressed in Eq. (6.61). The wave vector expression is then substituted into Eq. (6.59) and gives:

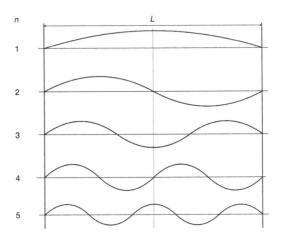

FIGURE 6.4 Lowest lying standing spin wave modes for the case of perfect boundary pinning. The modal numbers are indicated left, n.

$$H_n = \frac{\omega}{\gamma} + H_K + \mu_0 M - D\left(\frac{n\pi}{L}\right)^2 = H^{FMR} - D\left(\frac{n\pi}{L}\right)^2 \qquad (6.62)$$

showing the displacement of the spin wave mode with respect to the uniform FMR mode. From this expression, we note that all spin wave modes are excited for fields below the uniform mode, which is also evident from Eq. (6.59). For frequency-swept measurements, the higher order modes occur at higher frequencies than that of the uniform mode, as seen from Eq. (6.60). The field separation between successive modes varies as n^2 and for the field-swept measurement can be expressed as:

$$\delta H_n = H_{n-1} - H_n = D\left(\frac{\pi}{L}\right)^2 (2n-1) \qquad (6.63)$$

We will now briefly show the case for perfect freedom, where the surface spins have the same freedom as bulk spins. In Figure 6.5, we show the corresponding modal patterns for the first few orders. It is tempting to designate the first mode as $n = 0$, since then we could use the same expressions for the wavelength and wave vectors as give in Eq. (6.61). However, it is more convenient to maintain the order of numbers commencing with 1. In this case, we write the relations for the wavelengths and the wave vectors of the spin-wave modes as:

$$\lambda_n = \frac{2L}{(n-1)}; \qquad \text{and} \qquad k_n = \frac{2\pi}{\lambda_n} = \frac{(n-1)\pi}{L} \qquad (6.64)$$

The only difference between the two cases being the "-1" in the case of perfect freedom. While this may seem to show, that the cases of perfect pinning and perfect freedom are very similar, this is very far from being the case. Firstly, the sequence of resonance fields will be different since the latter case has the zero-order mode and secondly, the intensities of the modes will be very different, as we shall formalize further on. Suffice it to say for the moment, that the mode intensity is proportional to the transverse magnetization. This means, that for the case of perfect freedom only the $n = 0$ mode will have a net contribution, with all higher-order modes having zero intensity. This is a special case and only the uniform mode is excited. For the case of perfect pinning, only the odd modes are excited, with decreasing intensity with increasing mode number.

In the two cases, we have so far considered, the pinning conditions are symmetric, i.e., they are the same for both boundaries, $p_1 = p_2 = p$. There is no physical reason why this is necessarily the case in reality. For example, in thin films, one surface is generally bound by the substrate while the other is a free surface or may have a capping layer. In many practical cases, they will be different, giving asymmetric boundary conditions, $p_1 \neq p_2$. In this case, even order modes can be observed, though they still generally tend to be weaker modes appearing between the odd-order modes. The intermediate case of partial pinning is illustrated in Figure 6.6.

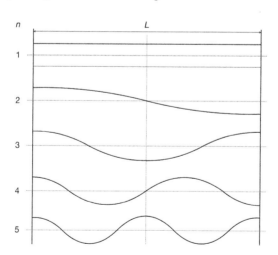

FIGURE 6.5 Lowest lying standing spin-wave modes for the case of free pinning or perfect freedom. The modal numbers are indicated left, n. In this case, the first mode corresponds to the case of all spins precessing in phase with no variation of amplitude across the film, which is another way of saying the uniform mode.

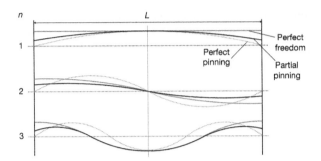

FIGURE 6.6   Lowest lying standing spin-wave modes for the case of partial pinning. We show the modal patterns also for the cases of perfect pinning (dashed gray lines) and perfect freedom (solid gray lines) for comparison.

Considering the modal patterns for the partial pinning case, it is not so simple to find the analytical expression for the wavelengths and wave vectors, since this will depend on the pinning parameter. We can introduce a new pinning factor, δ, which we can used to write a general expression for the wave vectors in the form:

$$k_n = \frac{2\pi}{\lambda_n} = \frac{(n-\delta)\pi}{L} \tag{6.65}$$

where we impose the condition $0 < \delta < 1$. In this case, we state that for perfect freedom $\delta = 1$, while for perfect pinning, we have $\delta = 0$. In the case of asymmetric pinning, we can modify Eq. (6.65) to read:

$$k_n = \frac{2\pi}{\lambda_n} = \frac{(n-\delta_1-\delta_2)\pi}{L} \tag{6.66}$$

where now we have $0 < \delta_{1,2} < 0.5$. Symmetric pinning in this instance will be found for $\delta_1 = \delta_2 = \delta/2$ and Eq. (6.65) is re-established.

We can also consider the case of underpinning, Figure 6.7, where the surface spins have more freedom than those in the bulk. This produces surface or interfaces localized modes, of which there will only be two, all other modes will be bulk volume modes. As we mentioned previously, these modes have imaginary wave vectors such that the surface modes adhere to the relation:

$$H_n^{surf} = \frac{\omega}{\gamma} + H_K + \mu_0 M - D(i\kappa_n)^2 = H^{FMR} + D\kappa_n^2 \tag{6.67}$$

Therefore, any modes appearing at field above H^{FMR} can be designated as surface localized modes. Once again, pinning conditions can be asymmetric, meaning that the degree of localization can be different at the two interfaces.

For further details on these issues, the reader is referred to the review by Puszkarski (1979).

Mode intensities for the various types of standing spin-wave resonances are related to the transversal component of the magnetization across the film. This is a complex calculation requiring the integration of the transverse magnetization profile across the film. This can be expressed in normalized form as:

$$I_n = \frac{1}{L} \frac{\left| \int_0^L m(y)dy \right|^2}{\int_0^L m^2(y)dy} \tag{6.68}$$

where we consider the case where the film plane coincides with that of the $x - z$ plane. Regarding the cases we considered above, for the perfect symmetric pinning model, only odd modes will have a net transverse moment, and using the sinusoidal profile, as indicated above, the intensities drop off as $1/n^2$. In Figure 6.8, we show an early example of a measurement of the spin-wave resonance in a permalloy thin film. We note the diminishing mode intensity as the mode number increases and that the separation between the spin-wave resonance modes also increases for higher mode numbers, in conformity with expressions (6.68) and (6.67), respectively. We can also note from Figure 6.8, that the appearance of small peaks between these odd-order modes illustrates that even order modes are present, though at much-reduced intensity when compared to the odd modes. This is in agreement with the

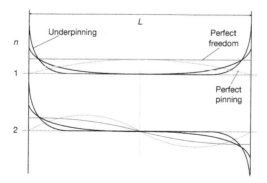

FIGURE 6.7 Lowest lying standing spin-wave modes for the case of underpinning. We show the modal patterns also for the cases of perfect pinning (dashed gray lines) and perfect freedom (solid gray lines) for comparison. As the underpinning increases, the modes become more and more localized.

above discussion and would indicate, that the pinning conditions appear to be slightly asymmetric.

6.5 Damping and Relaxation

The relaxation of the magnetization back to an equilibrium position has been treated by a number of approaches and we will here outline the more usual forms of treatment. One of the earliest attempts to do this, was that by Landau and Lifshitz in 1935, who added a term to the equation of motion, Eq. (6.16), that can be expressed in the form:

$$\frac{\partial \mathbf{M}}{\partial t} = -\gamma(\mathbf{M} \times \mathbf{H}_{eff}) - \frac{\lambda}{M^2}[\mathbf{M} \times (\mathbf{M} \times \mathbf{H}_{eff})] \qquad (6.69)$$

The vector diagram for this equation is shown in Figure 6.9. The vector products $\mathbf{M} \times \mathbf{H}_{eff}$ and $\mathbf{M} \times (\mathbf{M} \times \mathbf{H}_{eff})$ define the precession of the magnetization vector and its relaxation back to equilibrium, respectively. The vector sum of these two, as expressed in Eq. (6.69), describes a spiralling motion of the magnetization around the effective field. The rate, at which this occurs, is governed by the damping parameter λ. A slightly different form of this relaxation process was introduced by T. L. Gilbert in 1955 and is given as:

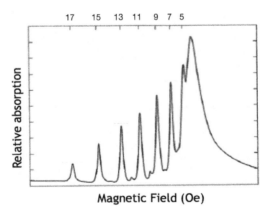

FIGURE 6.8   Spin wave resonance spectrum for a 390 nm thick permalloy film. The spin-wave mode numbers are indicated at the top, where modes 1 and 3 are overlapped and appear as the largest peak in the spectrum. Reprinted figure with permission from M. H. Seavey, Jr. & P. E. Tannenwald, (1958). *Phys. Rev Lett., 1*, 168, ©(1958) by the American Physical Society.

$$\frac{\partial \mathbf{M}}{\partial t} = -\gamma(\mathbf{M} \times \mathbf{H}_{\text{eff}}) + \frac{\alpha}{M}\left(\mathbf{M} \times \frac{\partial \mathbf{M}}{\partial t}\right) \tag{6.70}$$

This form of the phenomenological relaxation is probably the most commonly used form for spin dynamics in ferromagnetic materials. It is often referred to as the Landau–Lifshitz equation with Gilbert damping (or LLG, for short). In this formulation, the relaxation rate is governed by the parameter α, the Gilbert damping parameter. The interpretation is very much the same as that for the Landau–Lifshitz equation, where the vector $\mathbf{M} \times \partial\mathbf{M}/\partial t$ replaces that of $\mathbf{M} \times (\mathbf{M} \times \mathbf{H}_{\text{eff}})$. We can show the correspondence between these equations more formally using the following manipulations. We first substitute Eq. (6.70) inside itself for the $\partial\mathbf{M}/\partial t$ term on the RHS, from which we can write:

$$\frac{\partial \mathbf{M}}{\partial t} = -\gamma(\mathbf{M} \times \mathbf{H}_{\text{eff}}) + \frac{\alpha}{M}\left\{\mathbf{M} \times \left[-\gamma(\mathbf{M} \times \mathbf{H}_{\text{eff}}) + \frac{\alpha}{M}\left(\mathbf{M} \times \frac{\partial \mathbf{M}}{\partial t}\right)\right]\right\} \tag{6.71}$$

This equation is clearly non-linear. The degree of non-linearity depends on the strength of α. In general, this parameter is quite small, typically $< 10^{-2}$, which means that higher order terms can be considered as negligible. The leading term in α allows us then to approximate the LLG equation to that of the original LL form. In fact, this approximation in tantamount to considering the precessional amplitude as being very small, which for most practical

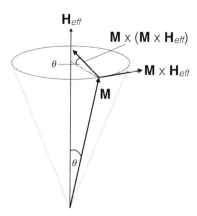

FIGURE 6.9 Geometric construction of the vector diagram for the Landau–Lifshitz equation with damping. We note, that the vector triplet of $\mathbf{M}$, $\mathbf{M} \times \mathbf{H}_{\text{eff}}$ and $\mathbf{M} \times (\mathbf{M} \times \mathbf{H}_{\text{eff}})$ are mutually perpendicular vectors.

cases of ferromagnetic resonance is true. Rewriting the previous equation, we have:

$$\frac{\partial \mathbf{M}}{\partial t} = -\gamma(\mathbf{M} \times \mathbf{H}_{\text{eff}}) + \frac{\alpha}{M}\left\{-\gamma[\mathbf{M} \times (\mathbf{M} \times \mathbf{H}_{\text{eff}})] + \frac{\alpha}{M}\left[\mathbf{M} \times \left(\mathbf{M} \times \frac{\partial \mathbf{M}}{\partial t}\right)\right]\right\} \tag{6.72}$$

The triple product in the final term of this equation can be evaluated as:

$$\mathbf{M} \times \left(\mathbf{M} \times \frac{\partial \mathbf{M}}{\partial t}\right) = \left(\mathbf{M} \cdot \frac{\partial \mathbf{M}}{\partial t}\right)\mathbf{M} - M^2\frac{\partial \mathbf{M}}{\partial t} = -M^2\frac{\partial \mathbf{M}}{\partial t}$$

We now obtain:

$$\frac{\partial \mathbf{M}}{\partial t} = -\gamma(\mathbf{M} \times \mathbf{H}_{\text{eff}}) - \frac{\gamma\alpha}{M}[\mathbf{M} \times (\mathbf{M} \times \mathbf{H}_{\text{eff}})] + \frac{\alpha^2}{M^2}\left[-M^2\frac{\partial \mathbf{M}}{\partial t}\right]$$

or more simply:

$$(1 + \alpha^2)\frac{\partial \mathbf{M}}{\partial t} = -\gamma(\mathbf{M} \times \mathbf{H}_{\text{eff}}) - \frac{\gamma\alpha}{M}[\mathbf{M} \times (\mathbf{M} \times \mathbf{H}_{\text{eff}})] \tag{6.73}$$

Comparing Eqs. (6.69) and (6.73) shows, that they are identical with the following substitutions:

$$\alpha = \frac{\lambda}{\gamma M}, \quad \text{and} \quad \gamma_G = \frac{\gamma}{(1 + \alpha^2)} \tag{6.74}$$

A further form of the equation of motion with relaxation was proposed Bloch in 1946 with subsequent modifications by Bloembergen in 1950 and can be expressed in component form as:

$$\frac{\partial M_x}{\partial t} = -\gamma(\mathbf{M} \times \mathbf{H}_{\text{eff}})_x - \frac{M_x(t)}{T_2} \tag{6.75}$$

$$\frac{\partial M_y}{\partial t} = -\gamma(\mathbf{M} \times \mathbf{H}_{\text{eff}})_y - \frac{M_y(t)}{T_2} \tag{6.76}$$

$$\frac{\partial M_z}{\partial t} = -\gamma(\mathbf{M} \times \mathbf{H}_{\text{eff}})_z - \frac{M_z(t) - M_0}{T_1} \tag{6.77}$$

Here, we note that there are two fundamentally different relaxations time; the longitudinal relaxation time, T_1, and the transverse relaxation time, T_2. These are used to explain the observed line shape of a resonance absorption curve where it is assumed that:

$$\frac{1}{T_2} = \frac{1}{2T_1} + \frac{1}{T_2^*} \tag{6.78}$$

where T_1 is the spin–lattice relaxation time and T_2^* is the spin–spin relaxation time. From Eqs. (6.75) - (6.77), it is possible to determine the temporal dependences of the different Cartesian components of the magnetization, which we can express as:

$$M_{x,y}(t) = M_{x,y}(0)e^{-t/T_2} \qquad (6.79)$$

$$M_z(t) = M_{z,eq} - [M_{z,eq} - M_z(0)]e^{-t/T_1} \qquad (6.80)$$

The Bloch–Bloembergen formalism is rarely used in ferromagnetic resonance, though it is more commonly used in the interpretation of nuclear magnetic resonance experiments.

The study of relaxation processes in ferromagnetic materials almost invariably is treated with respect to the LL and LLG forms of relaxation. In general, the linewidth of the resonance curve allows us a direct measurement of the relaxation. As we noted for the resonance field, the linewidth is also angle-dependent, where the intrinsic value can be expressed in terms of the free energy of the system as:

$$\Delta H_0 = \frac{\alpha}{M} \left(\frac{\partial^2 E}{\partial \theta^2} + \frac{1}{\sin^2 \theta} \frac{\partial^2 E}{\partial \phi^2} \right) \qquad (6.81)$$

A consideration of the LLG equation can also be shown to give the peak-to-peak linewidth as a function of the damping parameter and frequency:

$$\Delta H_G = 1.16 \frac{\alpha \omega}{\gamma} = 1.16 \left(\frac{\lambda}{\gamma M} \right) \frac{\omega}{\gamma} \qquad (6.82)$$

The factor of 1.16 in this expression arises from the inflection points in the Lorentzian line shape. The characteristic feature of the Gilbert damping is its linear dependence on the excitation frequency. Eq. (6.82) also shows, that the linewidth is directly proportional to the damping parameters α and λ. Clearly, with increased damping, the damping torque term will become increasingly important and dominate the RHS of the LL and LLG equations and $\partial M / \partial t \to 0$. In this case, the magnetic system will have a sluggish response and slowly approach the equilibrium state.

Some of the specific causes of intrinsic line width in metallic systems can arise from various mechanisms such as eddy currents, phonon drag, and spin-orbit relaxation, see Heinrich (2005). Defect scattering of magnons can also occur in magnetically inhomogeneous samples, where the uniform mode ($\mathbf{k} = 0$) is scattered into non-uniform modes, $\mathbf{k} \neq 0$ magnons. This process is referred to as *two-magnon* scattering. An additional damping mechanism can be observed in magnetic multilayers due to the spin-pumping effect, which occurs via the spin angular momentum transfer between magnetic layers, see Section 13.4. This produces an additional line-width of approximately:

$$\Delta H_{SP} = \Delta \mu + \hbar \omega \qquad (6.83)$$

where $\Delta \mu = \mu_\uparrow - \mu_\downarrow$ is the difference in the chemical potentials of the spin-up and spin-down electrons.

In most systems, additional contributions to the measured line-wdith cannot be excluded, with inhomogeneous (extrinsic) broadening mechanisms playing an important role. The degree, to which this is important, is intimately related to the nature of the sample. In a generic manner, we can express the various contributions as follows:

$$\Delta H = \Delta H_0 + \left(\frac{\partial H}{\partial \phi} \right) \Delta \phi + \left(\frac{\partial H}{\partial H_i} \right) \Delta H_i + \left(\frac{\partial H}{\partial V} \right) \Delta V + \left(\frac{\partial H}{\partial S} \right) \Delta S \quad (6.84)$$

The first term on the RHS of this expression refers to the intrinsic linewidth, the second term accounts for the spread in crystalline axes, the third term results from magnetic homogeneities in the sample, while the fourth term expresses the variation of sizes for particle systems and the final term accounts for differences in the surface spin contribution. In bulk samples, the final two terms are neglected. In Figure 6.10, we show a set of data for linewidths of $Co_{90}Zr_{10}$ thin films as a function of frequency. The positive slopes are related to the Gilbert damping parameter. It is also important to note, that the offsets indicate an increase of the extrinsic broadening for the thinner films. This is

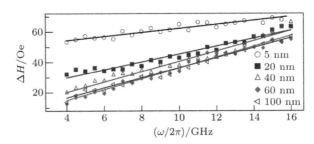

FIGURE 6.10   Frequency dependent FMR linewidth, ΔH, for a series $Co_{90}Zr_{10}$ thin films with different thicknesses. (Chang-Jun et al., 2015).

related to the surface contribution, probably related to the surface roughness, which decreases as a function of layer thickness.

6.6 Experimental Techniques

Magnetization dynamics can be observed using different experimental methods. In the following, we will give a brief outline of some of the most frequently used techniques. Traditionally, FMR has been studied experimentally using a fixed frequency microwave spectrometer, though, in recent decades, a number of new techniques have been developed, which allow a more versatile measurement and are not limited in frequency. The most common technique being the use of broadband methods and notably using a vector network analyzer (VNA). Ultrafast optical methods using a pump-probe methodology have also been extensively applied to the study of magnetization dynamics and can be used not only to probe the precessional dynamics related to ferromagnetic and spin-wave resonance, but also the dynamics related to ultrafast demagnetization due to the interaction with a high powered ultrafast laser pulse. We will outline the principal characteristics of these three techniques in this section. While other techniques are also available for the study of spin dynamics, we will limit our discussion to these, since they are the most commonly used and provide a substantial amount of relevant information that can be obtained experimentally.

6.6.1 The Microwave Spectrometer

Microwave spectroscopy is a general method for the measurement of various types of resonances such as FMR, ESR, and cyclotron resonance. The technique uses a microwave circuit constructed of waveguides, cavities, and other components. The microwaves are typically generated by a source such as a klystron or, more commonly now, with a solid-state diode. The microwave radiation is then directed via the waveguide system to a cavity, via a coupling hole or iris, which is so designed to support a standing wave mode of the incident electromagnetic radiation. Its size and geometry are matched to the wavelength of the radiation and the specific standing wave mode of the cavity. There are many possible designs of microwave spectrometer and we will not go into the specifics here. The interested reader can find more information in books on the subject, see, for example, Poole (1983). An example of a typical spectrometer is illustrated in Figure 6.11.

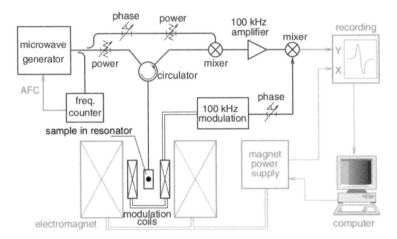

FIGURE 6.11   Schematic illustration of a microwave spectrometer. In red are the main components of the microwave bridge circuit, including the source or generator, the circulator, and the microwave cavity resonator, where the sample is housed. This cavity is placed between the poles of an electromagnetic and modulation coils. The signal recovery system uses the phase-sensitive detection method with a lock-in amplifier and data acquisition is made via a computer.

Of critical importance to the sensitivity of the method is the coupling from waveguide to the cavity. The sample to be studied is placed inside the microwave cavity in a position of the maximum oscillating magnetic field. It is this field, which is the driving force of the resonance. In the conventional FMR experiment, the cavity mode will determine the frequency of excitation, i.e., that of the microwave source, and the ferromagnetic resonance condition is found by sweeping an externally applied magnetic field. This field can typically be rotated with respect to the sample, allowing angular studies of the resonance field to be plotted, thus permitting studies of magnetic anisotropies in the sample under study. In the off-resonance state, the spectrometer will usually be set up such that the detector gives a null signal, typically using a bridge system. The resonance state corresponds to the maximum deviation of the magnetization precession angle with respect to its equilibrium orientation. As the magnetic field is varied, the magnetization of the sample under study will begin to precess at the microwave frequency and the precessional angle increases to a maximum and then decreases again as the field is further increased. The applied field corresponding to the maximum of the precession angle is defined as the resonance field. In terms of the experiment, at this point the sample absorbs the maximum energy from the microwave field, thus changing the cavity conditions and as a consequence the reflected signal from the cavity itself. This is what is measured in the typical FMR experiment. It is common to use field or frequency modulation techniques in conjunction with phase-sensitive detection. The results are then displayed as a derivative of the microwave power absorption as a function of the applied magnetic field. This form of signal recovery increases the experimental sensitivity and displays the derivative spectrum of the absorption. This method, along with many variants, has been used extensively to probe and measure the physical properties and dynamic effects in all forms of magnetic materials, from bulk crystalline to amorphous, from thin films to magnetic nanoparticles, including both metallic samples and magnetic oxides.

6.6.2   The Vector Network Analyzer Method

In the preceding discussion, we described the traditional method for measuring FMR. In recent years, a number of alternative methods have been developed, which have adapted the basic principles of the FMR experiment, making it more suitable for the measurement of nanostructured materials and nanoparticles. Of the methods available, the use of micro-resonators and stripline technologies in tandem with the vector network analyzer (VNA) is extremely promising and has now developed into a well-established method of performing ferromagnetic resonance (VNA-FMR) on thin films and low-dimensional structures. In this technique, the VNA acts as both source and detector, in which the two-port VNA device is connected, via high-frequency cables, to a coplanar waveguide (CPW) or stripline. The use of a planar micro-resonator (PMR) can also increase the sensitivity of the measurement, though it limits measurement to a fixed frequency, as we will discuss shortly. For the coplanar stripline, there is no resonant cavity, which means that measurements can be made over a broad range of frequencies (commonly referred to as a broadband FMR measurement). In this case, the measurement can be made continuously up to tens of gigahertz. The two-port VNA is connected via high-frequency cables to the CPW, through which a high-frequency electrical signal is passed from the VNA. The detection is made by measuring the four scattering or S-parameters; these consist of the two transmitted signals (port 1–port 2, S_{12}, port 2–port 1, S_{21}) and the two reflected signals (port 1–port 1, S_{11}, port 2–port 2, S_{22}). These four parameters make up the elements of the S matrix. Since the CPW is impedance matched (50 Ω) to the VNA output, this will maximize the transmitted signal, which makes the technique very sensitive to changes in the line impedance. The method requires a full two-port calibration to be implemented to remove background reflections from the cable/waveguide system.

The formal description of the signal obtained from the VNA-FMR method is based on the transmission and reflection coefficients, which are given in the form of the scattering or S-parameters and take into account the line impedance including the sample. These are expressed as lumped elements with the effective inductance (L), series resistance (R), shunt conductance (G) and capacitance (C), and can be expressed as (Maksymov and Kostylev, 2015) (Ding et al., 2004):

$$S_{11} = \frac{i\omega L + R + Z_0/[1 + Z_0(G + i\omega C)] - Z_0}{i\omega L + R + Z_0/[1 + Z_0(G + i\omega C)] + Z_0} \qquad (6.85)$$

and

$$S_{21} = \frac{2Z_0/[1 + Z_0(G + i\omega C)]}{i\omega L + R + Z_0/[1 + Z_0(G + i\omega C)] + Z_0} \qquad (6.86)$$

where Z_0 is the characteristic impedance of the stripline and ω is the microwave angular frequency. For a symmetric set-up, we would expect $S_{11} = S_{22}$ and $S_{21} = S_{12}$. In terms of the complex reflection coefficient, we can write:

$$\Gamma = \frac{Z - Z_0}{Z + Z_0} \qquad (6.87)$$

or alternatively:

$$\frac{Z}{Z_0} = \frac{1 + \Gamma}{1 - \Gamma} \qquad (6.88)$$

where Z is the impedance of the (sample) loaded stripline.

It should be noted, that the sensitivity of this method can be limited by the quality of the cables and connectors. Often poor-quality components will introduce further reflections, thus limiting the transmission characteristics of the high-frequency signals. This is particularly true of measurements made at the high-frequency end and above around 40 GHz in general. The magnetic sample, usually in the thin-film form, is placed (face-down) on top of the waveguide and located inside the poles of an electromagnet, whose field direction should be ideally parallel to the stripline. Placing the sample on the stripline changes the characteristic impedance of the waveguide.

The signal-to-noise ratio is improved by covering as much of the stripline as possible. This can be important for broadband measurements, where there is no signal amplification due to Q-factors. The measurement of the FMR spectrum can then proceed in one of two methods: (i) field sweep at fixed frequency or (ii) frequency sweep with a fixed static magnetic field. The VNA provides a measurement of the line impedance via transmission and reflection coefficients, which are related to the various S parameters. It should be noted, that the electrical signal, which passes through the CPW, will produce a small oscillating magnetic field around the CPW. It is this high-frequency magnetic field, which is the driving field for the resonance measurement. As the field or frequency is swept through the resonance of the ferromagnetic sample placed on the CPW, the line impedance will change, hence altering the S parameters, providing the measurement of the resonance itself. Figure 6.12 shows a schematic representation of the VNA setup.

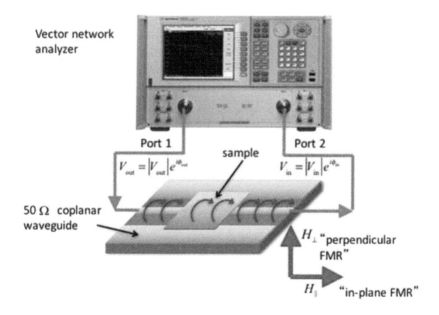

FIGURE 6.12 Schematic diagram of the VNA-FMR set-up. The high-frequency signal is generated in the VNA and sent through a coplanar waveguide, from one port and detected in the second port, as illustrated. Changes in the line impedance are detected as the magnetic sample passes through resonance. This can be performed by sweeping the applied magnetic field (in either the perpendicular or parallel configuration) with a fixed frequency or by sweeping the signal frequency at a constant applied field. (Figure credit NIST: https://www.nist.gov/programs-projects/spin-transport (accessed on 30 March 2020)).

A limitation of the traditional FMR experiment resides in the fact, that it must be, by its very nature, a fixed frequency measurement. The VNA-FMR technique, however, overcomes this problem, since it does not require a cavity and broadband measurements are possible. This, therefore, allows for direct measurement of the frequency-field dispersion relation for a magnetic sample. Excellent agreement with theory is found using this technique, as illustrated in the example of the dispersion relation for a thin permalloy film, shown in Figure 6.13.

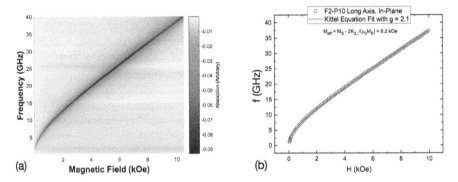

FIGURE 6.13 (a) VNA data for a 50 nm permalloy thin film showing the frequency–field characteristics. The color variation shows the absorption intensity. (b) Corresponding fit of the experimental data (points) and the theory (line) for the so-called Kittel equation: $\omega = \gamma\mu_0\sqrt{H(H+M_{\text{eff}})}$. [Data: D. Markó and D. S. Schmool, unpublished, (2018)].

Other derivatives of the coplanar waveguide method are also available using fixed frequency and variable frequency microwave generators. For the case of a planar micro-resonator (PMR), a one-port setup can be used, where the measurement is analogous to that of the traditional FMR method. Here, the PMR acts as a cavity and the VNA or microwave generator is set to its resonance frequency. The advantage of this technique is the improvement of the sensitivity due to the quality factor of the resonator, though this is typically around 50 and much lower than that of the normal microwave cavity, typically of the order of 10^4. Another important parameter is the filling factor of the sample. In a conventional FMR measurement, the filling factor is quite small ($\sim 10^{-8}$), since the cavity is quite large with respect to the sample dimensions. The combination of filling factor and Q-factor plays an important role in the practical sense of resonance-type measurements. The increased effective filling factors are of great importance in VNA-type measurements, being one of the main reasons, why its sensitivity makes it a viable alternative to conventional FMR techniques. The micro-resonator improves significantly the sensitivity in this respect, where filling factors can be several orders of magnitude larger. Since the resonators can be relatively small, smaller samples can be measured. Banholzer et al. (2011) and Schoeppner et al. (2014) have reported on the measurement of micron-sized ferromagnetic elements using this method. In Figure 6.14, measured FMR spectra are shown for a single and multilayer CoFe/Ni thin films. Real and imaginary parts of the FMR spectra are illustrated in Figure 6.14(b).

The angular dependence of the FMR for this sample is illustrated in Figure 6.15.

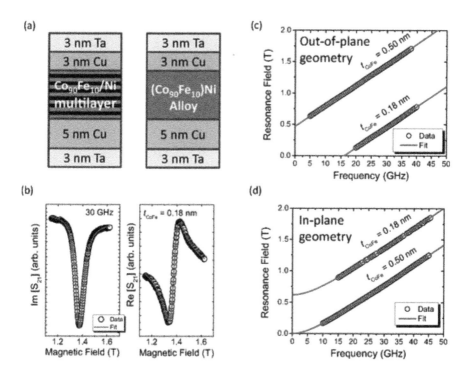

FIGURE 6.14 (a) Schematic diagram of the multilayer CoFe/Ni and alloy CoFeNi sample structures. (b) Examples of imaginary and real parts of the FMR spectra taken at 30 GHz for the $t_{CoFe} = 0.18$ nm multilayer sample. Example Kittel plots of the resonance field versus frequency for the (c) out-of-plane and (d) in-plane geometries are also illustrated. Reprinted figure with permission from: Justin M. Shaw, Hans T. Nembach, & T. J. Silva, (2013). *Physical Review B, 87,* 054416. ©(2013) by the American Physical Society.

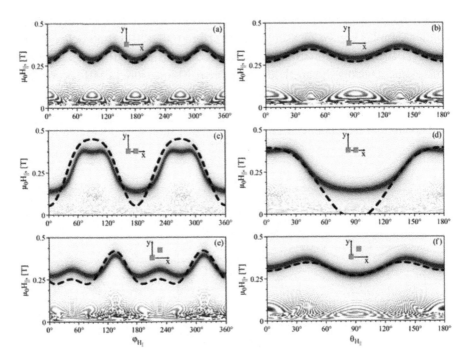

FIGURE 6.15 Angular dependence of FMR spectra for the system of a single magnetic moment with (a) azimuthal and (b) polar angle variation. The case of two nanocubes placed along the x-axis is shown for (c) ϕ- and (d) θ-angle variations of the static magnetic field. Variation of the angles for a diagonal alignment of the nanocubes is shown in panels (e) and (f), respectively. Dashed black curves correspond to analytical solutions. The darker tone marks higher magnetic susceptibility value χ_{Im}. Filled squares illustrate the schematic view of the system studied as seen from above. Reprinted by permission from Springer Nature, P. P. Horley, A. Sukhov, J. Berakdar, & L. G. Trápaga-Martínez, (2015). *The European Physical Journal B, 88*, 165, ©(2015).

6.6.3 Pump-Probe Spectroscopy

Another experimental technique, that has found widespread application, is the use of pump-probe spectroscopies, which employ ultrafast optical methods to obtain measurements of magnetization dynamics with temporal resolutions in the 10s of femtoseconds (Kirilyuk et al., 2010) and more recently in the sub-10 femtosecond range (Gonçalves et al., 2016). The main principles of the technique reside in the use of ultrafast lasers, typically a Ti:sapphire laser, whose beam is split into two (not necessarily equal) beams, which serve as the pump beam and the probe beam. The ultimate temporal resolution of the measurement derives from the pulse length of the laser itself. Much work has been performed to control the pulse width of laser systems, a discussion here is beyond the scope of this work, but the interested reader is referred to the following references (Gonçalves et al., 2016; Silva et al., 2014; Rothhardt et al., 2012) and references therein. In the following, we will give a brief overview of the main working principles of the pump-probe technique as applied to the study of magnetization dynamics.

Once the main beam has been split into the two components, these are each directed towards the sample under study, which is usually a thin film. The two beams must overlap at the sample surface. The pump acts as the excitation for the magnetization dynamics and the probe beam detects the magnetic state of the sample via the Kerr effect. The temporal variation of the magnetization is measured by varying the temporal delay between the pump and probe beam arrival times at the sample via a delay stage mounted into the pump beamline. The apparatus is illustrated schematically in Figure 6.16.

Signal treatment using lock-in and boxcar amplifiers is standard for such systems. The detected signal can show a number of effects such as ultrafast demagnetization and the subsequent relaxation of the sample magnetization, precessional dynamics are seen as an oscillating component of the magnetization, depending on which of the Kerr effects is used. The damping can also be readily measured with this method. It is common to perform a Fourier transform of the data to allow the results to be presented in the frequency domain. Research using ultrafast measurements has become a well-established field attracting much interest in current research in magnetism in low dimensional systems.

6.7 Spin Dynamics in Nanometric Systems

Research into the spin dynamics in thin films, multilayers, and nanostructured materials is a vast area. Here, we shall outline some of the principal themes and give some representative results. For a more detailed study, the reader is referred to the review by Schmool (2010).

By far the most studied system is that of the thin magnetic film, which has been the subject of uncountable studies since the 1950s. The early work by Kittel and co-workers led to the development of the analysis for thin films and the so-called Kittel equation. A consideration of the sample geometry and boundary conditions, much as we outlined in Section 4.7.4, allows us to determine the general resonance equation. The dipole exchange spin waves

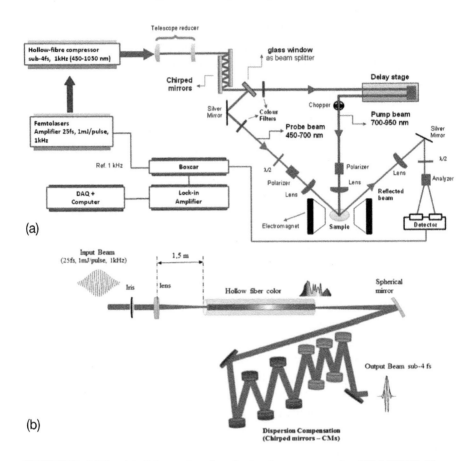

FIGURE 6.16 (a) Schematic of a dual-color pump-probe TR-MOKE (time-resolved magneto-optic Kerr effect) apparatus. (b) Simplified set-up of a hollow fiber and chirped mirror compressor. Although the chirped mirrors are identical, different colors are used to note, that they should be placed at two different angles (5° and 19°). Reprinted from C. S. Gonçalves, A.S. Vieira, D. Navas, M. Miranda, F. Silva, H. Crespo, & D. S. Schmool, (2016). *Sci. Rep.*, 6, 22872, under a Creative Commons Attribution 4.0 International License.

for an infinite ferromagnetic medium was given by the Herring-Kittel formula:

$$\left(\frac{\omega}{\gamma}\right)^2 = \mu_0^2 (H + Dk^2)(H + Dk^2 + M \sin^2 \theta_k) \qquad (6.89)$$

where θ_k defines the angle between the directions of the wave vector and the static magnetization.

For the case, where the film is in the $x - y$ plane and the field applied along the x-axis, the resonance equation can be expressed in the form:

$$\left(\frac{\omega}{\gamma}\right)^2 = \mu_0^2 (H + Dk^2)[H + Dk^2 + MF_{qq}(k_{\parallel}L)] \tag{6.90}$$

Here, $F_{qq}(k_{\parallel}L)$ is the matrix element of the magnetic dipole interaction. In the case of free and perfect pinning, the wave vectors can be expressed in the form:

$$k^2 = k_x^2 + k_y^2 + \left(\frac{n\pi}{L}\right)^2 = k_{\parallel}^2 + \left(\frac{n\pi}{L}\right)^2 \tag{6.91}$$

The quantization, as we saw in Section 4.7.4, for the spin waves has the same general pattern of modal number separated by 1 for the two cases. However, for an arbitrary angle between $\mathbf{k}_{\parallel}$ and $\mathbf{M}$, the matrix elements of the dipole interaction are expressed in a modified form as:

$$F_{qq}(k_x, k_y) = 1 + P_{qq}(k)[1 - P_{qq}(k)]\left(\frac{\mu_0 M}{H + Dk^2}\right)\left(\frac{k_y^2}{k^2}\right) - P_{qq}(k)\left(\frac{k_x^2}{k^2}\right) \tag{6.92}$$

If the spin wave propagates in the plane of the film, but perpendicular to the external magnetic field ($k_z = 0, k_y = k_{\parallel}$), the expression for $F_{qq}(k_{\parallel}L)$ takes the form:

$$F_{qq} = 1 + P_{qq}(k)[1 - P_{qq}(k)]\left(\frac{\mu_0 M}{H + Dk^2}\right) \tag{6.93}$$

For the lowest value mode, $q = 0$, the function $P_{qq}(k)$ takes the form:

$$P_{00}(k) = 1 + \frac{1 - e^{-k_{\parallel}L}}{k_{\parallel}L} \tag{6.94}$$

More complex function forms of $P_{qq}(k)$ exist for higher mode numbers. If we neglect the exchange, the dispersion relation for the lowest modes results in the so-called Damon-Eshbach (DE) surface magnetostatic modes, which is expressed in the form:

$$\left(\frac{\omega_{DE}}{\gamma}\right)^2 = \mu_0^2[H(H + M) + M^2(1 - e^{-2k_{\parallel}L})/4] \tag{6.95}$$

When the film is magnetized in the plane with $\mathbf{k}_{\parallel} \perp \mathbf{M}$, spin wave modes can be divided into dipole dominated modes ($k = 0$), with frequencies expressed in Eq. (6.95), and exchange dominated modes $(k > 0)$, with frequencies given by the perpendicular standing spin wave modes (PSSW):

$$\left(\frac{\omega}{\gamma}\right)^2 = \mu_0^2 \left\{ H + D\left[k_\parallel^2 + \left(\frac{n\pi}{L}\right)^2\right]\right\}\left\{ H + \left[D + H\left(\frac{M/H}{n\pi/L}\right)\right]k_\parallel^2 + D\left(\frac{n\pi}{L}\right)^2 + M\right\}$$

$$(6.96)$$

The above situation pertains to the thin film case in the low amplitude limit. For larger amplitude oscillations, the spin waves become non-linear and the spectrum will also be amplitude-dependent.

For the ferromagnetic resonance mode in thin films, the resonance condition can be expressed in the simple Kittel form as:

$$\left(\frac{\omega}{\gamma}\right)^2 = \mu_0^2[H\cos(\theta - \Theta_H) + M_{\text{eff}}\cos^2\theta][H\cos(\theta - \Theta_H) + M_{\text{eff}}\cos 2\theta]$$

$$(6.97)$$

From this, the resonance field is obtained, and therefore Eq. (6.97) can be used to fit angle-dependent FMR measurements, once the equilibrium conditions of the magnetization have been determined from the free energy of the system. For the special cases, where the external field is applied in the film plane, $\Theta_H = 90°$, which, in addition, is the easy axis of magnetization, $\theta = 90°$, Eq. (6.97) reduces to

$$\left(\frac{\omega}{\gamma}\right)^2 = \mu_0^2 H(H - M_{\text{eff}})$$

$$(6.98)$$

If the external field is applied along the film normal, $\Theta_H = 0°$, which is an easy axis, $\theta = 0°$, Eq. (6.97) yields:

$$\left(\frac{\omega}{\gamma}\right) = \mu_0(H + M_{\text{eff}})$$

$$(6.99)$$

In Figure 6.17, we show the variation of the resonance frequency as a function of the external field for a system with easy axis out-of-plane (left-hand panel) and one having an easy axis in-plane (right-hand panel) as expressed in Eq. (6.97). In the right-hand panel, we note that the easy axis corresponds to the case of $\Theta_H = 0°$, while the hard axis is for $\Theta_H = 90°$. We note, that the cusp of this latter occurs for the anisotropy field, $H_K = 2K_u/\mu_0 M$, where K_u is the uniaxial anisotropy strength. We also note, that the critical frequency at zero field is also related to the anisotropy field and can be expressed as: $\omega_K = \gamma\mu_0 H_K$. We see, therefore, that there is a strong orientational dependence of the resonance line as a function of the direction of the applied field

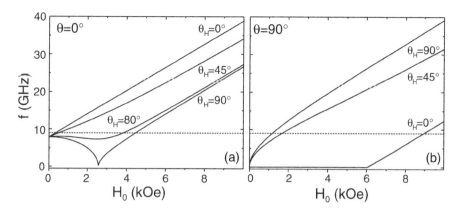

FIGURE 6.17 Resonance frequency as a function of the external field for a film having an easy axis out-of-plane (a) and an easy axis in-plane (b) for several orientations of the external magnetic field, Θ_H. Reprinted from J. Lindner, & K. Baberschke, (2003). *J. Phys.: Condens. Matter 15*, R193. Journal of Physics: Condensed Matter by American Institute of Physics; Institute of Physics (Great Britain) Reproduced with permission of IOP Publishing in the format Book via Copyright Clearance Center.

and is directly related to the sample magnetic anisotropies. It is frequent to perform angle-dependent studies at say a fixed frequency, which is one of the principal early uses of FMR, to elucidate the nature of the magnetic anisotropies in magnetic systems. Care must be taken that the frequency of measurement adheres to the condition $\omega > \omega_K$, since below this point, the full angular measurement will not yield the correct results. Indeed, for the hard axis, such a measurement would give two resonances one at low field and one at higher fields. The easy axis would not reveal any resonance at all. For the in-plane measurement, we can rotate the field within the plane to obtain the field values and there is no frequency requirement. In Figure 6.18, we show a good example of such measurements in a Fe_3Si thin film. Fitting to the experimental data then allows the extraction of the values of the anisotropy constants. In this case of a 4 nm film, the cubic anisotropy constants were found to be $K_4/M = 3.5$ mT, which is superimposed by a small uniaxial in-plane anisotropy field $K_2/M = 0.45$ mT, while for the 40 nm film, we have $K_4/M = 3.9$ mT and $K_2/M = 0.08$ mT.

The model of surface anisotropy developed by Puszkarski (1979) defines a surface pinning parameter defined as:

$$\mathscr{A} = 1 - \frac{L^2}{DM}\mathbf{H}_{surf} \cdot \mathbf{M} \tag{6.100}$$

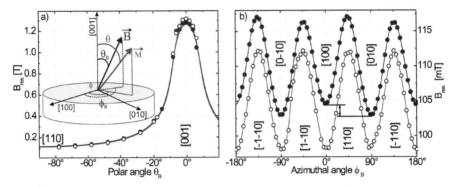

FIGURE 6.18   (a) Polar and (b) azimuthal angular dependencies of the resonance field of a 4 nm (closed symbol) and a 40 nm (open symbol) Fe_3Si film measured at a microwave frequency of 9.9 GHz. The solid lines are fits to the experimental data. Error bars are smaller than the symbol size. The inset shows the coordinate system used for the data analysis. Reprinted figure with permission from Kh. Zakeri, I. Barsukov, N. K. Utochkina, F. M. Römer, J. Lindner, R. Meckenstock, U. von Hörsten, H. Wende, W. Keune, M. Farle, S. S. Kalarickal, K. Lenz, & Z. Frait, (2007). *Phys. Rev B, 76*, 214421, ©(2007) by the American Physical Society.

where H_{surf} is an effective surface field for spin at the film boundary and differs from the bulk effective field. By considering the wave functions of the spin waves and their boundary conditions, it is possible to determine the spin wave mode profile and relative intensity as a function of the surface pinning parameter. Identification of the surface energy and its relation to the effective field allows this to be treated in the same framework as the Rado–Weertman equation, Eq. (6.46), and the boundary equation can be established in the form:

$$L\partial_n m_{surf} - (\mathscr{A} - 1)m_{surf} = 0 \qquad (6.101)$$

where m_{surf} is the amplitude of the transversal (dynamic) surface component of the magnetization. A calculation of the mode profiles and spectra, depicted as intensity vs. mode energy, is illustrated in Figure 6.19.

We will now consider the case of magnetic nanodots. As we discussed in Section 4.6.1, ferromagnetic nanodots have a limited spatial extent in the three spatial directions and provide further conditions for magnetic confinement, when compared to magnetic thin films, which have confinement in only one direction, i.e., that in the direction perpendicular to the film plane. In addition to this, the proximity of nanodots can also provide a further con-

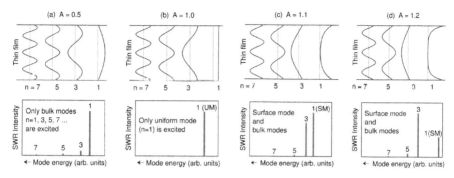

FIGURE 6.19 Profiles of the lowest spin-wave resonance modes (top) and the corresponding SWR spectra (bottom) depicted separately for different values of the surface pinning parameter, $\mathscr{A}$, taken along the in-plane pinning path. The spectra only exhibit peaks corresponding to odd symmetric modes, n = 1, 3, 5, 7. The calculations are performed for L = 75 layers in the film. UM denotes uniform mode, and SM surface mode. Reprinted from H. Puszkarski, & P. Tomczak, (2017). *Surf. Sci. Rep.*, 72, 351, ©(2017), with permission from Elsevier.

tribution to the free energy of the system in the form of dipolar interactions between the nanodot elements.

In the case of a circular magnetic nanodot, we previously noted, that the ground state frequently takes on the form of a zero magnetostatic energy vortex mode with a single vortex core. Such a system has been studied by many research groups and here, we show some representative results. In a system with a radius corresponding to $R = 51a$, where a is the simple cubic lattice constant, by taking into account exchange and dipolar interactions, which are characterized by the parameter:

$$d = \frac{(g\mu_B)^2 \mu_0}{8\pi a^3 J} \tag{6.102}$$

the low energy eigenmodes of the system can be evaluated as a function of the mode numbers. This is illustrated in Figure 6.20(a), where the modal patterns are illustrated for the lowest energy modes (m, n), for the mode labeling in terms of the radial (n) and azimuthal (m) orientations. We note, that the intensity profiles are non-uniform for the $(0, 0)$, $i = 8$ mode and are an artefact of the square lattice symmetry. In Figure 6.20(b), the frequency spectrum for the spin waves is illustrated as a function of the dipolar-to-exchange interaction ratio, as given in Eq. (6.102).

Perpendicular measurements have been performed on circular nanodot structures, in which additional SWR modes are observed with respect to the

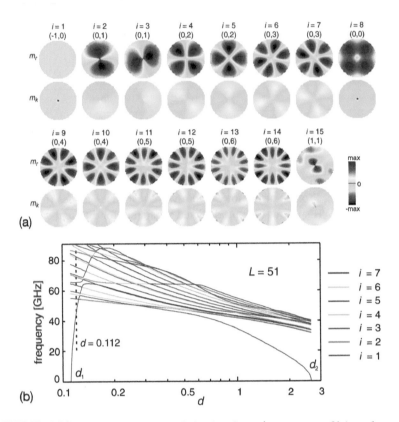

(a)

(b)

FIGURE 6.20   (a) Distribution of the in-plane (m_r, top profile) and out-of-plane (m_k, bottom profile) amplitude of precession of elementary magnetic moments for 15 lowest-frequency normal modes ($i = 1$ to 15) in a dot of size $L = 51$ in the in-plane vortex state for $d = 0.112$. (b) Spin-wave frequency of 24 lowest modes versus the dipolar-to-exchange interaction ratio d (in logarithmic scale) for a 2D circular dot of size $L = 51$. The elementary magnetic moments are assumed to form an in-plane vortex. The lack of zero-frequency modes is indicative of the stability of the assumed magnetic configuration. The dashed line marks the value of $d = 0.112$, for which the profiles of 15 lowest modes are presented (a). The color assignment of the first seven mode lines is indicated at the left; the colors repeat cyclically for successive modes. Reprinted from S. Mamica, J.-C. S. Lévy, & M. Krawczyk, (2014). *J. Phys. D: Appl. Phys., 47*, 015003. Journal of Physics D: Applied Physics by Institute of Physics and the Physical Society; Institute of Physics (Great Britain) Reproduced with permission of IOP Publishing in the format Book via Copyright Clearance Center.

continuous film of the same thickness in Ni and NiFe samples. In Figure 6.21, we show the SWR spectrum for a permalloy film, which has been patterned with circular dots of about 1 m. The additional modes arise from the confined

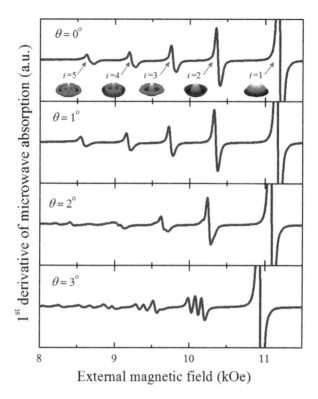

FIGURE 6.21 Spin wave resonance spectra of a patterned permalloy film. Evolution of the 1st derivative of the spin-wave spectra of an array of circular dots with an increase of angle θ. The main mode with index $i = 1$ was partially cut to make all other modes clearly visible. Inserts show zero-order Bessel-function profiles for corresponding circular Òdrumheadó modes. Reprinted from S. A. Bunyaev, V. O. Golub, O. Yu. Salyuk, E. V. Tartakovskaya, N. M. Santos, A. A. Timopheev, N. A. Sobolev, A. A. Serga, A. V. Chumak, B. Hillebrands, & G. N. Kakazei, (2015). *Scientific Reports, 5*, 18480, under the Creative Commons Attribution 4.0 International License.

in-plane geometry. It was found, that dot separation affects only absolute, but not relative positions of the resonance fields, implying that the DDI between dots creates an additional effective field perpendicular to the sample. The spectral separation between resonance modes is governed by the size, shape and magnetic properties of the dots. Using dipole-exchange theory, outlined above, the spectra are accounted for, where the in-plane wave vectors are quantized due to the lateral spatial confinement. For a perpendicularly magnetized film, the dispersion relation for spin waves takes the following form of Eq. (6.90). Due to lateral confinement, the in-plane wave vectors will now be quantized; $k \to k_m; m = 1, 2, 3, ...$, and an inhomogeneous demagnetizing field can be expected for a non-ellipsoidal shape, for which the internal field will be:

$$H_i = H_0 - MN(\rho) + H_{K\perp} \qquad (6.103)$$

where $N(\rho)$ is the effective demagnetizing factor and $H_{K\perp}$ is the perpendicular anisotropy field. For strong dipolar pinning at the edges, the eigenmodes of the disk-shaped dot will be expected to have the form of zeroth-order Bessel functions; that is, the mode profiles will be of the form: $\mu_m(\rho) = J_0(k_m\rho)$, where $k_m = \beta_m/R$, R is the dot radius and β_m are the roots of the zeroth-order Bessel function, $J_0(\beta_m) = 0$. Since the internal field is inhomogeneous, being dependent on the coordinate-dependent effective demagnetizing field, we can expect $H_i \to H_{i,m}$ to be quantized, where $N(\rho) \to N_m$ and

$$N_m = \frac{1}{A_m} \int_0^R N(\rho) J_0^2(k_m\rho) \rho \, d\rho \qquad (6.104)$$

with $A_m = R^2 J_1^2(k_m)/2$. With a variation of the orientation of the applied magnetic field the resonance-field values can be seen to shift and there is a splitting of the spin-wave modes in to subtends, as shown in Figure 6.22 (Bunyaev et al., 2015).

Antidot systems present a unique type of magnetic systems, in which long-range (dipolar interactions) and short-range (confinement) effects can be studied. Experimental results using FMR typically give multi-peaked resonance spectra with strong angular dependencies. In the study of Yu et al. (2003), samples with 1.5 m diameter holes in square and rectangular arrays separated by 3–7 m were measured. The resulting spectra were always double-peaked, with intensity and amplitude of the angular variation dependent on the separation, see Figure 6.23. The symmetries show an in-plane twofold symmetry, which is related to the symmetry of the arrays, however, the two resonance lines display an opposing phase in terms of the easy and hard magnetic axes. Simulations (OOMMF) of the antidot structure show, that there are two regions for the distribution of the demagnetizing dipolar fields, as shown in Figures 6.24(a) and 6.24(b) for a 3 × 4 m antidot mesh. When the magnetization is saturated along with the long axis (a), a larger portion of the film (region A) has an additional internal field which opposes the applied field causing the resonance field to shift to higher values, while the region between the short axis (region B) tend to align along the applied field giving a lower resonance field. When the field saturates the sample along the

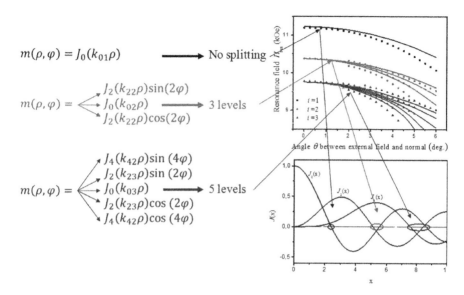

$$m(\rho, \varphi) = J_0(k_{01}\rho) \longrightarrow \text{No splitting}$$

$$m(\rho, \varphi) = \begin{cases} J_2(k_{22}\rho)\sin(2\varphi) \\ J_0(k_{02}\rho) \\ J_2(k_{22}\rho)\cos(2\varphi) \end{cases} \longrightarrow \text{3 levels}$$

$$m(\rho, \varphi) = \begin{cases} J_4(k_{42}\rho)\sin(4\varphi) \\ J_2(k_{23}\rho)\sin(2\varphi) \\ J_0(k_{03}\rho) \\ J_2(k_{23}\rho)\cos(2\varphi) \\ J_4(k_{42}\rho)\cos(4\varphi) \end{cases} \longrightarrow \text{5 levels}$$

FIGURE 6.22 Explanation of the dependence of the number of split peaks on the mode number. The left panel: analytical expressions of the profiles for three first modes. The lowest mode (black letters) is not degenerated in the case of perpendicular field. The second mode (red letters) contains the radially symmetrical Bessel eigenfunction ($n' = 0$) and two Bessel eigenfunction with sine and cosine angular dependence ($|n'| = 2$). Very close zeros of these Bessel functions (red oval in the right bottom panel) determine almost equal values of the radial wave vectors, which in turn leads to the almost equal frequencies, i.e., the second mode is degenerated three times. In the same way, the third mode (blue letters) contains five Bessel eigenfunctions with correspondent indices. The right top panel: the splitting of the degenerated modes described above in the slightly canted magnetic field. The right bottom panel: zeros of Bessel functions, which determine the radial wave vectors, are divided into groups (by black, red and blue ovals), corresponding to the three first split modes. Reprinted from S. A. Bunyaev, V. O. Golub, O. Yu. Salyuk, E. V. Tartakovskaya, N. M. Santos, A. A. Timopheev, N. A. Sobolev, A. A. Serga, A. V. Chumak, B. Hillebrands, & G. N. Kakazei, (2015). *Scientific Reports, 5,* 18480, under the Creative Commons Attribution 4.0 International License.

short axis direction (b), the smaller region along the short axis opposes the applied field, while the larger region also opposes the applied field, but with smaller intensity. The angular variation of the dipolar field, H_d, is shown in Figure 6.24(c) and displays a very similar variation to the resonance field. Given this variation, the authors associate the larger resonance field with the larger region (curve A) and the weaker resonance with the smaller region (curve B). The latter has a larger amplitude variation due to the closer prox-

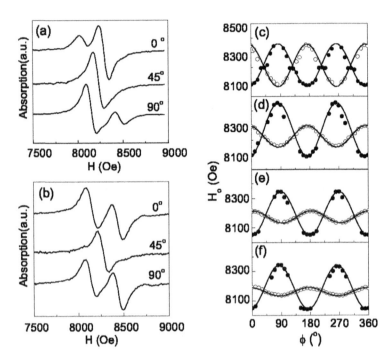

FIGURE 6.23 FMR spectra of (a) rectangular 3 × 4 m and (b) square 3 × 3 m antidot array with the applied in-plane field at various angles, ϕ, from the long axis; (c)-(f) the in-plane angular dependence of resonance fields extracted from FMR spectra for different hole mesh. The solid and open circles are the data from the weak and main resonance peaks, respectively, and the solid lines represent theoretical fits. Figure reprinted from M. J. Pechan, *Magnetic Nanostructures and Spintronic Materials*, Technical DOE Closeout Report (1986–2014), https://www.osti.gov/servlets/purl/1236143 (accessed on 30 March 2020).

imity of the antidots. As such, this explains also the origin of the orthogonal uniaxial anisotropies of the two resonances. The results for the square mesh will thus appear four-fold. The fact, that the resonances cross and are independent implies, that they are uncoupled modes and as such will resonate in their local dipole fields. This is borne out by the fact, that the square sample has two resonances of roughly equal intensity, while the rectangular arrays have different amplitudes, which arise from the disparity in the two different zones. The data were modeled using the LL equation and a free energy density of the form:

$$E = -\mu_0 HM \sin \theta \cos(\phi - \Phi_H) - K_\perp \sin^2 \theta - K_u \sin^2 \theta \cos^2 \phi \qquad (6.105)$$

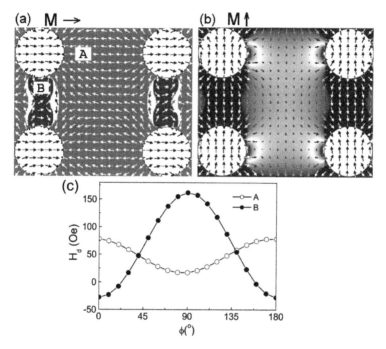

FIGURE 6.24 Micromagnetic simulation of the 3 × 4 m sample with magnetization along(a) long axis, $\phi = 0°$, and (b) short axis, $\phi = 90°$. Shading and arrow size indicates the induced dipole field strength. The antidot sample is approximately divided into two regions in terms of the demagnetization field orientation and amplitude. Figure reprinted from M. J. Pechan, *Magnetic Nanostructures and Spintronic Materials*, Technical DOE Closeout Report (1986–2014), https://www.osti.gov/servlets/purl/1236143 (accessed on 30 March 2020.

where all symbols have their usual meaning. The fits are shown in Figures 6.23(c) and 6.23(f). The short direction is unchanged in all samples measured and has a uniaxial-anisotropy field of $H_K = 2K_u/\mu_0 M = 190$ Oe. The second uniaxial component due to the larger spacing in the rectangular mesh has an anisotropy field, which varies as $1/r^3$, indicating the dipolar origin of this effect.

6.8 Magnonics

The periodic structure of the antidot array allows it to have particular properties, which depend on the materials and importantly the periodicity of the array. Indeed, the propagation of spin waves in such periodic systems is analogous to the propagation of light in photonic crystals. In the same manner, such systems are often referred to as *magnonic crystals* and the topic is called *magnonics*. In parallel with other forms of periodic structures, the magnetic crystal must be designed to perform specific functionalities related to the order of the antidot arrangement. To be able to realize magnetic waveguides and active spin-wave materials, it is important to control the interplay of localization and delocalization of the dynamic modes in the magnetic crystal. This will depend not only on the periodic array symmetry, but also on the filling fraction of the antidot lattice, i.e., their size. When this is large, spin waves can be trapped over a large field range at places, where the internal field is reduced. This can be seen as the parallel to the forbidden states, that are common to periodic structures.

The solutions to the LLG, Eq. (6.70), in the magnetostatic limit yield a manifold of dynamic solutions. Indeed, we saw earlier the solution for the case, where the wave vector has the geometry $\mathbf{k}_\parallel \perp \mathbf{M} \parallel \mathbf{H}$, we obtain the DE modes, as was indicated in Eq. (6.95). For the case of $\mathbf{k} \parallel \mathbf{H}$, the solution yields the so-called backward volume modes (BVM) or backward-volume magnetostatic wave (BVMSW). This will lead to a negative dispersion due to a phase that travels backward. Between these two limits, a spin wave manifold is observed due to the continuous change of angle from parallel to perpendicular, as illustrated in Figure 6.25. The unusual energy dependence on the angle is due to dipolar interactions. For wavelengths below the 1 m level, the two curves eventually merge and exchange interactions become dominant. In thin films, the magnetic confinement in the perpendicular direction is dominated by the exchange interaction, while in lateral directions, there is no such restriction, and this is where the dipolar magnetostatic spin waves (MSSW) are formed.

As we have seen, spin waves can have wavelengths ranging over several orders of magnitude, from tens of microns to less than 1 nm. This means, that their frequencies vary from the GHz regime (for dipolar interactions) to the THz range (in exchange dominated interactions). It will also be noted, that the frequency can be adjusted by an external magnetic field.

It is possible to construct the band diagram for magnons in periodic structures in an analogous manner to that illustrated in Section 3.3 for photonic crystals. The difference here is the nature of the interactions and the dimensions of the periodic structure. As we mentioned above, for larger dimensions in the micron range, these will be dominated by dipolar effects while for structures in the 10–100s of nm, exchange forces will be most important. A schematic illustration is illustrated in Figure 6.26.

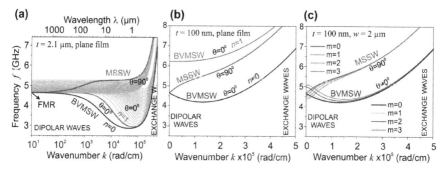

FIGURE 6.25 Spin wave dispersion characteristics for an infinite in-plane magnetized yttrium iron garnet (YIG) film of 2.1 m thickness (a), an infinite film of 100 nm thickness (b), and a 2 m wide spin-wave waveguide of 100 nm thickness (c). A magnetic field of 100 mT is applied in-plane, the saturation magnetization is 140 kA m^{-1}, the exchange constant is 3.6 pJ m^{-1}, and θ is the angle between the spin-wave propagation direction and the magnetization. (a) A logarithmic scale is used for the wavenumber. The blue line shows the lowest $n = 0$ thickness mode of a backward-volume magnetostatic wave (BVMSW) propagating along the magnetic field; the green line shows the first $n = 1$ BVMSW thickness mode and the red line shows the magnetostatic surface spin wave (MSSW) propagating transversely to the orientation of the magnetic field. Twentynine higher-order thickness modes are shown in gray. (b) A linear scale is used for the wavenumber. The dispersions for the MSSW, as well as zero $n = 0$ and first $n = 1$ BVMSW modes, are shown. (c) The dispersions for the zero thickness mode $n = 0$ with different width modes m are shown. Reprinted from A. V. Chumak, A. A. Serga & B. Hillebrands, (2017). *J. Phys. D: Appl. Phys., 50*, 244001. Institute of Physics and the Physical Society; Institute of Physics (Great Britain). Reproduced with permission of IOP Publishing in the format Book via Copyright Clearance Center.

In the case of magnetic crystals, the band structure can be obtained by solving the linearized Landau–Lifshitz equation of the form (Lenk et al., 2011):

$$i\Omega m_y - m_z + \frac{M}{H}h_z = 0 \tag{6.106}$$

$$i\Omega m_z + m_y \frac{M}{H}h_y = 0 \tag{6.107}$$

where $\Omega = \omega/(\gamma\mu_0 H)$. The dynamics components of the magnetization and magnetic field are, respectively:

$$m_y(x,y) = m_{0y}e^{i(k_x x + k_y y)} \quad \text{and} \quad m_z(x,y) = m_{0z}e^{i(k_x x + k_y y)} \tag{6.108}$$

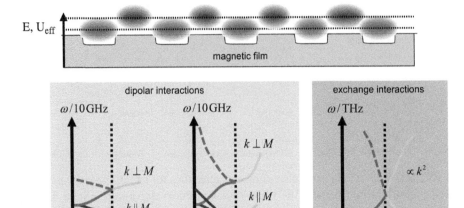

FIGURE 6.26   Schematic spin-wave dispersion for the different length scales in the periodic zone scheme. For a spin-wave Bloch state, prominent effects in the band structure are expected: 10 m range (left), 1 m (middle) and 100 nm (right). Reprinted from B. Lenk, H. Ulrichs, F. Garbs, & M. Münzenberg, (2011). *Phys. Rep.*, *507*, 107, ©(2011), with permission from Elsevier.

$$h_y(x,y) = -m_y \frac{kL}{2} e^{i(k_x x + k_y y)} \sin^2\theta \quad \text{and} \quad h_z(x,y) = -m_z \left(1 - \frac{kL}{2}\right) e^{i(k_x x + k_y y)}$$

$$(6.109)$$

The dispersion relation derived from this is given by:

$$\left(\frac{\omega}{\gamma}\right)^2 = \left(H + M\frac{kL}{2}\sin^2\theta\right)\left(H + M - M\frac{kL}{2}\right) \qquad (6.110)$$

where θ denotes the angle between the external field and the wave vector. It is then necessary to define the periodic modulation in the magnetization, which constitutes the periodic medium. This can be expressed in the form:

$$M(\mathbf{r}) = \sum_{\mathbf{G}} M(\mathbf{G}) e^{i\mathbf{G}\cdot\mathbf{r}} \qquad (6.111)$$

where $\mathbf{G}$ is a reciprocal lattice vector of the periodic structure; $\mathbf{G} = \mathbf{G}_{nm} = (2\pi/a)n\hat{\mathbf{x}} + (2\pi/b)m\hat{\mathbf{y}}$. With this definition, any two-dimensional lattice structure can be defined, much in the way we did for the 2D surface vectors in Section 3.3 of Volume 1. The generalization of this equation for a periodic material is a Bloch wave expansion of the form:

$$\mathbf{m}(\mathbf{r}) = \sum_{\mathbf{G}} \mathbf{m}_{\mathbf{k}}(\mathbf{G}) e^{i(\mathbf{k}+\mathbf{G})\cdot\mathbf{r}} \qquad (6.112)$$

The calculation of the band structure proceeds as with other periodic structures to define the allowed frequencies as a function of the wave vectors. Therefore, the nature of the material and its periodic structure are required. An example is illustrated in Figure 6.27.

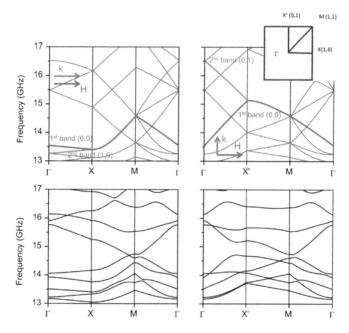

FIGURE 6.27 Band structure calculation for a CoFeB film for a two-dimensional square lattice. Top: Free spin wave band structure with first and second band marked in red and orange, respectively. Bottom: Solved by a set of Bloch states for a film thickness of $L = 50$ nm, hole distance $a = 3.5$ m and hole diameter $d = 1$ m. The splitting at the high symmetry points X, X' at the zone boundary is marked with the shaded yellow area. Reprinted from B. Lenk, H. Ulrichs, F. Garbs, & M. Münzenberg, (2011). *Phys. Rep., 507*, 107, ©(2011), with permission from Elsevier.

The experimental measurement of the magnon dispersion relation is frequently performed using the Brillouin Light Scattering (BLS) method, also in conjunction with VNA-FMR. In Figure 6.28, we show an example of a square antidot lattice (ADL) of permalloy coupled to an AFM layer of FeMn.

The antidot structure is formed by the partial etching of periodic holes in the permalloy.

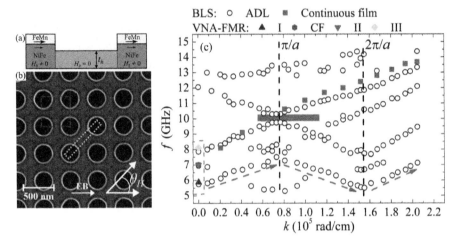

FIGURE 6.28   (a) Schematic of the cross-section view of the partially etched ADL. The ADL consists of a fully etched FeMn layer adjacent to a partially etched NiFe layer. The thickness of the remaining NiFe is labeled as t_h. A nonzero exchange bias field, H_b, is expected right underneath the FeMn layer, while in the regions from which the FeMn has been removed, the expected value for H_b is zero. (b) SEM image of the NiFe(20 nm)/FeMn(10 nm) ADL fabricated. The red dashed square illustrates the ADL unitary cell, whose side is 420 nm long. The hole diameter is 280 nm. (c) BLS measurements at $\theta_H = 0°$, while in the presence of an external field $\mu_0 H = 50$ mT, of the ADL (open circles) and the continuous film (red squares). For $k = 0$, the resonances match the FMR data. The periodicity of the lattice causes a change in the slope of the dispersion modes at $k = \pi/a$ and $2\pi/a$ as highlighted by the red dashed arrows. The magnonic band gap is highlighted by the gray shaded area at $(\pi/a$ rad cm^{-1}, ~ 10 GHz). The spin-wave vector is perpendicular to the applied field direction (DE configuration). Reprinted figure with permission from F. J. T. Gonçalves, G. W. Paterson, R. L. Stamps, S. O'Reilly, R. Bowman, G. Gubbiotti, & D. S. Schmool, (2016). *Phys. Rev. B, 94*, 054417, ©(2016) by the American Physical Society.

Many examples of the application of magnonics can be found in the literature, see for example the review article by Chumak et al. (2017). As an example, we show the case for a 1D dynamic magnonic crystal, constructed from a YIG strip, which acts as a spin-wave waveguide. Onto this is deposited a meandering wire, which, when traversed with a current, will produce a magnetic field. This magnetic field will have the structure of the wire and interfere with the effective field inside the YIG waveguide. By controlling the current in the

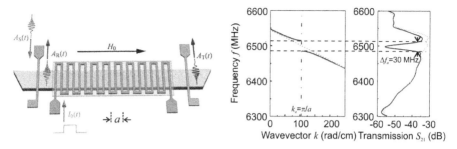

FIGURE 6.29 Schematic of the dynamic magnonic crystal comprising a planar current-carrying meander structure with 20 periods with a lattice constant $a = 300$ m (10 shown), positioned close to the surface of the YIG spin-wave waveguide. Right panel: Dispersion characteristic and measured spin-wave transmission characteristic of this dynamic magnonic crystal for 1 A current sent through the meander structure. Reprinted from A. V. Chumak, V. S. Tiberkevich, A.D. Karenowska, A. A. Serga, J. F. Gregg, A. N. Slavin, & B. Hillebrands, (2010). *Nat. Comm., 1*, 141, under a Creative Commons license.

wire, the potential of the periodic field in the waveguide can be altered, hence it being nominated as a dynamic magnetic crystal. In Figure 6.29, we show a schematic illustration of the structure and the dispersion relation as measured using a VNA shows a rejection band, which corresponds to the band-gap in the magnetic structure. The position of this rejection band can be tuned using the current and can also be switched on and off very rapidly ($<$ 10 ns). Another form of a magnonic crystal can be induced acoustically using a surface acoustic wave (SAW), which interferes with the magnons via the Doppler effect, causing a shift in their frequencies.

6.9 Ferromagnetic Resonance in Magnetic Nanoparticle Assemblies

There are many experimental studies of nanosystems by ferromagnetic resonance. In this section, we aim to give some representative results for a variety of magnetic nanoparticle systems.

It is well-known, that fine particle systems suffer from superparamagnetic effects. Such behavior is characterized by the thermal fluctuation of the spontaneous magnetization. As temperature increases the thermal energy, $k_B T$, becomes sufficient to overcome the energy barrier, E_B, defined by the magnetic anisotropies and particle size. These thermal excitations induce rapid fluctuations of the particle magnetic moment and, in the simplest case, are described by the Arrhenius law:

$$\tau = \tau_0 \exp\left(\frac{E_B}{K_B T}\right) \tag{6.113}$$

Typically E_B is taken to be $K_{eff}V$, where K_{eff} is an effective anisotropy and V the particle volume. The characteristic time, τ_0, depends on various parameters; temperature gyromagnetic ratio, magnetization, energy barrier, the direction of the applied field and damping constant. Most authors take τ_0 to be constant for simplicity (with a value of 10^{-11} s), where this assumption along with Eq. (6.113) is known as the Néel–Brown model. If the time window of experimental measurement, τ_m, is shorter than τ at a fixed temperature, the particle's magnetic moment remains blocked during the observation. This is the so-called blocked regime, where an assembly of identical non-interacting particles would appear ferromagnetic. Since this regime is temperature-dependent, a blocking temperature, T_B, is defined as that temperature below which the sample appears ferromagnetic, and will be specific to the measurement technique. At this temperature, $\tau = \tau_m$ and the blocking temperature can be defined from Eq. (6.113) as:

$$T_B = \frac{E_B}{k_B}\frac{1}{\ln(\tau_m/\tau_0)} \sim 0.43\frac{E_B}{k_B} \tag{6.114}$$

where we have used $\tau_m = \tau_{FMR} \approx 10^{-10}$ s. In Figure 6.30, the effect of size distribution is illustrated, where a distribution of blocking temperatures is predicted for ferromagnetic resonance and SQUID magnetometry (Antoniak et al., 2005). It will be noted, that this distribution arises from the volume distribution given for example by a log-normal distribution and that $T_B = T_{max}$. Above the blocking temperature, the rapid fluctuations produced by thermal excitations will mean, that the particle's magnetic moment reverses between local minima so rapidly, that its behavior mimics atomic paramagnetism and the particle is said to be in the superparamagnetic (SPM) state.

Introducing interactions between the particles can be shown to modify Eq. (6.113) to give the form of a Vogel–Fulcher law (Dormann et al., 1988): $\tau = \tau_0 \exp[E_B/k_B(T_B - T_0)]$, in which T_0 is an effective temperature proportional to H_i^2, where H_i is the effective interaction field. T_0 increases with interaction strength. This approach is valid in the weak interaction limit. Models of FMR in nanoparticle assemblies by de Biasi and Devezas (1978) and Berger et al. (1997), which are valid for strong fields or weak anisotropies

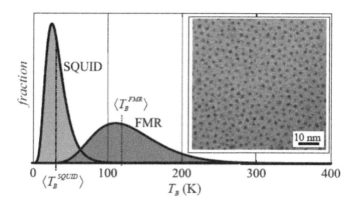

FIGURE 6.30 The effect of size distribution is illustrated where a distribution of blocking temperatures is predicted for FMR and SQUID magnetometry. The calculations use $\tau_{SQUID} \sim 10^2$ s, $\tau_{FMR} \sim 10^{-10}$ s, where the inset shows a transmission electron microscope (TEM) image of the FePt NPs, from which the size distribution is determined. Reprinted from C. Antoniak, J. Lindner, & M. Farle, (2005). *Europhys. Lett., 70*(2), 250–256. ©(2000) EDP Sciences.

(magnetic moment parallel to the applied field, H) in spherical nanoparticle systems, represent the resonance condition in the form:

$$\frac{\omega}{\gamma} = H + H_A(\psi) \tag{6.115}$$

with the anisotropy field being given as $H_A(\psi) = (K/M)\left[3\cos^2\psi - 1\right]$, ψ being the angle between the anisotropy axis and the magnetization. In the SPM regime, thermal fluctuations produce a dynamic narrowing of the resonance spectra, in which

$$H_A \rightarrow H_A^{SPM} = H_A(\psi) < P_2(\cos\vartheta) >= H_A(\psi)\frac{[1 - 3L(x)/x]}{L(x)} \tag{6.116}$$

$\langle P_2(\cos\vartheta)\rangle$ is the 2nd order Legendre polynomial and $L(x)$ is the Langevin function: $L(x) = \coth(x) - 1/x$, where $x = MVH/k_BT = mH/k_BT$. A realiztic model of the SPM regime requires that the magnetization of the NP assembly be defined so as to take into account the effects of size distributions such that:

$$M \rightarrow M_{SPM} = M_s^{bulk}\int_0^\infty L(x)P(m)dm \tag{6.117}$$

$P(m)$ follows the log-normal variation for the particle assembly and $L(x)$ is the Langevin function. The model of Kliava and Berger (1999) uses the spectral function for an assembly of SPM nanoparticles of the form:

$$\frac{dP}{dH} \int_{\varphi} \int_{\vartheta} \int_{V} F\left[H - H_r(V, \vartheta, \varphi), \triangle H\right] P(V) \sin \vartheta \, dV d\vartheta d\varphi \qquad (6.118)$$

The lineshape function, $F[H - H_r(V, \vartheta, \varphi), \triangle H]$, can be Gaussian or Lorentzian, with H_r being the resonance field and $\triangle H$ the individual linewidth for a particle of a given size. In Eq. (6.118), P represents the resonant absorption of a particular line, whose intensity can be expressed as $I_n = \int P_n dH$. However, a more practical form of the intensity can be expressed in the empirical form: $I_n = I_{pp}^{(n)} \left(\triangle H_{pp}^{(n)}\right)^2$, where "pp" are the peak-to-peak extrema in the derivative absorption spectra (dP/dH vs. H). Figure 6.31 shows some predicted and experimental curves. The spectra in general display a sharp and a broad resonance, which are understood to arise from SPM and blocked particles in polydisperse samples.

In Figure 6.32, the temperature variation of the FMR resonance lines and linewidth, along with some representative spectra, are shown for $\gamma - Fe_2O_3$ nanoparticles with an average diameter of 6.4 nm (Dutta et al., 2004). Multi-peaked spectra are common in FMR studies of NP systems and have been observed by many authors. The high and low field resonance lines have typically very different temperature dependencies, where the high field lines are generally much less sensitive to changes in temperature. Despite this, the

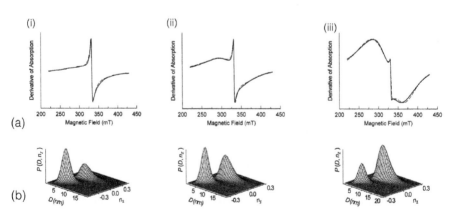

FIGURE 6.31 Computer fit (full line) to the experimental X-band SPR spectrum (dashed line) of the borate glass (a): (i) annealed at 465 °C, (ii) annealed at 475 °C, and (iii) annealed at 480 °C. In all cases, the reconstruction of the best-fit joint distribution density of diameters and demagnetizing factors are shown (b). Reprinted from J. Kliava, & R. Berger, (1999). *J. Magn. Magn. Mater.* 205, 328, ©(1999), with permission from Elsevier.

linewidth variations are quite similar. It is noted, that the transition through the blocking temperature ($T_B \sim 100$ K) does not show any marked behavior in the resonance field, though the linewidth does noticeably increase below this transition. For higher temperatures, the two resonance lines tend to merge, producing strongly overlapping lines. Such results are typical of NP assemblies and it is frequently necessary to use a fitting procedure to separate the various contributions. Further resonances can also be observed in NP FMR experiments and are usually due to surface effects and inhomogeneities in the assembly. The temperature dependence of FMR can be used to show the variation of the effective anisotropy of the particles, which show a strong enhancement at low temperatures (Antoniak et al. 2005; Schmool and Schmalzl 2007).

In experimental measurements of nanoparticle systems, the resonance linewidths are typically very large ($\sim$ 1-2 kOe). The extensive linewidth broadening is principally due to extrinsic effects, being mainly due to the spread of crystalline and consequently anisotropy axes. The characteristic line narrowing observed at temperatures above the blocking temperature is due to the alignment of the magnetic moment (or magnetization vector) of the individual particles under the influence of the external magnetic field, where the thermal energy is sufficient to overcome the anisotropy energy barrier.

Taking into account the dipolar interactions between magnetic nanoparticles in multilayered discontinuous films, Schmool et al. (2007) obtain the resonance condition given by:

$$\left(\frac{\omega}{\gamma}\right)^2 = \frac{\sin\theta\cos(\theta - \vartheta_0)}{\sin\vartheta_0}H_a^2 - 2C\frac{\sin\theta\cos 2\vartheta}{\sin\vartheta_0}H_a \qquad (6.119)$$

where the angular variation of the resonance field depends on the constant C, which is a geometric factor, that can be expressed in terms of the shape factors of the nanoparticles and more importantly, the in-plane and out-of-plane interactions. The constant C, which also depends on sample material constants: magnetization, volume fraction and the particle shape factor, can be expressed as:

$$C = 2\pi\rho M_s\left(1 - \rho\right)\left(N_\| - N_\perp\right) + \frac{1}{M}\left[n\Gamma^{IP} - (n-1)\Gamma^{OP}\right] \qquad (6.120)$$

where $N_\|$ and $N_\perp$ are the demagnetizing factors of the nanoparticle in the parallel and perpendicular orientations, respectively, while Γ^{IP} and Γ^{OP} denote the in-plane and out-of-plane averaged dipolar interactions. Using the same

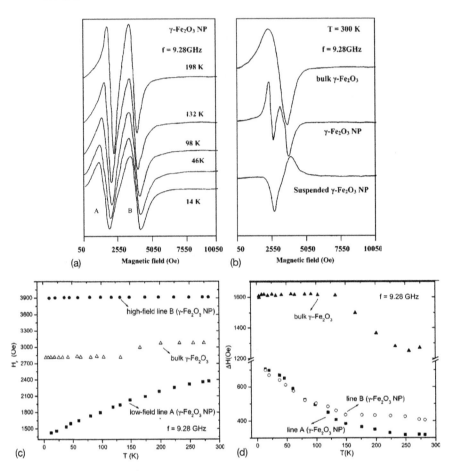

FIGURE 6.32 (a) Absorption spectra for γ-Fe$_2$O$_3$ NP at various temperatures; (b) γ-Fe$_2$O$_3$ bulk, NP, and suspended NP spectra at room temperature; (c) temperature variation of the FMR resonance lines (NP and bulk γ-Fe$_2$O$_3$); and (d) temperature dependence of linewidth. Reprinted figure with permission from P. Dutta, A. Manivannan, M. S. Seehra, N. Shah, & G. P. Huffman, (2004). *Phys. Rev. B, 70*(17), 174428, ©(2004) by the American Physical Society.

approach, Schmool and Schmalzl (2009) found, that for three dimensional arrays of NPs the C constant can be expressed as:

$$C = \frac{\pi \langle r \rangle^3 M_s}{6\rho} V_{mag} = \frac{\pi}{6} \langle r \rangle^3 M_s V \qquad (6.121)$$

ρ is the volume fraction of particles, which is defined as; $\rho = V_{mag}/V$, V being the total volume of the sample and $\langle r \rangle$ is the average particle radius. For non-spherical particles, this constant will have an additional term related to the shape anisotropy. The temperature variation of the resonance is a reflection of the variation of magnetization M in the superparamagnetic regime, which can be expressed using a weighted Langevin function:

$$M = M_s \int L \left(\frac{HMV_{mag}}{k_B T} \right) P(V) dV \qquad (6.122)$$

$P(V)$ represents the log-normal distribution. The comparison of the Langevin function with the experimental data is shown in Figure 6.33 for these γ-Fe_2O_3 samples with $\langle D \rangle = 4.6$ nm. In low temperature measurements for the magnetic nanoparticles in the discontinuous multilayered films, Schmool et al. (2006) observed an enhanced resonance field, the size of which scaled as the inverse of the effective size of the particles. This was taken as an indication of the existence of enhanced surface anisotropy in this system at low temperatures.

Monodisperse fcc Co arrays were studied by FMR with in-plane (azimuthal) and out-of-plane (polar) angular measurements to the effective magnetization and in-plane anisotropy field by Spasova et al. (2002). Regular arrays of Co particles of about 12 nm were obtained by drying a solution of the NP in an applied field of 0.35 T on a grid. The resulting assembly consisted of stripes of regular triangular Co nanocrystals with a width of around 200–250 nm. The lowest resonance field was obtained when the external field was applied along the direction of the stripes; $(H_{res})_{min} = 0.233$ T, which is lower than the EPR field of $\omega / \gamma = 0.3085$ T, showing that an additional intrinsic

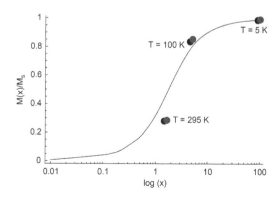

FIGURE 6.33 Experimental data points at the specific temperatures with a fit using the Langevin function (line) from Eq. (6.122). Reprinted from D. S. Schmool, & M. Schmalzl, (2007). *J. Non-Cryst. Solids, 353*, 738, ©(2007), with permission from Elsevier.

magnetic field due to an effective magnetization and an easy-axis magnetization in the film plane is evident. Three considerations are made to the
interpretation of the resonance field: (i) shape anisotropy due to stripes, (ii)
magnetic anisotropy due to interparticle magnetostatic coupling in an fcc-
like lattice inside the stripes, and (iii) the effective magnetic anisotropy of
the individual particles (including shape, volume and surface anisotropy contributions). Assuming, as a first approximation, that all anisotropies due to
spin-orbit coupling vanish, with only that for shape being present, a resonance equation is obtained as given by:

$$\left(\frac{\omega}{\gamma}\right)^2 = [H_{res} + (N_x - N_z)4\pi M_s][H_{res} + (N_y - N_z)4\pi M_s] \qquad (6.123)$$

This function is not sufficient to explain the angular dependence of the FMR
and further contributions are required. By introducing a cubic and uniaxial
symmetry, the in-plane resonance field could be obtained as:

$$\left(\frac{\omega}{\gamma}\right)^2 = \left[H_{res} + 2H_{an}^{4\parallel}\cos 4\varphi - H_{an}^{2\parallel}\cos 2(\varphi-\phi)\right]$$
$$\times \left[H_{res} + H_{eff} + 2H_{an}^{4\parallel}(2-\sin^2 2\varphi) - H_{an}^{2\parallel}\cos^2(\varphi-\phi)\right] \qquad (6.124)$$

where ϕ is the angle of the applied field with the axis of the uniaxial
anisotropy, $H_{an}^{2\parallel}$ and $H_{an}^{4\parallel}$ are the effective uniaxial and four-fold anisotropy
an an fields and the effective field is given by $H_{eff} = -2K_{2\perp}/M_s + 4\pi\rho M_s$,
where ρ is the volume fraction. The fit, shown in Figure 6.34(a), shows good
agreement with experiment where the following values were obtained: $H_{an}^{2\parallel}$
$= 0.037$ T and $H_{an}^{4\parallel} = 0$ and $H_{eff} = 0.127$ T, that is only uniaxial anisotropy is
observed. For the polar dependence, the following resonance equation was
used:

$$\left(\frac{\omega}{\gamma}\right)^2 = [H_{res}\cos(\vartheta - \theta) + H_{eff}\cos 2\vartheta]\left[H_{res}\cos(\vartheta - \theta) - H_{eff}\cos^2\vartheta + H_{an}^{2\parallel}\right]$$
$$\qquad (6.125)$$

The fit, Figure 6.34(b), yields $H_{eff} = 0.13$ T and $H_{an}^{2\parallel} = 0.037$ T, which is
in good agreement with the previous fit and experiment. The difference between the expected $4\pi\rho M_s = 0.222$ T for $\rho = 0.31$ and H_{eff} is accounted for
using the perpendicular anisotropy field $2K_{2\perp}/Ms$ and/or the possible existence of an antiferromagnetic CoO outer layer, which would reduce M_s. Such
core-shell Co-CoO nanoparticles have been further studied by Wiedwald et
al. (2003) using FMR and X-ray magnetic circular dichroism (XMCD) tech-

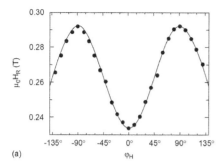

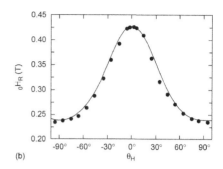

(a)

(b)

FIGURE 6.34 Dependence of the resonance field on the direction of the external magnetic field: (a) in-plane ϕ and (b) out-of-plane θ (measured from the normal to the sample). Reprinted from M. Spasova, U. Wiedwald, R. Ramchal, M. Farle, M. Hilgendorff, & M. Giersig, (2002). *J. Magn. Magn. Mater.* 240, 40–43, ©(2002), with permission from Elsevier.

niques to study the ratio of orbital-to-spin magnetic moment. Essentially, XMCD yields a value of $\mu_L/\mu_S^{eff} = 0.24 \pm 0.06$, which is over three times the value obtained from g-factor analysis of $g = 2.15 \pm 0.015$ corresponding to $\mu_L/\mu_S^{eff} = 0.075 \pm 0.008$. The difference is explained as being due to the presence of uncompensated Co magnetic moments at Co-CoO core-shell interface. High-resolution TEM corroborates the existence of the CoO shell.

The existence of oxide shells in the FMR in nanoparticle systems is demonstrated in Figure 6.35 on Fe nanocubes, where a plasma treatment is used to remove the oxide shell in a vacuum, with FMR being performed in-situ, Antoniak C. et al. (2012), see also Trunova et al. (2008). The explanation of the lineshapes can be explained as arising from a consideration of the distribution of particle axes and summing the various contributions in the final spectrum. The fit shows excellent agreement with the model. Tomita et al. (2004) have studied the FMR in Fe nanogranular films, where Fe NPs are dispersed in a SiO_2 matrix with variable concentrations. For concentrations of 5%, no angular dependence was observed for the FMR, while at 15%, a shift of around 600 Oe at 9.1 GHz is observed. This could be attributed to shape effects or, more likely, to interparticle interactions, which would be expected to be significant for the more concentrated sample. The temperature dependence of the FMR in these two samples were also studied, where for the 5% sample no significant variation was observed as opposed to the 15% sample, which for the perpendicular resonance showed an increase in value with reduced temperatures, while the parallel resonance shows a small downward shift with decreasing temperature. The magnetic percolation is

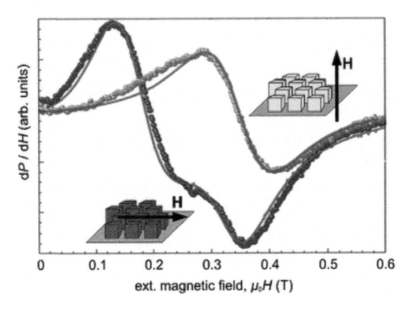

FIGURE 6.35 Experimental FMR spectra (symbols) of plasma cleaned and capped Fe-nanocubes at room temperature for different measurement geometries and simulations (lines). Reprinted by permission from C. Antoniak, et al. (2012) Intrinsic Magnetism and Collective Magnetic Properties of Size-Selected Nanoparticles. In: Lorke A., Winterer M., Schmechel R., & Schulz C. (eds.). *Nanoparticles from the Gas-phase. NanoScience and Technology.* Springer, Berlin, Heidelberg, ©(2012). Springer Nature.

expected with increased concentration and the authors use the Kittel equations to study the more concentrated sample. The shift in the FMR for this sample with decreasing temperature fits to an increase in $4\pi M$ as measured by SQUID.

The temperature dependence of the magnetic anisotropy is seen to be an important parameter in determining the magnetic behavior of nanoparticle assemblies. In the study of Antoniak et al. (2005), FMR and SQUID data are jointly used in the evaluation of τ_0 and K_{eff} in FePt nanoparticles from the two difference blocking temperatures for these two methods. From the analysis of the FMR using the Kittel equation:

$$\left(\frac{\omega}{\gamma}\right)^2 = \left[H_{res}\cos\left(\vartheta - \theta\right) + H_A\cos 2\vartheta\right]\left[H_{res}\cos\left(\vartheta - \theta\right) - H_A\cos^2\vartheta\right]$$

(6.126)

$H_A = 2K_{\text{eff}}/\mu_0 M$ is the effective anisotropy field, which is the uniaxial contribution due to small deviations from the spherical shape, as well as surface and step anisotropies at the particle surface, which are not averaged out. Averaging over the angles θ_H of the external magnetic field gives the numerical relation:

$$H_{res} = H_{res}^0 \left[1 - \left(\frac{H_A}{H_{res}^0} \right)^{1.25} \right]^{0.44} \tag{6.127}$$

here, $H_{res}^0 = \hbar\omega/g\mu_0\mu_B$ and g is the g-factor, which is obtained as 2.054 ± 0.010 from frequency-dependent measurements. Since the intensity of the FMR line is proportional to the magnetization, the blocking temperature was evaluated analyzing the intensity versus temperature. This will give a higher value than SQUID measurements since the time windows for the two methods are very different; $\tau_{FMR} \approx 10^{-10}$ s and $\tau_{SQUID} \approx 10^2$ s. By comparing the two blocking temperatures, see Figures 6.36 and 6.30(a), and using the Arrhenius relationship, of the form of Eq.4.95 , the effective anisotropy constant can be written as:

$$K_{\text{eff}}\left(\left\langle T_B^{SQUID}\right\rangle\right) \approx \frac{27k_B}{V_m} \left[\frac{1}{\left\langle T_B^{SQUID}\right\rangle} - \frac{\alpha}{\left\langle T_B^{FMR}\right\rangle} \right]^{-1} \tag{6.128}$$

In Eq. (6.128), $\alpha = K_{\text{eff}}\left(\left\langle T_B^{FMR}\right\rangle\right)/K_{\text{eff}}\left(\left\langle T_B^{SQUID}\right\rangle\right) = H_A\left(\left\langle T_B^{FMR}\right\rangle\right)/H_A\left(\left\langle T_B^{SQUID}\right\rangle\right)$ with a temperature–dependent K_{eff}, $\alpha = 1$ and V_m as the mean volume. A small deviation of T_B from T_{max} arises from a distribution of sizes. From the FMR data, $\left\langle T_B^{FMR}\right\rangle \approx 110$ K, which is about five times that from the magnetization measurements. The anisotropy constant is shown in Figure 6.36(b) as a function of temperature, with $\alpha = 0.8$ and using Eq. (31), $K_{\text{eff}} = (8.4\pm0.9) \times 10^5$ J/m^3 and from Eq. (6.128), $\tau_0 \approx 1.7 \times 10^{-12}$ s. The experimental values of K_{eff} are found to follow a Bloch law-like dependence, with a power of 2.1; that is $K_{\text{eff}} \propto \left[M_s(1 - T/T_B)^{3/2}\right]^{2.1}$ (Antoniak et al., 2006). The anisotropy was found to be about an order of magnitude higher than in the bulk.

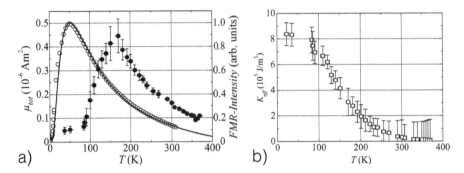

FIGURE 6.36 (a) ZFC measurements of the total magnetic moment using the SQUID (open circles) and FMR (full circles) technique. (b) Temperature-dependent anisotropy constant of $Fe_{70}Pt_{30}$ nanoparticles. Reprinted from C. Antoniak, J. Lindner, & M. Farle, (2005). *Europhys. Lett., 70*(2), 250–256. ©(2000) EDP Sciences.

6.10 Summary

Spin dynamics covers a range of phenomena, which includes ultrafast switching processes and resonance behavior, and generally considers only those processes that are on a short time scale. Typically resonances for ferromagnetic materials are in the range of GHz and ultrafast demagnetization effects can occur in the femtosecond regime. Processes such as domain wall motion and magnetic viscous effects are usually much slower, with domain wall resonances occurring in the 100s of MHz range.

Ferromagnetic resonance (FMR) refers to the precessional motion of the magnetization vector around the effective magnetic field. This process can be described as a classical phenomenon, where the magnetization vector is considered as a normal classical vector processing about a magnetic field under the action of a driving field, which is applied in the transverse direction. This description was initially proposed by Landau and Lifshitz as a cross product of the magnetic field and the magnetization vectors, also referred to as the magnetic torque of the system. The Landau–Lifshitz equation forms the basis of much of the theoretical descriptions of the dynamics in ferromagnetic systems. The relaxation process can also be described in a phenomenological manner using viscous damping terms, such as those introduced by Landau and Lifshitz and later by Gilbert. These approaches have been extremely successful in describing the dynamical behavior of a vast range of materials.

The uniform precession of the magnetic moments in ferromagnetic resonance can be seen as a fundamental mode of excitation of a ferromagnetic system. Higher-order excitations are also possible and can be readily observed in confined magnetic structures. These are described as standing spin-wave modes and give rise to multi-peaked spectra in a resonance experiment, often referred to as spin-wave resonance. A number of techniques are available for the measurement of dynamical effects in magnetism. The specific excitations of standing spin waves depend explicitly on the physical dimensions of the magnetic body as well as the boundary conditions which will determine the values of the allowed wave vectors of the excitations. Surface localized modes can also be observed, depending on the specific boundary conditions, which are related to the surface anisotropy of the magnetic system. The ferromagnetic resonance method has been widely applied to magnetic systems and in particular to the study of magnetic anisotropies and magnetic interactions. Measurements being made as a function of the orientation of the applied magnetic field and the sample temperature.

FMR was traditionally observed using microwave spectroscopy, though in more recent times the vector network analyzer (VNA) has become the preferred method of observation since it allows the manipulation of both the applied magnetic field as well as the frequency of excitation. VNA apparatus can typically function in the 10s of GHz, though more expensive devices can work above 100 GHz. Another increasingly popular experimental method employs ultrashort laser pulses to directly interfere with the magnetic system on very fast time scales. This coupled with pump-probe methodologies allows the direct observation of the variation of the magnetization as a function of time. Many systems function in the 100 fs range, though well-controlled optical systems can be employed with temporal resolutions well below 10 fs. The excitations on the ultrashort time scale concern what is referred to as ultrafast demagnetization. Here the laser pulse rapidly transfers energy to the electron system which reduces the magnetization on the femtosecond timescale.

6.11 Problems

(1) Derive the result that the spin wave energy for a spin wave of wave vector q has the form:

$$\hbar\omega_q = 2J[1 - \cos(qa)]$$

(2) Consider the spin wave spectrum for a 50 Å Fe thin film. Evaluate the expected frequency for the first spin wave mode. Use $D = 280 \, \text{meV Å}^2$.

(3) The dynamics of the motion of the magnetization vector is generally described by one of the following equations:

$$\frac{\partial \mathbf{M}}{\partial t} = -\gamma(\mathbf{M} \times \mathbf{H}_{\text{eff}}) - \frac{\lambda}{M^2} [\mathbf{M} \times (\mathbf{M} \times \mathbf{H}_{\text{eff}})] \quad \text{LL equation}$$

$$\frac{\partial \mathbf{M}}{\partial t} = -\gamma(\mathbf{M} \times \mathbf{H}_{\text{eff}}) + \frac{\alpha}{M} \left(\mathbf{M} \times \frac{\partial \mathbf{M}}{\partial t} \right) \quad \text{LLG equation}$$

Use vector diagrams to explain the various terms of these equations and show under what conditions they can be assumed to be equivalent.

(4) Explain the form of the Bloch–Bloembergen equation of motion and discuss how it fundamentally differs from the phenomenological descriptions given by the equations of motion given in the previous question.

(5) Show that the damping term of the Landau–Lifshitz equation can be expressed in the following form:

$$-\mathbf{M} \times (\mathbf{M} \times \mathbf{H}) = \mathbf{M}^2 \cdot \mathbf{H} - [(\mathbf{M} \cdot \mathbf{H}) \cdot \mathbf{M}]$$

(6) Using only the precessional component of the equation of motion for the magnetization vector, show that the resonance condition for a ferromagnetic material can be expressed in terms of the so-called Smit–Beljers equation:

$$\left(\frac{\omega}{\gamma} \right)^2 = \frac{1}{M_0^2 \sin^2 \theta} \left[\frac{\partial^2 E}{\partial \theta^2} \frac{\partial^2 E}{\partial \phi^2} - \left(\frac{\partial^2 E}{\partial \phi \partial \theta} \right)^2 \right]$$

where E is the magnetic free energy density.

(7) Describe how the existence of a confined magnetic system can provide the necessary conditions for higher order excitations, otherwise known as *standing spin waves*.

(8) For a thin magnetic film with an applied field along the sample normal, it is possible to show that the resonance condition gives rise to the following expression for the resonance field of the nth standing spin wave mode:

$$H_n = \left(\frac{\omega}{\gamma} \right) + H_K + \mu_0 M - D k_n^2$$

Consider the case of perfect surface pinning and establish the relevant form of the above equation. N.B. H_K denotes the anisotropy field.

(9) The figure below shows a typical spectrum for a thin film of permalloy. Given that the exchange stiffness constant for this material is 1×10^{-6} erg cm^{-1}, estimate the thickness of the film. State any assumptions you make in this calculation. Comment on the existence of the small intermediate peaks evident in the spectrum.

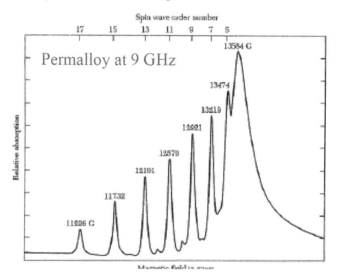

(10) Compare and contrast the experimental techniques of traditional ferromagnetic resonance with a microwave spectrometer and some of the more modern methods, such as the network analyzer and pump–probe techniques. State the advantages and disadvantages of each as well as discussing the similarities and differences.

(11) Recently, the inverse Faraday effect (IFE) has been found to produce ultrafast magnetization reversal in a non-precessional mechanism (see, for example, K. Vahaplar et al., Phys. Rev. Lett., **103**, 117201 (2009)). Explain the IFE process and how it can be exploited for device applications.

(12) Explain the phenomenon of *spin angular momentum transfer* and give an application of this effect in a physical device.

References and Further Reading

Antoniak, C., Lindner, J., & Farle, M., *Europhys. Lett., 70* (2), 250–256 (2005).

Antoniak, C., Lindner, J., Salgueirino-Maceira, V., & Farle, M., *Phys. Stat. Sol. (a), 203*, 2968 (2006).

Antoniak, C., Friedenberger, N., Trunova, A., Meckenstock, R., Kronast, F., Fauth, K., Farle, M., & Wende, H., (2012). "Intrinsic Magnetism and Collective Magnetic Properties of Size-Selected Nanoparticles." In: Lorke A., Winterer M., Schmechel R., Schulz C. (eds.). *Nanoparticles from the Gasphase*. NanoScience and Technology. Springer, Berlin, Heidelberg.

Banholzer, A., Narkowicz, R., Hassel, C., Meckenstock, R., Stienen, S., Posth, O., Suter, D., Farle, M., & Lindner, J., (2011). *Nanotechnology, 22*, 295713.

Berger, R., Kliava, J., Bissey, J.-C., & Baïetto, V. (2006). *J. Phys.: Condens. Matter, 10*, 8559.

Bunyaev, S. A., Golub, V. O., Salyuk, O. Yu., Tartakovskaya, E. V., Santos, N. M., Timopheev, A. A., Sobolev, N. A., Serga, A. A., Chumak, A. V., Hillebrands, B., & Kakazei, G. N., (2015). *Scientific Reports, 5*, 18480.

Chang-Jun, J., Fan, X.-L., & Xue, D.-S., (2015). High frequency magnetic properties of ferromagnetic thin films and magnetization dynamics of coherent precession. *Chinese Physics B, 24*, 057504.

Chumak, A. V., Neumann, T., Serga, A. A., Hillebrands, B., & Kostylev, M. P., (2009). *J. Phys. D: Appl. Phys. 42*, 205005.

Chumak, A. V., Tiberkevich, V. S., Karenowska, A. D., Serga, A. A., Gregg, J. F., Slavin, A. N., & Hillebrands, B., (2010). *Nat. Comm., 1*, 141.

Chumak, A. V., Serga, A. A., & Hillebrands, B., (2017). *J. Phys. D: Appl. Phys., 50*, 244001.

De Biasi, R. S., Devezas, T. C., (1978). *J. Appl. Phys. 49*, (4), 2466–2469.

Ding, Y., Klemmer, T. J., Crawford, T. M., (2004). *J. Appl. Phys., 96*, 2969.

Dormann, J. L., Fiorani, D., & Tronc, E., (1997). Magnetic relaxation in fine-particle systems, Prigogine, I., & S. A. Rice (Eds.). *Adv. Chem. Phys., Volume XCVIII*.

Dormann, J. L., Bessais, L., Fiorani, D., (1988). *J. Phys. C: Solid State Phys. 21*, 2015–2034.

Gonçalves, C. S., Vieira, A. S., Navas, D., Miranda, M., Silva, F., Crespo, H., & Schmool, D. S., (2016).*Sci. Rep., 6*, 22872.

Gonçalves, F. J. T., Paterson, G. W., Stamps, R. L., O'Reilly, S., Bowman, R., Gubbiotti, G., & Schmool, D. S., (2016). *Phys. Rev. B, 94*, 054417.

Heinrich, B. (2005). Bland, J. A. C., & Heinrich, B. (Eds.), *Ultrathin Magnetic Structures*, volume III, Springer, Berlin Heidelberg.

Horley, P. P., Sukhov, A., Berakdar, J., & Trápaga-Martínez, L. G., (2015). *The European Physical Journal B, 88*, 165.

Kakazei, G. N., Wigen, P. E., Guslienko, K. Y., Novosad, V., Slavin, A. N., Golub, V. O., Lesnik, N. A., & Otani, Y., (2004). *Appl. Phys. Lett., 85*, 443.

Kirilyuk, A., Kimel, A. V., & Rasing, Th. (2010). *Rev. Mod. Phys., 82*, 2731–2784.

Kliava, J., & Berger, R., (1999). *J. Magn. Magn. Mater. 205*, 328.

Lenk, B., Ulrichs, H., Garbs, F., & Münzenberg, M. (2011). *Phys. Rep., 507*, 107.

Lindner, J., & Baberschke, K., (2003). *J. Phys.: Condens. Matter, 15*, R193.

Maksymov, I. S., & Kostylev, M., (2015). *Physica E, 69*, 253–293.

Mamica, S., Lévy, J.-C. S., & Krawczyk, M., (2014). *J. Phys. D: Appl. Phys., 47*, 015003.

Markó, D., Valdéz-Bango García, F., Quirós, C. F., Hierro-Rodriguez, A., Fraga Vélez, M., Carbajo Martín, J. I., Maestro Alameda, J. M., Schmool, D. S., & Álvarez-Prado, L. M. (2019). *Appl. Phys. Lett., 115*, 082401.

Poole, C. P., (1983). *Electron Spin Resonance: A Comprehensive Treatise on Experimental Techniques*, Wiley-Interscience, New York.

Puszkarski, H., (1979). *Prog. Surf. Sci., 9*, 191.

Puszkarski, H., & Tomczak, P., (2017). *Surf. Sci. Rep., 72*, 351.

Rothhardt, J., Demmler, S., Hadrich, S., Limpert, J., & Tunnermann, A., (2012). Opt. Express, *20*, 10870–10878.

Schmool, D. S., Rocha, R., Sousa, J. B., Santos, J. A. M., Kakazei, G. N., (2006). *J. Magn. Magn. Mater., 300*, e331–e334.

Schmool, D. S., Rocha, R., Sousa, J. B., Santos, J. A. M., Kakazei, G. N., Garitaonandia, J. S., & Lezama, L., (2007). *J. Appl. Phys. 101* (10), 103907.

Schmool, D. S., & Schmalzl, M., (2007). *J. Non-Cryst. Solids 353*, 738.

Schmool, D. S., & Schmalzl, M., (2009). *Advances in Nanomagnetism*, B. Aktaş, & F. Mikailov (Eds.), Springer, 321–326.

Schmool, D. S., (2010). Spin Dynamics in Nanometric Magnetic Systems, *Handbook of Magnetic Materials*, J. Buschow (Ed.), pp. 111–346, Elsevier.

Schoeppner, C., Wagner, K., Stienen, S., Meckenstock, R., Farle, M., Narkowicz, R., Suter, D., Lindner, J., (2014). *J. Appl. Phys. 116*, 033913.

Shaw, J. M., Nembach, H. T., & Silva, T. J., (2013). *Physical Review B, 87,* 054416.

Silva, F., Miranda, M., Alonso, B., Rauschenberger, J., Pervak, V., & Crespo, H., (2014). *Opt. Express, 22,* 10181–10191.

Spasova, M., Wiedwald, U., Ramchal, R., Farle, M., Hilgendorff, M., Giersig, M., (2002). *J. Magn. Magn. Mater. 240,* 40–43.

Tamaru, S., Tsunegi, S., Kubota, H., & Yuasa, S., (2018). *Rev. Sci. Instrum., 89,* 053901.

Tannenwald, P. E., & Seavey, M. H., (1957). *Phys. Rev., 105,* 377.

Tomita, S., Hagiwara, M., Kashiwagi, T., Tsuruta, C., Matsui, Y., Fujii, M., Hayashi, S., (2004). *J. Appl. Phys. 95,* 8194.

Trunova, A. V., R. Meckenstock , Barsukov, I., Hasse, C., Margeat, O., Spasova, M., Lindner, J., Farle, M., (2008). *J. Appl. Phys. 104,* 093904.

Wiedwald, U., Spasova, M., Salabas, E. L., Ulmeanu, M., Farle, M., Frait, Z., Fraile-Rodriguez, A., Arvanitis, D., Sobal, N. S., Hilgendorff, M., Giersig, M., (2003). *Phys. Rev. B, 68* 064424.

Yu, C., Pechan, M. J., & Mankey, G. J., (2003). *Appl. Phys. Lett., 83,* 3948.

Yu, X. Z., Mostovoy, M., Tokunaga, Y., Zhang, W., Kimoto, K., Matsui, Y., Kaneko, Y., Nagaosa, N., & Tokura, Y. (2012). *Proc. Natl. Acad. Sci., 109,* 8856–8860.

Zakeri, Kh., Barsukov, I., Utochkina, N. K., Römer, F. M., Lindner, J., Meckenstock, R., von Hörsten, U., Wende, H., Keune, W., Farle, M., Kalarickal, S. S., Lenz, K., & Frait, Z., (2007). *Phys. Rev B, 76,* 214421.

Index

Q

Q-factor, 450, 452
Quadrant detector, 23
Quadratic
    equation, 433
    term, 154
Quadrupole, 321
Quantitative description, 379
Quantization, 107, 108, 111, 115, 116, 151,
    176, 187, 188, 271, 412, 457
Quantum
    confinement, 115, 116, 151, 188, 217,
      218
    dot, 116, 125–127, 130, 131, 145,
      151–153, 156, 164–169, 171–177,
      191, 217, 218, 263, 418
    effects, 105, 116, 263
    efficiency, 216, 218
    electrodynamics, 197
    field theory, 344
    fluctuations, 156
    interference effects, 141
    mechanical
      calculation, 115
      current density, 143
      description, 142
      phenomenon, 285
      representation, 111
    mechanics, 111, 116, 188
    membrane (QM), 215
    numbers, 111, 126, 134, 212, 218
    of conductance, 133, 189
    point, 131, 135
      contact (QPC), 125, 130–138,
        171, 172, 189
    state, 111, 206
    structure, 147
    theory, 206, 424
    transport, 131, 132, 136, 137, 189
    well, 116–118, 122, 124–126, 129, 131,
      134, 145, 147, 148, 171, 188, 189,
      191, 210–218, 221, 260, 263, 285,
      379–382, 408, 411, 412, 416
      region (GaAs), 145
      resonant states, 380
    wire (QWR), 114, 116, 124, 126, 130,
      131, 189

Quasi
    1D system, 185
    band gap, 86
    bound
      level, 146, 148
      state, 145–147
    static
      DC stator voltages, 90
      limit, 238
      regime, 237
    uniform leaf state, 328
Quenched orbital motion, 272

R

Radiation
    pressure, 85, 260
    tension, 260
Radiative
    recombination, 216, 408
    transition, 204, 205
Radio-frequency, 38, 39, 64–69, 261, 311
    applications, 95
Radiopharmaceuticals, 323
Rado-Weertman equation, 434, 460
Raman spectroscopy techniques, 199
Rashba effect, 409, 412, 414
Reciprocal
    lattice, 180, 181, 251, 470
      vectors, 180, 251
    space, 111, 181
Reconfigurability, 61, 64
Reconfigurable filters, 64
Rectangular planar geometries, 259
Red-shifted spectra, 249
Refraction, 198, 200, 239, 254, 256–258,
    262
Refractive
    index, 60, 198, 200, 201, 203, 221,
      225, 240, 252, 255–258, 260, 262,
      353–356
      changes, 356
      materials, 252
      medium, 256
    indices, 255, 260
Relaxation process, 322, 423, 442, 445, 484
Resistivity, 98, 105, 108, 371, 385, 388

Printed and bound by CPI Group (UK) Ltd, Croydon, CR0 4YY

23/10/2024

01777705-0015